Statistical Modeling in Machine Learning

Concepts and Applications

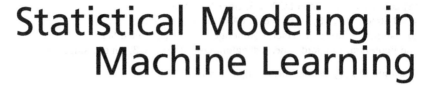

Statistical Modeling in Machine Learning

Concepts and Applications

Edited by

Tilottama Goswami

Professor
Department of Information Technology,
Vasavi College of Engineering, Hyderabad, Telangana, India

G.R. Sinha

Adjunct Professor
International Institute of Information Technology Bengaluru (IIITB),
Bangalore, Karnataka, India

ELSEVIER

ACADEMIC PRESS
An imprint of Elsevier

15.   Impact of Midday Meal Scheme in primary schools in India using exploratory data analysis and data visualization

Sonal Mobar Roy, Tilottama Goswami and Charan Kumar Nara

16.   Nonlinear system identification of environmental pollutants using recurrent neural networks and Global Sensitivity Analysis

Srinivas S. Miriyala, Ravikiran Inapakurthi and Kishalay Mitra

Contributors

Raja Banerjee Department of Mechanical and Aerospace Engineering, Indian Institute of Technology Hyderabad, Hyderabad, Telangana, India

Haider Banka Department of Computer Science and Engineering, IIT (ISM) Dhanbad, Dhanbad, Jharkhand, India

Barbara Cardone Università degli Studi di Napoli Federico II — Dipartimento di Architettura, Napoli, Italy

Bhanu Chander Department of Computer Science and Engineering, Indian Institute of Information Technology, Kottayam, Kerala, India

Ferdinando Di Martino Università degli Studi di Napoli Federico II — Dipartimento di Architettura, Napoli, Italy; Centro Interdipartimentale di Ricerca "Alberto Calza Bini", Università degli Studi di Napoli Federico II, Napoli, Italy

Divy Dwivedi Medibuddy, Bengaluru, Karnataka, India

Ranjana Dwivedi Department of Electronics and Communication Engineering, Motilal Nehru National Institute of Technology Allahabad, Prayagraj, Uttar Pradesh, India

Ashutosh Ganguly Emalpha, Mumbai, India

Kumaravelan Gopalakrishnan Department of Computer Science and Engineering, Pondicherry University, Pondicherry, India

Tilottama Goswami Department of Information Technology, Vasavi College of Engineering, Hyderabad, Telangana, India

V.V. Haragopal Aizenalgo Private Limited, Kompally, Hyderabad, India

Ravikiran Inapakurthi Global Optimization and Knowledge Unearthing Laboratory, Department of Chemical Engineering, Indian Institute of Technology Hyderabad, Sangareddy, Hyderabad, India

Pramod D. Jadhav   Department of Mechanical and Aerospace Engineering, Indian Institute of Technology Hyderabad, Hyderabad, Telangana, India

Sapna Singh Kshatri   Shri Shankaracharya Institute of Professional Management and Technology, Raipur, India

Sumit Kumar   Jio, Bangalore, Karnataka, India

Shailesh Kumar   Department of Electronics and Communication Engineering, Motilal Nehru National Institute of Technology Allahabad, Prayagraj, Uttar Prdesh, India

Basant Kumar   Department of Electronics and Communication Engineering, Motilal Nehru National Institute of Technology Allahabad, Prayagraj, Uttar Prdesh, India

G. Madhuri   OCR Lab, SCIS, University of Hyderabad, Hyderabad, Telangana, India

T. Manimozhi   Francis Xavier Engineering College, Tirunelveli, Tamilnadu, India

Dyna Marneni   Department of Computer Science and Engineering, Maturi Venkata Subba Rao Engineering College, Hyderabad, Telangana, India; Research Scholar, JNTUH, Hyderabad, India

Srinivas Soumitri Miriyala   Global Optimization and Knowledge Unearthing Laboratory, Department of Chemical Engineering, Indian Institute of Technology Hyderabad, Hyderabad, Telangana, India

Dhanajay Mishra   KarmaAI Life, Bangalore, Karnataka

Kishalay Mitra   Department of Chemical Engineering, Indian Institute of Technology Hyderabad, Hyderabad, Telangana, India; Global Optimization and Knowledge Unearthing Laboratory, Department of Chemical Engineering, Indian Institute of Technology Hyderabad, Hyderabad, Telangana, India

K. Mohanasundaram   Alliance School of Business, Alliance University, Bangalore, India

Charan Kumar Nara   Cognizant Technology Solutions India Private Limited

Atul Negi   OCR Lab, SCIS, University of Hyderabad, Hyderabad, Telangana, India; School of Computer and Information Sciences, University of Hyderabad, Hyderabad, Telangana, India

Priyanka D. Pantula   Global Optimization and Knowledge Unearthing Laboratory, Department of Chemical Engineering, Indian Institute of Technology Hyderabad, Hyderabad, Telangana, India

V. Pothyachi   Economics PGT, Sri S Badal Chand Sugan Chorida Vivekananda Vidayalaya, Chennai, Tamilnadu, India

NagaSree Keerthi Pujari   Department of Chemical Engineering, Indian Institute of Technology Hyderabad, Hyderabad, Telangana, India

Sonal Mobar Roy   Centre for Post Graduate Studies & Distance Education, National Institute of Rural Development and PR, Hyderabad, Telangana, India

Uponika Barman Roy   Tata Consultancy Services Limited, Bangalore, Karnataka, India

Tapan Kumar Sahoo   Department of Computer Science and Engineering, IIIT Bhubaneswar, Bhubaneswar, Odisha, India

Sabrina Senatore   Università degli Studi di Salerno, Dipartimento di Ingegneria dell'Informazione ed Elettrica e Matematica Applicata, Salerno, Italy

Deepak Singh   NIT, Raipur, Chhattisgarh, India

G.R. Sinha   International Institute of Information Technology Bengaluru (IIITB), Bangalore, Karnataka, India

Rohini Srivastava   Department of Electronics and Communication Engineering, Motilal Nehru National Institute of Technology Allahabad, Prayagraj, Uttar Prdesh, India

Vinay Kumar Srivastava   Department of Electronics and Communication Engineering, Motilal Nehru National Institute of Technology Allahabad, Prayagraj, Uttar Pradesh, India

Sridhar Vemula   Department of Computer Science and Engineering, Maturi Venkata Subba Rao Engineering College, Hyderabad, Telangana, India; Research Scholar, Osmania University, Hyderabad, India

K.A. Venkatesh   Professor & Dean, School of Mathematics and Natural Sciences, Chanakya University, Bangalore, Karnataka, India

Editors' biographies

Tilottama Goswami has received a BE degree with Honors in Computer Science and Engineering from the National Institute of Technology, Durgapur; and an MS degree in Computer Science (High Distinction) from Rivier University, Nashua, New Hampshire, United States. She was awarded a PhD in Computer Science from the University of Hyderabad. Presently, Dr. Goswami is Professor in the Department of Information Technology, Vasavi College of Engineering, Hyderabad, India. She has, overall, 23 years of experience in academia, research, and the IT industry. Her research interests are computer vision, machine learning, and image processing. She has been granted an Australian patent for her research work.

Dr. Goswami has been conferred with the Distinguished Scientist Award by IJIEMR-Elsevier SSRN, Vijayawada, India. She is also a recipient of the Women Researcher Award, awarded by the REST Society for Research International, India. Dr. Goswami is the recipient of University Grants Commission-Basic Scientific Research (UGC-BSR) Fellowship (under the Government of India). She has been awarded the Star Team Award for developing efficient software for GeoMedia (GIS), leading to complete customer satisfaction at Hexagon (Intergraph), Hyderabad. She is Editorial Board Member of two international journals and has contributed editorial articles and chapters in Scopus-indexed books.

Dr. Goswami is an IEEE Senior Member in IEEE CIS/GRSS Chapter Hyderabad Section. She is presently serving as Chairperson of ACM Hyderabad Deccan Professional Chapter and has also served as ACM-W Chair. Dr. Goswami is an active researcher and contributes to society by delivering workshops and guest lectures, participating as technical program committee, tutorial chair, and reviewer in international conferences and journals. She has delivered more than 20 invited talks in the area of artificial intelligence, machine learning algorithms, statistical methods, computer vision, and color image processing. Dr. Goswami has been convener for international events such as Distinguished Lecture Series, AI Webinar Series, Workshops, and Conclaves. Dr. Goswami actively maintains her industry engagement through industry exchange program and project consultancy on applications of AI in various problem domains.

Prof. G. R. Sinha (Fellow IETE, Fellow ISTE, SMIEEE) is working as Adjunct Professor at International Institute of Information Technology Bengaluru (IIITB), India. Prior to IIITB, he was working as Professor at IIITB-mentored Myanmar Institute of Information Technology (MIIT) Mandalay Myanmar. He has been Visiting Professor (Online) in National Chung Hsing University Taiwan, University of Sannio Italy and Visiting Professor (Honorary) in Sri Lanka Technological Campus Colombo.

He has published 293 research papers, book chapters and books at International and National levels; and edited 20 books in the field of Cognitive Science, Biomedical Signal Processing, Biometrics, Optimization Techniques, Sensors, Outcome based Education, Data Deduplication with Internationally reputed publishers Elsevier, IOP, Springer, Taylor & Francis, IGI. He owns two Australian patents. He is Associate Editor of five SCI/Scopus indexed journals and has been Guest Editor in various SCI journals.

Dr Sinha has been ACM Distinguished Speaker in the field of DSP (2017–2021). He has been Expert Member for Vocational Training Program by Tata Institute of Social Sciences (TISS) for Two Years. He has been contributing CSI Distinguished Speaker in the field of Image Processing since 2015. He also has served as Distinguished IEEE Lecturer in IEEE India council for Bombay section. He has received more than 12 National and International level Awards and Recognitions. He has delivered more than 60 Keynote/Invited Talks and Chaired many Technical Sessions in International Conferences across the world. He has been Vice President of Computer Society of India for Bhilai Chapter for two years. He is regular Referee of Project Grants under DST-EMR scheme of Govt. of India. He has been Expert Member of Professor promotion committee of GermanJordanian university Jordan and Project Proposal evaluation committee of UK-Israel Research Grants.

Dr Sinha has supervised 08 PhD Scholars, 15 M. Tech. Scholars, 100 UG level students and has been Supervising 01 more PhD Scholar. His research interest includes Biometrics, Medical/Biomedical Image Processing & Cognitive Science applications, Computer Vision, Outcome based Education (OBE) and Assessment of Student Learning Outcomes.

Preface

Machine learning aims to find a pattern from the inputs to the system by transforming data using models to decouple and capture maximum information, mainly with the help of mathematics, statistics, and learning theory. The data can be from any application domain such as environmental science, social science, behavioral science, industrial operations, medical science, etc. Statistics have ruled in almost all research domains for knowledge extraction, and therefore the intricacies, theory, and applications of statistics need to be highlighted and elaborated to help numerous academicians and researchers from various fields. *Statistical Modelling in Machine Learning: Concepts and Applications* covers theoretical background and importance of statistics; sampling theory, encoding and scaling process for data processing; statistics for evaluation and prediction; statistics for classification, regression and clustering; role of statistics in data preparation and data analytics in environmental science, social science, education, music, medical science, and aeronautics. The book is unique because it caters to basic concepts and applications of the role of statistics, exploratory data analysis, and machine learning.

The knowledge of statistics is considered as prerequisite for in-depth understanding of machine learning. The existing books on statistics most of the time cater to readers from mathematics and statistics backgrounds. The theories, notations, and proofs are of not much interest and use to the programming community and machine learning practitioners. This book will be useful to statisticians, programmers, machine learning practitioners, and all those who apply machine learning to the benefit of innovating and automating to solve various machine learning tasks such as classification, predictive analytics, regression, clustering, recommending, etc. The book attempts to explain the concepts in a very lucid manner with appropriate case studies and simple mathematical illustrations wherever possible. Machine learning techniques are growing rapidly, and researchers are developing new algorithms and techniques to maximize the model performance. The new techniques for evaluation and validation, etc., are covered. This book takes a much-needed holistic approach—putting all together with an in-depth treatise of a multidisciplinary applications of machine learning. The book covers a comprehensive overview of the state-of-the-art with help of real-life problems and applications.

This book includes 17 chapters and the chapter descriptions of the book is as follows. Chapter 1 presents an introduction to statistical modeling in machine learning with the aim to provide a deep overview of the major machine learning techniques and algorithms with prediction of prison overcrowding. While using statistics to construct a data representation, we have to infer connections between variables to find insights. Machine learning is the process of gaining a comprehensive understanding of data via mathematical and/or statistical models to make predictions. The chapter also emphasizes on the advantages of machine learning algorithms from an application viewpoint to help in making an informed decision on implementing various learning algorithms. In Chapter 2, a data collection technique is discussed. The preparation stage of data is the most fundamental block to dive into the world of artificial intelligence. The magic lies in the efficient usage of clean data from a heap of raw information. Data collection makes the initial contribution in the lifecycle of data. There are various effective ways to collect data worldwide. This chapter represents a familiar technique to obtain data from heap which is often known as web scrapping, which is an automated system that extricates data from the internet. The authors also discuss how the Python language, with its vast library support, implements web scrapping to collect data from the internet. Chapter 3 presents an analysis of COVID-19 using machine learning techniques. COVID-19 is caused by a newly detected coronavirus, and its proper analysis is very much needed. This chapter presents an analysis on the symptoms of disease and identifies significant symptoms that impact the cause of the illness. Machine learning techniques such as multiple regression, SVM, decision tree, random forest, and logistic regression are applied to understand the evaluation with respect to the measures like coefficient of determination and mean-squared error. Hypothesis testing is used to determine if at least one of the features is useful in the diagnosis of the disease. Further feature selection process is used to identify the most significant symptoms that will cause the virus. Different visualization methods are used to figure the substantial reasoning from the model's prediction and perform analysis on the results obtained.

Chapter 4 studies discriminative dictionary learning based on statistical methods. A brief review of statistical techniques applied in discriminative dictionary learning is provided. The main objective of the methods described in this chapter is to improve classification using sparse representation. In this chapter, a hybrid approach is also described, where sparse coefficients of input data are used to train a simple multilayer perceptron with back-propagation. The classification results on the test data are comparable with other computation-intensive methods. In Chapter 5, artificial intelligence—based uncertainty quantification technique for external flow CFD simulations are discussed. In this work, authors propose a novel multiobjective evolutionary optimization approach that aims to achieve optimal estimation of such hyperparameters: architecture, sample size for training, and choice of activation, simultaneously at the time of building ANN surrogates. The data for training the ANN models is obtained from the high-fidelity time-expensive CFD simulations for modeling the supersonic flow over a cruciform missile system. Chapter 6 presents contrast between simple and complex classification algorithms. The chapter addresses Music Information Retrieval (MIR) and sheds light on the features involved in audio signal processing, its importance, and ways to model it. The chapter discusses in detail about the comparative study of results obtained after fitting the data with various classification models: K nearest-neighbors, Fisher linear discriminant analysis, quadratic discriminant analysis, and feed-forward neural networks. In Chapter 7, classification model for medical data analysis is presented. This chapter provides a broad aspect of all types of classification models such as logistic regression, decision tree, random forest, ANN, SVM, radial basis function neural network, and deep neural net classification models for medical data analysis as well as medical image analysis. In Chapter 8, regression models for machine learning are discussed. This chapter introduces the theoretical aspects of regression from simple to multilinear models. The chapter deals with statistical modeling via data visualization and showcases implementation in R programming. Chapter 9 presents model selection and regularization, introducing some fitting methods than least squares so that the linear model (regression) improves in terms of its accuracy and the model interpretability. It helps in dealing with the problem of multicollinearity between the independent variables. Ridge regression can be used to reduce the complexity of the model, which in turn results in a decrease in the overfitting problem.

Chapter 10 discusses data clustering using unsupervised machine learning. The chapter elaborates unsupervised-based clustering approaches along with cluster evolution criteria in terms of distance measurements and clustering loss functions. The chapter also highlights some interesting challenges and future outlooks in unsupervised deep clustering. In Chapter 11, Emotion-based classification through fuzzy entropy enhanced FCM clustering is described. The chapter proposes a novel approach to the emotion-based classification of microblogging messages such as Twitter. The classification method is unsupervised and exploits the well-known fuzzy c-means (FCM) clustering algorithm, proposing an enhanced version called entropy-weighted FCM (EwFCM) that overcomes the main drawback of the FCM, viz., the sensitivity to the random cluster initialization by leveraging a fuzzy measure to evaluate the entropy in the data distribution. The proposed method converges faster and provides promising classification performance, as evaluated by common metrics such as accuracy, precision, and F1-score. Chapter 12 presents fundamental optimization methods for machine learning. The modern prevalent fundamental optimization methods are discussed from the perspective of gradient information including the first-order methods, high-order methods, and derivative-free optimization algorithms. The chapter discusses issues and challenges in the field of deep neural networks and various machine-learning optimization methods. Chapter 13 presents stochastic optimization of industrial grinding operation through data-driven robust optimization (RO). A new data-based sampling technique for RO is presented, which utilizes unsupervised machine learning and novel generative modeling framework for identifying the intended space more accurately and sampling in the desired regions of uncertainty. Chapter 14 discusses about dimensionality reduction using PCAs in feature partitioning framework. This chapter presents variants of one-dimensional principal component analysis (PCA) in feature partitioning frameworks, namely, subpattern principal component analysis (SpPCA), cross-correlation subpattern principal component analysis (SubXPCA), extended subpattern principal component analysis (ESpPCA), and extended cross-correlation subpattern principal component analysis (ESubXPCA). The issues such as summarization of variance, space and time complexities of the above feature partitioning methods are addressed theoretically.

Chapter 15 reports the impact of the mid-day meal scheme in primary schools in India using exploratory data analysis and data visualization. Malnutrition has emerged as a serious issue over the last few years. In this chapter, the authors have made an attempt to look at various schemes, especially the mid-day meal scheme and study its impact in primary schools of states across India. Statistical analysis, especially exploratory data analysis (EDA) has been done by the authors. Secondary data was used from census records and government websites for the three parameters: Anemia, Stunt Growth and Enrollment of children. EDA and visualization techniques such as stacked-bar plots, box plots, violin plots and scatter plots for the three parameters is done across states, to understand their role in affecting the mid-day meal scheme. Chapter 16 presents nonlinear system identification of environmental pollutants using recurrent neural networks and global sensitivity analysis. In this work, optimally designed recurrent neural networks (RNNs) are utilized to capture the non-linearities of fifteen pollutants measured in Taiwan. A novel evolutionary based neural architecture search algorithm balancing the variance-bias trade-off is proposed. To identify the most potent features effecting concentration of pollutants, Monte Carlo−based global sensitivity analysis using the optimally designed RNNs is performed. Finally, Chapter 17 presents a comparative study of automated deep learning techniques for wind time series forecasting. The authors propose a novel and generic automated machine learning strategy to design them optimally under the framework of multiobjective optimization solved by NSGA-II. The study in this work demonstrates the importance of forecasting and its impact in wind farm design and control.

Scientists, researchers, academicians, research scholars, economists, social science enthusiasts working in multidisciplinary fields for predictive analytics using machine learning and statistics can benefit from concepts and case studies of real time applications, as depicted in the book.

Acknowledgments

Dr. Tilottama Goswami is indebted to her parents for instilling good values and providing her with a quest for knowledge. She would like to extend her profound gratitude to her husband Samir, daughters Prakriti and Sanskriti, and various well-wishers for their kind cooperation and encouragement.

Dr. Sinha expresses his gratitude and sincere thanks to his wife Shubhra, daughter Samprati, his parents, and teachers.

We would like to thank all our friends, well-wishers, and those who keep us motivated in doing more and more, better and better. We sincerely thank all contributors for writing the relevant theoretical backgrounds, concepts, and real-time applications of statistical modeling in machine learning.

We express our humble thanks to the editorial team, commissioning editor, and editorial staff at Elsevier for their great support, necessary help, appreciation, and quick responses. Finally, we want to thank everyone in one way or another, who helped us in editing this book.

Last but not least we would also like to thank God for showering us his blessings and strength to do this type of novel and quality work.

Tilottama Goswami
G. R. Sinha

1

Introduction to statistical modeling in machine learning: a case study

Sapna Singh Kshatri[1], Deepak Singh[2], Tilottama Goswami[3], G.R. Sinha[4]

[1]SHRI SHANKARACHARYA INSTITUTE OF PROFESSIONAL MANAGEMENT AND TECHNOLOGY, RAIPUR, INDIA; [2]NIT, RAIPUR, CHHATTISGARH, INDIA; [3]DEPARTMENT OF INFORMATION TECHNOLOGY, VASAVI COLLEGE OF ENGINEERING, HYDERABAD, TELANGANA, INDIA; [4]INTERNATIONAL INSTITUTE OF INFORMATION TECHNOLOGY BENGALURU (IIITB), BANGALORE, KARNATAKA, INDIA

1.1 Introduction

Following the "AI winter" of the 1980 and 1990s, interest in data-driven artificial intelligence (AI) approaches in a range of technological disciplines, such as speech and image analysis [1] and communications [2], progressively rose. Unlike prior AI research, which was dominated by logic-based expert systems, the success of machine learning-based pattern recognition tools has encouraged growing trust in data-driven approaches. These tools combine decades-old methods like back-propagation, the Expectation-Maximization (EM) algorithm, and Q-learning [3] with revolutionary regularization approaches and variable learning rate schedules. Their success is due to the unprecedented availability of data and computing resources across a wide range of engineering fields. It's a science study of algorithms models that computers use to accomplish a job without explicitly programming it. Algorithms for learning are used in a variety of applications that we use on a regular basis. Machine learning is a subset of Artificial Intelligence and has been a critical component of digitalization solutions that have garnered significant attention in the digital realm.

The area of machine learning, which can summarize as enabling computers to make accurate predictions based on prior experiences, has seen significant growth in recent years, owing to the fast rise in computers' data storage and processing power. Machine-learning techniques have been widely applied in bioinformatics and a variety of other fields. Due to the challenges and costs associated with biological studies, advanced machine-learning techniques are developed for this application area. We begin this chapter by reviewing core machine-learning concepts such as feature evaluation, unsupervised versus supervised learning, and classification types. Essentially, we are pursuing supervised machine learning, as shown in Fig. 1.1. Then, we'll address the crucial obstacles inherent in developing and evaluating machine-learning research.

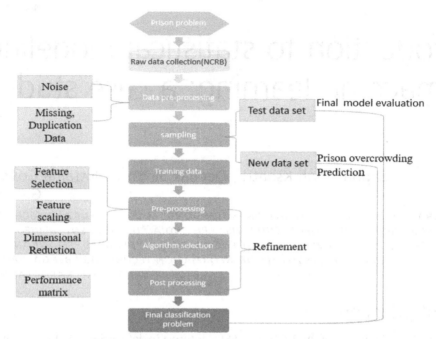

FIGURE 1.1 Supervised machine-learning flow diagram.

As a result, the first step is to put together the data set. Following that, it is necessary to establish which attributes/fields/features are the most relevant, which is accomplished via a process called as feature extraction. The simplest method is "machine learning," which examines everything available and derives the most relevant/informative attributes from those measurements. Additionally, a technique known as feature subset selection is used in order to discover and eliminate as many duplicates, irrelevant, and redundant attributes as is possible from the data set. Second, it is important to do data preparation, noise (outliers), missing feature values, and categories that must be converted to dummy variables are all common features in data sets. There are other ways for dealing with missing data that may be used, and they should all be explored. A comprehensive data preparation is required as a consequence of this.

1.1.1   Machine-learning research in early age

Presents the brief information describes previously used methods for crime investigation, tracking, hotspot detection, and crime estimation; it's a quick way to implement new algorithms and models from 1992 to now, with an overview of various studies in machine-learning techniques from old to new. This research is one of only a few that allows a machine to learn directly from unlabelled and unstructured voice streams, similar to how human infants learn. The success of the tests was aided by IHDR. Using the AUC (area under the receiver operating characteristic [ROC] curve) as a presenting metric for AI algorithms. Six machine-learning algorithms (C4.5, Multiscale Classifier, Perceptron, Multilayer Perceptron, k-Nearest Neighbors, and a Quadratic Discriminate Function) are tested on six "real-world" clinical diagnostics informational collections as part of a contextual inquiry [4]. According to Tong, presented another algorithm for performing dynamic learning with SVMs. By exploiting the duality between boundary space and highlight space, we showed up at three algorithms that endeavor to decrease adaptation space however much as could be expected at each inquiry [5]. Strano and Marco represent a brief conversation of criminal profiling followed by a prologue to the Italian Neural Network for Psychological Criminal Profiling (NNPCP) venture. This task, given an alleged neural organization and information mining [6], rendered to violations submitted in a "virtual" situation. Meaningfully, the assessment presumes that, albeit a portion of the hypothesis's central ideas can without a doubt be applied to cybercrime, there stay significant contrasts among "virtual" and "earthly" universes that limit the hypothesis's helpfulness [7]. The iterated detainee's problem is utilized to represent and demonstrate the marvels in financial aspects, humanism, and brain research, just as in the organic sciences, for example, transformative science. The disclosure and stream-lining of IPD techniques in open applications require adaptable procedure portrayal. The correlation of deterministic and nondeterministic limited state machines as the portrayals of strategies for the iterated detainee's predicament is introduced [8]. This chapter outlines the capacity of a Native American Indian ethnic character poll in catching personality information from a detainee populace. Data were gathered during the Southern Ute Indian Community Safety Survey (SUICSS), a US Division of Justice, Bureau of Justice Statistics-supported investigation of wrongdoing and savagery on the Southern Ute Indian reservation. The SUICSS had three prongs: (a) dispersion of a re-view poll, (b) individual meetings, and (c) an assessment of the Tribal Code [9].

If we consider machine learning in conjunction with other technologies, we can see how advancements in sensor technology, IoT, and machine-learning approaches have transformed environmental monitoring into a really intelligent monitoring system. The structure for robust ML methods, denoising techniques, and the establishment of appropriate standards for wireless sensor networks (WSNs) has been proposed [10]. On the other hand, the misrepresentation ratio (MRR) is applied to the input healthcare text data and models the PE criteria for hypothesis validation. Additionally, such a revolu-tionary method enables the amalgamation of numerous ML system variables, such as data size, classifier type, partitioning protocol, and % MRR [11].

1.1.2 Ensemble machine-learning technique

The base selected for conventional classification rankings and bagging tries to increase the accuracy of the basis by concatenating the learned classifier's predictions into a single prediction via the construction of a composite classifier. Bagging employs the voting mechanism. The Bagging classifier is a technique for constructing a community of learners suggested by Leo Breiman in 1994 [12]. Apart from deaths and property damage, arson may have substantial societal consequences and instill psychological fear in the populace [13]. This issue discusses a hybrid method that combines techniques from ensemble learning and intelligent optimization. Create a feature selection technique based on recursive feature elimination (RFE) to eliminate redundant features. Second, we pick the optimum data imbalance processing algorithm from a set of 18 candidates. Support vector machines (SVMs) are an exciting machine-learning tool that has demonstrated superior performance in most prediction problems [14].

1.2 Classification of algorithms in machine learning

Artificial Intelligence aims to train computers to use model information or previous experience to solve a problem [15,16]. Machine learning has been extensively applied to problems in predicating crime [17]. This application was made as a rule by different names inside notable logical teaches, for example, signal processing, data hypothesis, coding hypothesis, and so on. Instances of this remember measurable models for linear and nonlinear regression, utilization of compression methods. Can be seen, and so on. This application has commonly been restricted to exemplary machine-learning procedures. Nonetheless, numerous ongoing advances in machine learning have been predicted in different territories (for example, picture, discourse, and video processing, standard language processing.). It was found in the most refreshed surveys on the utilization of machine learning for prediction.

To demonstrate the statistical model's concepts and applications in machine learning, "Prediction of Prison Overcrowding" is an example of a problem statement. NCRB (National Crime Record Bureau) data in its raw form. A model for predicting jail overcrowding is built using prison data. Crime prediction is a significant step in crime analysis as the success of the prediction depends on the data type and accuracy of the model. The prediction of crime is an important step in crime analysis because the success of the prediction depends on the data type and accuracy of the model [18]. Preliminary studies say this analytical, statistics-based approach is working in a language of machine learning called matrix-based; Critics, however, warn that predictive policing can open up a trickle of performance problems. Here are four possible predictors of possible policing.

An algorithm can be understood as high error and low accuracy when we choose faulty modules. Furthermore, data sets without any preprocessing or detailed data cause difficulty in their implementation, which leads to the presence of defective modules in prediction. Many missing values, failure of the developed model, and reliability of the forecast are affected. That Mistakes can be found in prophecy in any of the following ways: The first cause of a crime prediction defect is the presence of data fault, incorrect classification, or sometimes hominoid mistakes. Thus, the need to predict crime has emerged as a busy challenge. Most crime prediction strategies aim to increase the efficiency of crime prediction through improving the performance matrix or by

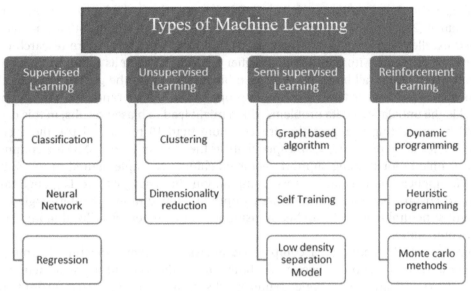

FIGURE 1.2 Types of machine learning.

determining faulty modules. As mentioned below Fig. 1.2, machine learning approaches are classified into three broad categories.

Some single classifiers are Artificial Decision Trees, Neural Networks, Fuzzy Systems, Naive Bayes, SVM, K-Nest Neighbors, etc. [19]. Most predictive models have been employed to accept the best and effective classifier for predicting faults by creating different data sets. Data sets with other numbers are taken and subjected to a classification that employs training and testing data with a data set of variable instances. Data sets with varying crime numbers are classified based on crime, which uses training and testing with a data set of inconsistent cases. Classifiers are examined and categorized utilizing the presentation criteria of individual classifiers to plan their adequacy. Characterization is classified by relying on parameters, for example, accuracy, FP rate, TN rate F-measure, etc., that determine the nature of data. The classifier representing the most extreme estimate of the exposition metric is chosen as the best classifier and appointed as the best classifier for the programming deficiencies available because of accessible faults because the data contains many improvements. They are caused by mistakes made during the process. The best classifier used to predict programming flaws means that crime shortcomings are evaluated with high accuracy and precision [20].

- Supervised learning

The Random Forest method is a supervised learning technique that generates a forest with many trees. The Random Forest method is a supervised learning technique that generates a forest with many trees. It is used to train multiple decision trees for regression, classification, and other tasks by creating numerous decision trees during training. Using Random Forest, it is possible to rank the variables according to their importance [21]. SVMs are a promising machine-learning technique that has demonstrated excellent performance in most prediction problems [14]. Future research could involve experiments with additional classifier algorithms, such as Random Forest [22], XGBoost, and others, all based on the ensemble method, with the goal of boosting their performance as much as feasible. Another proposal is to utilize random under sampling to tackle the unbalanced data problem, which might lead to biased results; in this case, it would be interesting to repeat the process more than 10 times and use the average, standard deviation of accuracy, and specificity as the outcomes [23]. SMO is a commonly used technique for training support vector machines are implemented in the famous LIBSVM application. It is an iterative computation that uses just two Lagrange multipliers to ensure that each advancement's optimal configuration is intermingled. The relevant structural type of the arched nonsmooth improvement is utilized to initiate the SMO computation.

Hazwani et al. conducted a comparative analysis of several machine-learning approaches, including support vector machines, fuzzy theory, and artificial neural networks. The multivariate time series report results from a thorough assessment of crime prediction methodologies. The future scope still revealed the limitations of current approaches for optimizing and modifying parameters to acquire more accurate findings and improve performance [24].

The Naive Bayes method is popular owing to its simplicity and utility since models can be formed easily, and predictions can be produced quickly. The process determines the probability that an item with specific characteristics belongs to a particular category of class, and a probabilistic classifier assumes that the class attributes are independent [25].

- Networks of neurons: The most well-known supervised learning technology is the neural network without such a question. They are fundamental approximators of nonlinear functions, and numerous studies have been undertaken to establish their utility. Assert that a sufficiently large and dense network can approximate any function. Recent research has demonstrated that sparsely connected; deep neural networks are the best nonlinear approximators for a wide variety of processes and systems. The strength and flexibility of NNs stem from its modular structure, which is based on the neuron as a fundamental building component, a caricature of neurons in the human brain. Each neuron receives an input, processes it using an activation function, and generates an output. Numerous neurons can be connected to create various architectures that convey knowledge about the issue and data type. Feedforward networks are an extremely prevalent network architectural type. They are composed of layers of neurons, each receiving an input from the preceding layer via a weighted output. An architecture of a neural network is composed of an input layer that absorbs data and an output layer that provides predictions. The network weights are determined using nonlinear optimization approaches such as back-propagation (Rumelhart et al. 1986) to decrease the error between the prediction and labeled training data. Deep neural networks employ multiple layers and a range of nonlinear activation functions.

- Classification: Generally speaking, classification is the process of dividing data into a predetermined number of classes. The basic goal of a classification challenge is to classify a group or a class of people or things. Classification aids in finding a series of templates that can be used to anticipate future unidentified class labels by allowing for the prediction of future unknown class labels. It is necessary to utilize the training data set to calculate the model's anticipated accuracy [26,27].
- Regression: Machine learning in regression analysis [28] comprises several artificial intelligence approaches that foresee a constant outcome variable (y) while simultaneously evaluating one or more predictor variables (x). Regression models are designed to establish mathematical conditions in which y may be represented as a component of the x-factors, which is the ultimate goal of the model. Using machine-learning techniques in regression may lawfully extract information from data even when there is no stated previous programming objective. For subjects that are too unexpected to be fully specified or for things that cannot be precisely defined, reversal with its ability to learn from data is especially suitable. The prediction algorithms for regression and classification are shown in Fig. 1.3.

These are, precisely, the sort of issues that emerge in constant developing values. Many machine-learning algorithms apply to the various problems that start in data networks, for example, Random Forest, Gradient Boosting Machine (GBM) [29], SVM [30], Logistic Regression, Multinomial Logistic Regression [31], Multilayer Perceptron (MLP) [32], K-Nearest Neighbors (KNN) [33], Principal Component Analysis [34], K-Means [35], Naïve Bayes [35], and many more.

1. Semisupervised learning [36]: Semisupervised learning techniques are used when working with poorly labeled training data or with extra corrective information from the environment. Two algorithms fit this category: generative adversarial networks (GANs) and reinforcement learning methods (RL). When the LM is (self-)trained in either case, it follows a game-like approach discussed in further detail below.

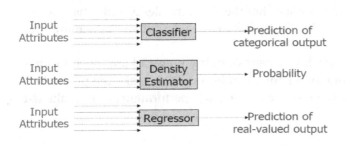

FIGURE 1.3 Types of crime prediction.

2. Unsupervised learning [37]: An extracting feature from data using global criteria established in advance is required for this learning task. There is no necessity for supervision or a ground-truth label for the results in this learning job. A wide range of topics, including dimension reduction, quantization, and clustering are treated in this chapter.

Vector clustering [32] and quantization are two important techniques. Clustering is an unsupervised learning strategy that may be used to discover groupings of data that are similar to the original data. For example, k-means clustering splits data into k clusters, with each observation being allocated to the cluster with the closest centroid. This results in a Voronoi cell partition of the data space, which may be seen as a Voronoi cell partition of the data space. Vector quantizers discover representative points in data sets that may be partitioned into a certain number of clusters using a predetermined number of points. These points may then be used in lieu of the whole data set to estimate future samples, saving time and money. The vector quantizer (x, w) creates a connection between the two x and the coordinates of the cluster centers by calculating the correlation coefficient. According to standard definitions, the loss function is the squared distortion of a data set from cluster centers, which must be decreased in order to establish the settings for the quantizer.

1.3 Regression algorithms in machine learning

The term regression is characterized as dissecting or estimating the connection between a reliant variable Furthermore, at least one independent variable. Regression can be characterized by two kinds of linear regression and calculated regression. Logistic regression is the speculation of linear regression [38]. Regression evolution is a statistical method for evaluating the association among factors that have cause and final product connection. Principle cognizance of invariant relapse is examining the relationship between an established variable and an unbiased variable and defines the linear relation condition among dependent and independents variables [39].

- Simple linear regression is exactly what it sounds like: it is a very simple linear method for predicting the future value of a quantitative response Y using a single regression predictor variable X. Simple linear regression is also known as simple linear regression approach. It is assumed that the two variables X and Y have a linear connection in most cases. This linear link may be described mathematically as follows:
- Multiple Linear Regression Simple linear regression is a powerful approach for predicting a response when just one predictor variable is used to make the prediction. On artificial and natural tasks, the nonlinear strategies' performances are evaluated [40].

Regression evolution is a statistical method for evaluating the association among factors that have cause and final product connection. Principle cognizance of invariant relapse is examining the relationship between an established variable and an unbiased variable and defines the linear relation condition among dependent and independents variables follow regression model formulated as follow,

$$y = \rho_0 + \rho_1 x_1 \ldots\ldots \rho_n x_n + \text{\texteuro}$$

Y = dependent variable.
x_1 = independent variable.
ρ_1 = parameter.
€ = error.

The inconspicuous error portion of € indicates the failure of data on the straight line to lie on and refers to the difference between the actual and observed realization of y. Behind such a distinction, there might be a few reasons, such as the influence of all erased variables in the model, subjective variables, natural arbitrariness of perceptions, and so on.

We should presume that €with mean zero and consistent variance 1 € is used as an isolated and indistinguishably distributed arbitrary variable. Subsequently, we would also expect that euros are generally distributed.

The self-sufficient elements are considered by the experimenter to be compelled, so it is regarded as nonstochastic, whereas y is considered to be a discretionary variable with

$$A(y) = \rho_0 + \rho_1 x$$

and

$$\text{Var}(y) = \theta^2$$

To a great extent, X, can in like manner, be any subjective variable. Taking everything into account, as opposed to the model mean and test contrast of y, we consider the prohibitive mean of y given as

$$X = x$$

For the uninitiated, regression analysis is a type of prediction model approach that looks at the relationship between a dependent (target) variable and an independent (s) variable (indicator). Factors that contribute to relapse can be divided into two categories. The first is a class of variables or suppressors that can operate independently of the rest of the program. Exploration and learning are used to identify independent variables. The significant variable, also known as the response variable, is the second variable to be considered.

It is via relapse that the relationship between the factors may be demonstrated and examined. The errors are proportionally independent and normally distributed with a mean of 0 and variance σ. By decreasing the error or residual sums of squares, the βs are estimated:

$$S(\beta_0, \beta1, \ldots \ldots \beta m) = \sum_{i=1}^{n} \left(Y_i - \left(\beta_0 + \sum_{j=1}^{k} \beta_j X_{ij} \right) \right)^2 \tag{1.1}$$

To locate the base of (2) regarding β, the subsidiary of the capacity in (2), as for each of the βs, is set to zero and tackled. This gives the accompanying condition:

$$\frac{\delta s|}{\delta \beta | \hat{\beta}_0 \hat{\beta}_1 \ldots \hat{\beta}_m,} = -2 \sum_{i=1}^{n} \left(Y_i - \left(\hat{\beta}_0 + \sum_{j=1}^{k} \hat{\beta}_j X_{ij} \right) \right) = 0, j = 0, 1, 2 \ldots k, \tag{1.2}$$

and

$$\frac{\delta s|}{\delta \beta | \hat{\beta}_0, \hat{\beta}_1 \ldots \hat{\beta}_m} = -2 \sum_{i=1}^{n} \left(Y_i - \left(\hat{\beta}_0 + \sum_{j=1}^{k} \hat{\beta}_j X_{ij} \right) \right) X_{ij} = 0, j = 1, 2 \ldots k, \tag{1.3}$$

The ^βs, the answers for (3) and (4), are the least-squares appraisals of the βs.

It is helpful to communicate both the n conditions in (1) and the k+1 condition in Eqs. (1.4) and (1.5) (which depend on straight capacity of the βs) in lattice structure. The model (1) can be communicated as

$$y = X\beta + \varepsilon \tag{1.4}$$

Where y is the nx1 vector of perception, X is a nx(k+1) network of autonomous factors (and an additional section of 1s for the intercept $\hat{\beta}_0$,β is a (k+1) X_i vector of coefficients and ε is a X_i vector of free and indistinguishably circulated mistakes related with (1).

So as to discover the B̀, the (k+1) X_i vector of β̀s and the gauge of β that limits the blunder, (2) in framework structure is

$$\begin{aligned} S(\beta) &= (y - X \beta)^T (y - X\beta) \\ &= y^T y - \beta^T X^T y - y^T X\beta + \beta^T X^T X\beta \\ &= y^T y - 2\beta^T X^T y + \beta^T X^T X\beta \end{aligned} \tag{1.5}$$

With a superscript "*T*" meaning the render of a network or vector. The articulation β TX Ty is a scalar. Along these lines, the least-squares estimator must fulfill the (k+1) Eqs. (1.3) and (1.4) written in matrix structure as

$$\frac{\delta s|}{\delta \beta|_{\hat{\beta}}} = -2X^T y + 2X^T X\hat{\beta} = 0 \tag{1.6}$$

Where 0 is the (k+1)×1 vector of 0's. This condition can be rearranged to

$$X^T X\beta = X^T Y \tag{1.7}$$

Under fitting conditions for example T X where X is not solitary, this equation will at long last net the following least-squares coefficients:

$$\frac{\tau}{\beta} = (X^T X) X^T Y^{-1} \tag{1.8}$$

These coefficients could then be used for anticipating or evaluating the usual ward variable for estimating free variables that should not be used in the example used to estimate β. While forecasting a multiple regression model to select goal overcrowding, which is a predictor variable, it may not be a good idea to use the same model to choose just one algorithm based on 15 years of the data-dependent variable. This question can be answered a little cumbersome, but it could be essential.

In this instance, advertising expenses are considered input variables, while sales input is considered output. Typically, the input variables are denoted by the variable output variable sign X, followed by a subscript to distinguish them from one another. For example, X1 might represent the television budget, X2 could represent the radio budget, and X3 could represent the newspaper budget. It can refer to the inputs in several ways, including predictors and features, independent variables, independent variable features, and occasionally just straight-forward variables. It is frequently referred to as the response or dependent variable, and the letter Y typically represents it. The output variable, in this case, is sales. Throughout this book, we will interchangeably use all of these terms.

1.4 Case study: prison crowding prediction

ML is a relatively new topic of study, there are much more learning algorithms than we can discuss in this introduction. I've chosen to outline six techniques that we commonly employ while tackling data analysis problems (usually classification). The first four approaches are established strategies that have been widely utilized in the past and perform well when evaluating low-dimensional data sets with a sufficient number of labeled training samples [41]. In the second section, I will quickly discuss two approaches (SVMs and Boosting) that have lately garnered considerable attention in the machine-learning field. They are capable of solving complex issues using a small number of instances (e.g., 50) fairly precisely and efficiently.

1.4.1 Methods and material

To characterize the learning issue in more detail, defined as the process of inferring correlations between a system's inputs, outputs, and parameters from a limited amount of data Cherkassky and Mulier in 2007. An LM is a sample generator (also known as the system in question), and as given in Fig. 1.4, we differentiate between the two. Our main point is that LMs are inherently stochastic in their estimates, and their learning process may be thought of as the reduction of a risk functional:

$$R(w) = \int B[y, \varphi(i, o, w)] \, t(o, y) \, dpdy,$$

The structure of the LM is defined by (i, y, w), the parameters of the LM are defined by (i, y, w), and the loss function B balances the multiple learning goals (for example, unsupervised learning, accuracy, simplicity, smoothness, and so on). where the data I (inputs) and o (outputs) are sampling from a probability distribution t, the structure of

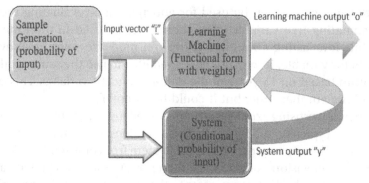

FIGURE 1.4 With the use of data from a sample producer and observations from the system, a learning machine approximates the output of a machine-learning algorithm.

the LM is defined by (I, o, w), the As previously stated, the risk functional is weighted by a probability distribution t(o, y), which restricts the LM's capacity to foresee. In Fig. 1.4, it can be seen that the many different types of learning algorithms may be divided into three basic categories: unsupervised, supervised, and semisupervised learning algorithms. These discrepancies demonstrate the degree to which the LM has access to external supervisory data provided by an expert in the field. Fig. 1.4 has shown the proposed machine-learning output in conditional probability of input.

We begin with a basic example to serve as a springboard for our statistical learning study. Consider the following scenario: we are using 60,000 data by a client to provide recommendations on enhancing NCRB of a violent crime. The prison data set includes Indian violent crime in 28 distinct state and FIR expenditures for the class in each of those crime across three different algorithms: SMO, multilinear and SMO regression. The crime data set includes violent crime 28 distinct states and union tertiaries.

1.4.2   Data collection

For the period 2001−18, unstructured crime data (as text or pdf) was acquired from the official website of India's crime records, the NCRB, as well as additional Public Domain Data sets available on ZIPNet Delhi Police, data.gov.in, and censusindia.gov.in. The prison crime data set includes statistics on several types of violent crimes, crime rates, and criminal characteristics. The data set collected from NCRB of violent crime. The imported data set is pictured with the class attribute being STATE/UT. The representation diagram shows the distribution of attribute STATE/UT with different attributes in the data set, each shade in the perception graph represents a specific state. The imported data set is pictured in WEKA; the representation diagram shows the circulation of crime as one to five level specific attributes with class attributes, which are people captured during the year.

In Fig. 1.5 shown the blue area in the figure depicts high-level crime such as murder, whereas the pink site reflects low-level crime such as the abduction of a specific

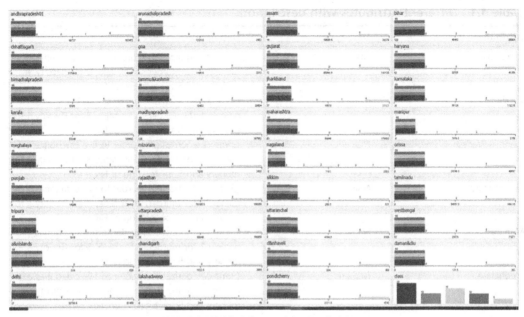

FIGURE 1.5 All violence crime visualization.

characteristic in the data set. According to police statistics, murder, attempted murder, and dowry death are classified as 1—the rape, 2—the attempt to rape, 3—dacoity, assembly to commit dacoity, and, similarly, up to 5.

1.4.3 Data preprocessing

Preprocessing data is a critical step in the in-data mining process. Data Field has various undesirable features that lead to an incorrect inspection. Preprocessing data has two main goals: (1) information challenges and (2) information examination framework [42] For instance, the data may contain invalid fields, and it might contain columns that are insignificant to the current investigation, etc. Data preprocessing procedures have been actualized in an adjusted form of WEKA [43] As a result, Consequently, accommodate the sort study, the data would be preprocessed when new states are generated, the data frequently diverges. The data is regularly divided when, leaving information blank or incorrect. Other characteristics and some missing values have been corrected. From 2001 to 2015, the data was finished using the new state names, which were incorrectly understood. This unit has been notified of the incorrect FIR: defined "total domestic violence" in our study as shown in Table 1.1 crime as different attributes.

The preprocessing module does it. Preliminarily, let's check out the Currently Two focal points that can be determined from this subwindow.

Table 1.1 Different attributes with description.

Attributes	Data-type	Description
Karnataka	Numeric	Crime categorization murder to five categories, namely 1 for murder, 2 for rape, 3 for dacoity, 4 for robbery and 5 for total no of FIR.
Kerala	Numeric	Crime categorization murder to five categories, namely 1 for murder, 2 for rape, 3 for Decioty,4 for robbery and 5 for total no of FIR.
Madhya Pradesh	Numeric	Crime categorization murder to five categories, namely 1 for murder, 2 for rape, 3 for Decioty,4 for robbery and 5 for total no of FIR.
Maharashtra	Numeric	Crime categorization murder to five categories, namely 1 for murder, 2 for rape, 3 for Decioty,4 for robbery and 5 for total no of FIR.
Manipur	Numeric	Crime categorization murder to five categories, namely 1 for murder, 2 for rape, 3 for Decioty,4 for robbery and 5 for total no of FIR.
Meghalaya	Numeric	Crime categorization murder to five categories, namely 1 for murder, 2 for rape, 3 for Decioty,4 for robbery and 5 for total no of FIR.
Mizoram	Numeric	Crime categorization murder to five categories, namely 1 for murder, 2 for rape, 3 for Decioty,4 for robbery and 5 for total no of FIR.
Nagaland	Numeric	Crime categorization murder to five categories, namely 1 for murder, 2 for rape, 3 for Decioty,4 for robbery and 5 for total no of FIR.
Orissa	Numeric	Crime categorization murder to five categories, namely 1 for murder, 2 for rape, 3 for Decioty,4 for robbery and 5 for total no of FIR.
Punjab	Numeric	Crime categorization murder to five categories, namely 1 for murder, 2 for rape, 3 for Decioty,4 for robbery and 5 for total no of FIR.
Rajasthan	Numeric	Crime categorization murder to five categories, namely 1 for murder, 2 for rape, 3 for Decioty,4 for robbery and 5 for total no of FIR.
Sikkim	Numeric	Crime categorization murder to five categories, namely 1 for murder, 2 for rape, 3 for Decioty,4 for robbery and 5 for total no of FIR.
Tamil Nadu	Numeric	Crime categorization murder to five categories, namely 1 for murder, 2 for rape, 3 for Decioty,4 for robbery and 5 for total no of FIR.
Tripura	Numeric	Crime categorization murder to five categories, namely 1 for murder, 2 for rape, 3 for Decioty,4 for robbery and 5 for total no of FIR.

Table 1.1 Different attributes with description.—cont'd

Attributes	Data-type	Description
Uttar Pradesh	Numeric	Crime categorization murder to five categories, namely 1 for murder, 2 for rape, 3 for Decioty,4 for robbery and 5 for total no of FIR.
Uttaranchal	Numeric	Crime categorization murder to five categories, namely 1 for murder, 2 for rape, 3 for Decioty,4 for robbery and 5 for total no of FIR.
West Bengal	Numeric	Crime categorization murder to five categories, namely 1 for murder, 2 for rape, 3 for Decioty,4 for robbery and 5 for total no of FIR.
A & N Islands	Numeric	Crime categorization murder to five categories, namely 1 for murder, 2 for rape, 3 for Decioty,4 for robbery and 5 for total no of FIR.
Chandigarh	Numeric	Crime categorization murder to five categories, namely 1 for murder, 2 for rape, 3 for Decioty,4 for robbery and 5 for total no of FIR.
D & N Haveli	Numeric	Crime categorization murder to five categories, namely 1 for murder, 2 for rape, 3 for Decioty,4 for robbery and 5 for total no of FIR.
Daman & Diu	Numeric	Crime categorization murder to five categories, namely 1 for murder, 2 for rape, 3 for Decioty,4 for robbery and 5 for total no of FIR.
Delhi	Numeric	Crime categorization murder to five categories, namely 1 for murder, 2 for rape, 3 for Decioty,4 for robbery and 5 for total no of FIR.
Lakshadweep	Numeric	Crime categorization murder to five categories, namely 1 for murder, 2 for rape, 3 for Decioty,4 for robbery and 5 for total no of FIR.
Pondicherry	Numeric	Crime categorization murder to five categories, namely 1 for murder, 2 for rape, 3 for Decioty,4 for robbery and 5 for total no of FIR.

1. The database includes Andhra Pradesh, Assam, Bihar, Chhattisgarh, and Goa, among other Indian states.
2. A 180-item data set was used to test our method

The collection contains seven association domains and 28 states and regions. So, the patterns in 35 attribute combinations, as illustrated in Table 1.2. The attribute "bars" (along the right side of the board) outline the strength of each characteristic separately. This board can also be opened in a separate window from the classifier board to visualize classifier predictions. When the class is discrete, the misclassified points appear as a case in the shade anticipated by the classifier; when the class is continuous, the size of each plotted point changes depending on the classifier's error.

Table 1.2   The performance of each model in comparison.

Algorithm	Correlation coeff.	Mean abs error	Root mean squared error	Root rel squared error	Total nom of instances	
Linear regression	−0.1234	0.2904	0.3645	100	100	100
Multi linear regression	0.99	0.29	0.3803	99.8443	104.33	100
SMO regression	0.0499	0.2601	0.3886	89.545	106.596	100

1.4.4   Proposed regression-based prison overcrowding prediction model (RBPOPM)

After research in machine learning, we have proposed the RBPOPM model for regression-based jail crowd prediction. To comprehend the present number of inmates, we must first understand the history. Continuing jail data shows that inmate numbers expand around 1.5 times faster than imprisonment rates. Regression analysis is used to find a connection between two variables. Using current and historical statistics, one may anticipate future jail swarms. It also enables us to look at the consequences of elements like unfavorable behavior changes and the number of proven criminals in jail. With the help of experts, data professionals, and data scientists, we can avoid and survey the best statical factors to use when building insightful models.

As seen in Fig. 1.6, the steps of data visualization are discussed; the first step is to collect raw data, separating it from the data utilized. Which information is gathered by data mining? Which necessitates the adoption of several data-mining methods? Preprocessing is performed on the data acquired in the second step, which eliminates noise and omissions of information. Following that third step of model building, the third section of data regression is model selection. Multiple classifiers are available, and it is critical to choose the optimal model for a given data type. Following the optimal model selection for our issue, the last step is prediction, followed by visualization.

1.5   Result and discussion

Regression modeling is one of the most significant statistical approaches used in analytical epidemiology. The influence of one or more independent variables (e.g., inputs, subject characteristics, and risk factors) on predictor variables such as strength or overcrowding can be examined using regression models. It delivers the mean absolute error of several linear regressions for SMO regression trees. SMO regression with gradient yields the best results, with a −0.2601 mean absolute error. Multilinear reversal is second with 0.29 MSE, followed by linear regression (0.2904 MAE). For example, the mean fundamental error of linear regression is 0.2904. The tree-based classifier uses classification or regression tree selection to help prediction (Fig. 1.7).

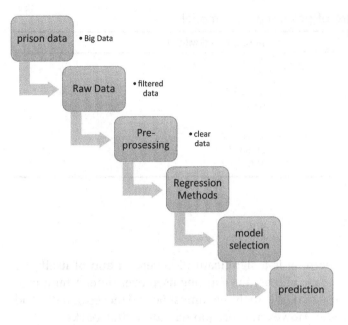

FIGURE 1.6 Overview of regression methods.

In this study, the training data set is sunspot numbers from 2001 to 2015, whereas the prediction set is from 2016 to 2018. Because sunspot activity studies demonstrate a 15-year periodicity, this research employs data from the preceding 3 years as the output vector and data from 15 years ahead as the input vector.

A linear, multilinear SMO approach is used to forecast time series in this study. The SMO technique is generally applicable to the linear and multilinear algorithms, based on the simulation findings of the three previous situations. That is faster than the SMO approach but slower than the multilinear algorithm.

Table 1.2, models are classified into three types: linear, multilinear, and SMO. By using the SMO method, predictive data and has higher predictive accuracy than linear and multilinear regression. The findings are compared using the mean absolute error as criteria.

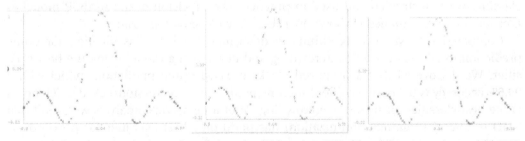

FIGURE 1.7 The curve of real value and predictive value (A) based on SMO, (B) multiliner regression, and (C) SMO regression.

Table 1.3 Performance measure of proposed SMO model.

Performance measures	Description (RMBSP)
Correctly classified instances	90
Incorrectly classified instances	10
Kappa statistics	0.9709
Mean absolute error	0.0392
Root mean squared error	0.1039
Relative absolute error	12.7991%
Root relative squared error	26.5817%
Total number of intense	100

1.6 Conclusion

Combining prison and machine learning is a significant discovery in and of itself; the work we conducted previously is a case of machine learning used over time. Attempts to advance have been made in our research. We analyzed almost 60,000 pieces of data. And build a well-defined standard model. However, every job has some drawbacks.

The study identified the following shortcomings.

1. A fundamental shortcoming of this effort is the difficulty of creating renationalized data.
2. Time required for data gathering and preprocessing; needed extensive training.
3. Determining crime's direct and indirect consequences on jail populations is a complex undertaking.
4. Machine learning and prediction cannot be applied directly to all crimes; the nature of some crimes differs from human perception.
5. There is a great deal of diversity in the data, which makes it challenging to work with a single algorithm; as a result, we had to evaluate the data against various methods to arrive at a more accurate conclusion.

We tested Several algorithms linear multilinear and SMO to find the best classifier for crime prediction to address the jail overcrowding problem. A model based on the best classifier was constructed and used in practice. The precision of the RMBSP model as measured by performance is shown in Table 1.3 with SMO regression.

Compared with various algorithms to determine the best classification for crime prediction to address prison overcrowding and developed a model using the best classifier. We discovered that a proposed stacking-based crime prediction model with a 99.5% accuracy is the best classifier for crime prediction. A regression-based SMO model is the best classifier for prison overcrowding. After an empirical analysis we found that SMO (sequential minimal optimization) has performed best classifier for prison overcrowding prediction with 90% accuracy.

A linear, multilinear SMO method is applied to predict time series in this study. From the simulation results for the preceding three cases, we may deduce that the SMO method has a higher degree of generalization than the linear and multilinear algorithms. The operation time is comparable to that of the multilinear algorithm and significantly shorter than that of the SMO method.

References

[1] G. Hinton, et al., Deep neural networks for acoustic modeling in speech recognition: the shared views of four research groups, IEEE Signal Process. Mag. 29 (6) (2012) 82−97, https://doi.org/10.1109/MSP.2012.2205597.

[2] M. Ibnkahla, Applications of neural networks to digital communications − a survey, Signal Process. 80 (7) (2000) 1185−1215, https://doi.org/10.1016/S0165-1684(00)00030-X.

[3] H. van Hasselt, A. Guez, D. Silver, Deep reinforcement learning with double Q-learning, in: Proceedings of the AAAI Conference on Artificial Intelligence, Vol. 30, No. 1 SE-Technical Papers: Machine Learning Methods, March, 2016.

[4] A.P. Bradley, The use of the area under the ROC curve in the evaluation of machine learning algorithms, Pattern Recogn. 30 (7) (1997) 1145−1159, https://doi.org/10.1016/S0031-3203(96)00142-2.

[5] S. Tong, D. Koller, Support vector machine active learning with applications to text classification, J. Mach. Learn. Res. (2001) 45−66, https://doi.org/10.1162/153244302760185243.

[6] M. Strano, A neural network applied to criminal psychological profiling: an Italian initiative, Int. J. Offender Ther. Comp. Criminol. 48 (4) (2004) 495−503, https://doi.org/10.1177/0306624X04265284.

[7] E. Jardine, Global cyberspace is safer than you think: real trends in cybercrime, in: Paper Series, no. 16, 2015.

[8] M. Yar, The novelty of 'cybercrime': an assessment in light of routine activity theory, Eur. J. Criminol. 2 (4) (2005) 407−427, https://doi.org/10.1177/147737080556056.

[9] A.-M. Cazan, L.E. Năstasă, Emotional intelligence, satisfaction with life and burnout among university students, Proc. Soc. Behav. Sci. 180 (November 2014) (2015) 1574−1578, https://doi.org/10.1016/j.sbspro.2015.02.309.

[10] S.L. Ullo, G.R. Sinha, Advances in smart environment monitoring systems using IoT and sensors, Sensors 20 (11) (2020). https://doi.org/10.3390/s20113113.

[11] S.K. Srivastava, S.K. Singh, J.S. Suri, Chapter 16 - a healthcare text classification system and its performance evaluation: a source of better intelligence by characterizing healthcare text, in: G.R. Sinha, J.S. Suri (Eds.), Cognitive Informatics, Computer Modelling, and Cognitive Science, Academic Press, 2020, pp. 319−369. https://doi.org/10.1016/B978-0-12-819445-4.00016-3.

[12] L.G.A. Alves, H.V. Ribeiro, F.A. Rodrigues, Crime prediction through urban metrics and statistical learning, Phys. Stat. Mech. Appl. 505 (2018) 435−443. https://doi.org/10.1016/j.physa.2018.03.084.

[13] N. Wang, S. Zhao, S. Cui, W. Fan, A hybrid ensemble learning method for the identification of gang-related arson cases, Knowl. Base Syst. 218 (2021) 106875. https://doi.org/10.1016/j.knosys.2021.106875.

[14] F. Anifowose, J. Labadin, A. Abdulraheem, Improving the prediction of petroleum reservoir characterization with a stacked generalization ensemble model of support vector machines, Appl. Soft Comput. 26 (2015) 483−496. https://doi.org/10.1016/j.asoc.2014.10.017.

[15] A.S. Lundervold, A. Lundervold, An overview of deep learning in medical imaging focusing on MRI, Zeitschrift fur Medizinische Physik 29 (2) (2019) 102−127. https://doi.org/10.1016/j.zemedi.2018. 11.002.

[16] T. Goswami, Chapter 16 - machine learning behind classification tasks in various engineering and science domains, in: G.R. Sinha, J.S. Suri (Eds.), Cognitive Informatics, Computer Modelling, and Cognitive Science, Academic Press, 2020, pp. 339−356. https://doi.org/10.1016/B978-0-12-819443-0.00016-7.

[17] R. Iqbal, M.A.A. Murad, A. Mustapha, P.H.S. Panahy, N. Khanahmadliravi, An experimental study of classification algorithms for crime prediction, Ind. J. Sci. Technol. 6 (3) (2013) 4219−4225. https:// doi.org/10.17485/ijst/2013/v6i3.6.

[18] S. Sathyadevan, M.S. Devan, S.S. Gangadharan, Crime analysis and prediction using data mining, in: 2014 First International Conference on Networks Soft Computing (ICNSC2014), August, 2014, pp. 406−412. https://doi.org/10.1109/CNSC.2014.6906719.

[19] E. Cantu-Paz, C. Kamath, Inducing oblique decision trees with evolutionary algorithms, IEEE Trans. Evol. Comput. 7 (1) (2003) 54−68. https://doi.org/10.1109/TEVC.2002.806857.

[20] Y. Shin, L. Williams, Can traditionally fault prediction models be used for vulnerability prediction? Empir. Software Eng. 18 (1) (2013) 25−59. https://doi.org/10.1007/s10664-011-9190-8.

[21] L.J. Muhammad, E.A. Algehyne, S.S. Usman, Predictive supervised machine learning models for diabetes mellitus, SN Comput. Sci. 1 (5) (2020) 1−10. https://doi.org/10.1007/s42979-020-00250-8.

[22] Z. Noshad, et al., Fault detection in wireless sensor networks through the random forest classifier, Sensors 19 (7) (April, 2019). https://doi.org/10.3390/s19071568.

[23] A. Fenerich, et al., Use of machine learning techniques in bank credit risk analysis, Rev. Int. Métodos Numéricos Cálculo Diseño Ing. 36 (3) (2020) 1−15. https://doi.org/10.23967/J.RIMNI. 2020.08.003.

[24] N.H.A. Halim, M.Y. Mashor, A.S.A. Nasir, R. Hassan, Performance Comparison between multilayer Perceptron and fuzzy ARTMAP networks for acute leukemia detection, Int. J. Res. Rev. Comput. Sci. 2 (5) (2011) 1−7.

[25] M. Zareapoor, R. SeejaK., M. Alam, Analysis on credit card fraud detection techniques: based on certain design criteria, Int. J. Comput. Appl. 52 (2012) 35−42.

[26] D.K. Renuka, T. Hamsapriya, M.R. Chakkaravarthi, P.L. Surya, Spam classification based on supervised learning using machine learning techniques, in: 2011 International Conference on Process Automation, Control and Computing, 2011, pp. 1−7. https://doi.org/10.1109/PACC.2011.5979035.

[27] S.S. Kshatri, B. Narain, Analytical study of some selected classification algorithms and crime prediction, Int. J. Eng. Adv. Technol. 9 (6) (2020) 241−247. https://doi.org/10.35940/ijeat.f1370.089620.

[28] L. Mcclendon, N. Meghanathan, Using machine learning algorithms to analyze crime data, Mach. Learn. Applicat.: Int. J. 2 (1) (2015) 1−12. https://doi.org/10.5121/mlaij.2015.2101.

[29] A. Natekin, A. Knoll, Gradient boosting machines, a tutorial, Front. Neurorob. 7 (2013) 21. https:// doi.org/10.3389/fnbot.2013.00021.

[30] F. Smach, C. Lemaître, J.-P. Gauthier, J. Miteran, M. Atri, Generalized fourier descriptors with applications to objects recognition in SVM context, J. Math. Imag. Vis. 30 (1) (2008) 43−71. https://doi. org/10.1007/s10851-007-0036-3.

[31] Z.Q. John Lu, The elements of statistical learning: data mining, inference, and prediction, J. Roy. Stat. Soc. 173 (3) (July, 2010) 693−694. https://doi.org/10.1111/j.1467-985X.2010.00646_6.x.

[32] D. Ruppert, The elements of statistical learning: data mining, inference, and prediction, J. Am. Stat. Assoc. 99 (466) (2004). https://doi.org/10.1198/jasa.2004.s339, 567−567.

[33] T. Hastie, R. Tibshirani, J. Friedman, The elements of statistical learning: data mining, inference and prediction probability theory: the logic of science the fundamentals of risk measurement

mathematicians, pure and applied, think there is something weirdly different about, Math. Intel. 27 (2) (2005) 83−85.

[34] J. Franklin, The elements of statistical learning: data mining, inference and prediction, Math. Intel. 27 (2) (2005) 83−85. https://doi.org/10.1007/BF02985802.

[35] K. Nordhausen, The elements of statistical learning: data mining, inference, and prediction, second edition by trevor hastie, robert tibshirani, jerome friedman, Int. Stat. Rev. 77 (3) (2009) 482. https://doi.org/10.1111/j.1751-5823.2009.00095_18.x.

[36] S. Yadav, M. Timbadia, A. Yadav, R. Vishwakarma, N. Yadav, Crime pattern detection, analysis & prediction, in: Proceedings of the International Conference on Electronics, Communication and Aerospace Technology, ICECA 2017, Vol. 2017-Janua, 2017, pp. 225−230. https://doi.org/10.1109/ICECA.2017.8203676.

[37] B. Widrow, Y. Kim, D. Park, The hebbian-LMS learning algorithm, IEEE Comput. Intell. Mag. 10 (4) (2015) 37−53. https://doi.org/10.1109/MCI.2015.2471216.

[38] H. Khodakarami, B.S.G. Pillai, W. Shieh, Quality of service provisioning and energy minimized scheduling in software defined flexible optical networks, J. Opt. Commun. Netw. 8 (2) (2016) 118. https://doi.org/10.1364/JOCN.8.000118.

[39] F.A. Anifowose, J. Labadin, A. Abdulraheem, Ensemble model of non-linear feature selection-based Extreme Learning Machine for improved natural gas reservoir characterization, J. Nat. Gas Sci. Eng. 26 (2015) 1561−1572. https://doi.org/10.1016/j.jngse.2015.02.012.

[40] S. Singh, Data analysis based on the visualization: a survey, IJCRT 6 (1) (2018) 1777−1785.

[41] S.S. Kshatri, D. Singh, B. Narain, S. Bhatia, M.T. Quasim, G.R. Sinha, An empirical analysis of machine learning algorithms for crime prediction using stacked generalization: an ensemble approach, IEEE Access 9 (2021) 67488−67500. https://doi.org/10.1109/ACCESS.2021.3075140.

[42] A. Famili, W.M. Shen, R. Weber, E. Simoudis, Data pre-processing and intelligent data analysis, Intell. Data Anal. 1 (1) (1997) 3−23. https://doi.org/10.3233/IDA-1997-1102.

[43] F. Kamiran, T. Calders, Data pre-processing techniques for classification without discrimination, Knowl. Inf. Syst. 33 (1) (2012) 1−33. https://doi.org/10.1007/s10115-011-0463-8.

2

A technique of data collection: web scraping with python

Sumit Kumar[1], Uponika Barman Roy[2]

[1]JIO, BANGALORE, KARNATAKA, INDIA; [2]TATA CONSULTANCY SERVICES LIMITED,
BANGALORE, KARNATAKA, INDIA

2.1 Introduction

The internet is a rich source of information. Every day we generate data in trillions of megabytes [1]. These highly populated heterogeneous data requires security as it carries indefinite personal information. Everything that we can see and listen on the internet is a data. It can be in any format such as text, audio, video, or an image. All the websites represent data to the users; thereby, they act as a data warehouse. While walking through the life cycle of data management, the first milestone which needs to be achieved is the data collection. Web scraping is a technique which mines the data available on the internet [2]. In the following section we will see the basic functionality of web scraping along with its significance and ethics to use. The third section will give a theoretical coverage of the elements involved in web scraping along with an architectural diagram. In the fourth section, a step-by-step guided walkthrough will help the readers to comprehend how to extract data from web. This section contains a worked-out example with each code snippet followed by the explanation. In addition, it will document on the stages of transformation of the data, and it will help the readers to visualize the difference in the processed data from the original form of the data. The fifth section will enlist few real-time web scrapings in distinguished domains. Essentially, that creates an overall understanding of the multiple wide benefits of web scraping. In the sixth section, the author will conclude the discussion summarizing the topics discussed followed by the reference area.

2.2 Basics of web scraping

2.2.1 Definition

To define, "A web scraping tool is a technology solution to extract data from web sites, in a quick, efficient and automated manner, offering data in a more structured and easier to use format, either for Business-to-Business (B2B) or for Business-to-Consumer (B2C) processes" [3]. This technique takes out huge amounts of unstructured data and stores it in an structured manner. The websites hold massive amounts of data for users. Let us see an example. Consider any big e-commerce website where the users buy and sell different number of items. A user wants to buy some fresh apples from any one of the platforms. To buy, the user will definitely visit the fruits section of the application and then proceed to the varieties of apples. These days companies are using machine learning techniques to bring up recommendations for users. The one who searches for apples also gets some choice to buy some grapes with extra discounts. This catches the user's attraction to buy grapes as an add-on with apples. But what exactly runs behind the story is a simple machine learning recommendation system. Now, the question is how the system knows to recommend grapes with apples but not a soap bar? The answer is the data that is collected from the user's choice undergoes analysis before recommendation. In simple terms, it is the data.

To develop a good machine learning model, it is a primary focus to provide the model with a correct form of authenticated data. It is very important what data is being collected and how it is being formatted for the utilization. The websites from where the data gets mined is in a raw unstructured format. Web scraping not only consumes those raw data but also transforms them into a usable structured layout to feed to the machine learning models [4]. It is obvious that to collect data from a website, the site's application programming interface (API) is to be accessed. Therefore, the authentication from the website plays a major role in the process. It is not mandated that all companies permit their data to be used for building any algorithm on it.

2.2.2 Why do we need to scrape data?

There are indefinite reasons why we data needs to be scraped. The information on the internet is of various forms like text, image, media, etc. Every individual organization has their own set of requirements of data for their development or operation. To understand the requirement of scraped data better, some examples can be enlisted, such as the following:

(1) **Research and Development**: Statistical data, survey information or sensor data are the biggest data sources in the field of research. To process any analysis or develop more into a domain, exploration and experiments are the backbone. So, the sector of research and development is highly reliable on data to carry out the experiments.

(2) **Business Marketing**: The companies stack up data (customers data like phone number, email ID, recently purchased products, choice of products, reviews, etc.) from the websites to enhance on better customer facility.

(3) **Social Media Technology**: The various giant media platforms like Twitter, YouTube, and Facebook bring the user an experience of trending streaming

media. They make use of very high-end search engines where the history data is the fundamental ingredient.

(4) Recommendation Websites: The admission portals or job listing websites provides a good example of application of web scraped data. These portals use the gathered data and recommend on the user's preference.

The above listed scopes are some of the high-end usages of web scraping but there are lot more other utilities at smaller scale where the data is gathered to fulfill organization level or personal level requirements [5].

2.2.3 Choice of programming language

Python language is deliberately chosen to discuss and demonstrate an example of web scraping. There are other helpful languages like C++, Ruby, Node.JS, R, Selenium, etc., which are used for the task [6]. All these mentioned languages have good support from their respective libraries to achieve web scraping. But to deal with internet data, the size of the data is noteworthy, and it is always big. To reach these website data, the API handling of the programming language should be flexible. So, any programming language chosen should support HTTP responses thoroughly. Even though Node.JS has an extensive HTTP library for API handling, it is limited in holding up to large amounts of data. Again, C++ sets off highly expensive to use for web related operations. Python is an easier selection due to its simplicity and vast functionalities. For any data-related programming, one of the significant aspects to keep in mind is the visualization. The data to play with is rigid and raw to analyze until they are text data. Visualization gives a better understanding of the data. There are distinct kinds of visualization libraries in Python like Matplotlib, seaborne, Pygal, etc.

The requirements have to be properly articulated to choose the best fit language for web scraping. To shortlist them as:

- Less complicated as a big size of data needs to be manipulated.
- Flexibility to float around various APIs to collect data.
- Faster and more efficient to utilize minimum CPU usage.
- Supports visualizations to data.
- Capable of generating end data in a structural format.
- Maintainability.
- Scalability.

To bring forth the above features in a coding language, Python is considered the most suitable one at the beginner's level.

2.2.4 Ethics behind web scraping

So far, what and why is discussed on web scraping. Now it's **time to start for How**!!

But let's pause a bit and rewind. There are lot of occasions where the reader is finding the buzz words like fetch the data from the internet, use a website's data, and so on. The question here is how ethical it is to pull out data from anywhere and why the websites would support it. Let us pick up the first part of the question.

The community of data scientists, data engineers, business analysts, data journalists, and everyone who relies on data, scrap them from internet. So, it is obviously not illegal, but everyone maintains basic ethics while scraping. Data scraping from the internet has some guidelines or legalities to follow, which is known as ethical web scraping [4].

Often an internet user ticks the privacy policy statement like "I accept the privacy and follow it" while web browsing. There is where he is abiding to the data policy of the website owner. Similarly, as an ethical scraper it is highly important to align to the honesty while taking out data [7,8].

- The information taken out should only be present with the scraper, not to be shared widely. The motive of web scraping should be genuine and only to bring out analysis from the data but not to prepare a duplicate copy of the collected data.
- The public APIs are the most reliable sites to scrap data from.
- The scraper should provide purpose to clear the intention of collecting data and own details to be communicated for any concern. There must be a collaborative attitude toward the website owner.
- The analysis brought up from the owner's data must end up benefiting him as well. It can help in raising the trafficking of his website. And this answers the latter half of the question.
- Likely, the website owner also to comply with certain ethics like encouraging the scrapers to utilize the data for mutual benefit.

To keep a healthy collaboration between the scraper and site owner helps to bring upon more innovation and profit to both ends.

2.3 Elements of web scraping

2.3.1 Architecture of web scraping

Fig. 2.1 gives a pictorial representation of scraping data from the web. In the next part of this section, the reader will get a detail understanding of each block of the architecture.

2.3.2 Components of web scraping

To start collecting data from websites, there are few prerequisites need to be followed. As discussed in last Section 2.2.4, data cannot be gathered from any random websites. It is always recommended to mine from public websites. From Fig. 2.1, it is observed that web scraping includes important techniques like website selection, inspection of data, etc. Let's see how these components contribute to the ultimate goal:

- **Website Selection**: The first component must be to understand the reason of data collection and then to know the nature of data required. Considering the use case

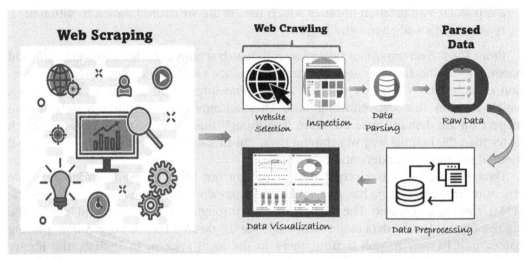

FIGURE 2.1 Architecture of web scraping.

to be price comparison of certain items among different online shopping websites, the motive should be to look for those commodities in different platforms and then gather their individual data.

- **Inspection of web data**: After selecting the website, the task is to inspect the data on the web page. While hovering around the data (required to be gathered), Ctrl + Shift + I option lets a Hypertext Markup Language (HTML) window pop at the right side of the page. This window demonstrates the HTML construction only for the inspected data.

- **Parsing**: Followed by the inspection of data, the most important part is to parse the data using the Python library. Python has a powerful library called Beautiful Soup, which parses the HTML data. By definition parsing refers to converting of machine language data to human readable data. In this case, the parsing will convert the HTML data into simple text data.

- **Data Preprocessing**: The parsing gives the data in a crude unstructured format. Python has extensive library support to format data in the desired form. This method of framing data is data preprocessing. This is a significant step as it generates a structure of the data which in the later gets used for analysis. More accurate the data is preprocessed better result can be obtained from analysis. Hence, this is considered the heart of any machine learning algorithms.

- **Data Visualization**: This is completely in the need of the user. It is a later stage operation which is done post–web scraping to understand the collected data better. Often while analysis it becomes easier to visualize the data in the form of graphs or charts. As data comprehension is an important task to make any system intelligent, so it is always suggested to be familiar with the data. In Python there are popular visualization libraries which results the structured data into different types of statistical charts and graphs.

From Fig. 2.1, an important take away is that web scraping includes web crawling and parsing. From the above bullets, the first two points, i.e., website selection and inspection of web data is together referred as web crawling. There are automated systems which perform this web crawling task. They automatically search the website and inspect the on-demand data out from it. Although this software techniques are much faster than the manual way of scraping data, but this section demonstrates the manual method for granular understanding purpose.

Data Parsing plays the actual role in pulling out the web content into readable structure. Every web page has its own HTML framework. While inspection of data, the HTML content is accessed. The target is to crawl through the HTML tags and identify the tag responsible for the data to get scraped. Once the desired HTML tag is found, parsing comes into its way to pull it from there to the local system. In Python, the library Beautiful Soup has an interesting function to crawl on the web data for the HTML tag identification and also for parsing it. The reader will get a hands-on experience of these concepts in the implementation walk through.

2.4 An implementation walkthrough

In this section we will go through a demonstration of a real-time scraping of web data. There are different libraries which are popularly used by the scrapers to mine data.

2.4.1 Libraries for web scraping

2.4.1.1 Scrapy

This framework is primarily used for web crawling. Scrapy builds very complex systems which is capable to scrape a high dimension of data. This tool consumes notably very less memory and CPU usage. The benefit of this tool is it is an open-source framework, and it has the powerful functions to carry out the HTTP responses and their parsing. Scrapy is a Python-supported tool.

An operating system or a program when performs several tasks at one single instant, it is referred as multithread or an asynchronous system. Scrapy framework is built on an asynchronous foundation, known as Twisted, hence the pace with which it operates the program to extract data, process them and to format them is exceedingly fast. That is the main reason of selecting Scrapy for mining large scale data. It generates the structured data in different forms like CSV, JSON, XML, etc. This tool basically collects data from the websites which are designed using Java scripts [9].

Scrapy has its own command-line interface known as "Scrapy Tool," it has different functions like fetch, crawl, check, parse, bench, etc. These are few command-line functions and alike there are other functions which help Scrapy to collect the data from the web and return into a structured format. After reading the high-end features offered by Scrapy, the readers may raise the query to justify the need of other scraping tools. To answer that, every system has their respective flaws. Scrapy is not a beginner's tool. It is used at high professional end to scrape high dimensional data from the very complex websites [10]. Due to this, the tool has configuration complexity at different operating systems.

2.4.1.2 Selenium

A web-based automation tool mainly used for developing testing scripts in the industry. Web Scraping is an add-on benefit of this tool. The basic steps to use Selenium is to install the package and import them followed by loading the browser (Google Chrome) driver. Once the configurations are satisfied, the website has to be accessed [11]. Selenium has different locators to navigate through the elements of the website. It uses elements as Tag name, Class name, IDs, XPath, CSS selectors, etc.

The above code snippet in Fig. 2.2, is sourced from GitHub [12]. In this it is seen that after loading the chrome driver, it's **get** () function finds the website to extract data from. The **find_elements_by_class_name** () detects the specific class (division, paragraph, table, etc.) of prime focus.

The popularity of Selenium as a web scraper is due to its ability to scrape data from the dynamic websites. There are two kinds of websites: static and dynamic. The static is the one where the data present on the website is constant whereas in the dynamic one the data keeps on changing like the geo location or navigation maps. Even Scrapy has a good potential to mine from dynamic web pages.

To compare performance between Scrapy and Selenium, it is obvious that both has individual features, but Selenium stays back of Scrapy to handle the large chunks of data. Even the multi-threading of Scrapy makes it faster. A website which has major JavaScript's depiction is better to be handled by Selenium as Scrapy suffers dependency issues to mine from those websites. Unlike Scrapy, Selenium is not a framework. It is a Python library, so it is super compatible to configure and use.

2.4.1.3 Beautiful Soup

A Python library to parse the unstructured web data to a formatted structured data. This package retrieves data out from the HTML or XML pages. Alone Beautiful Soup is not enough to access any website. It needs support exclusively from the request library of

```
from selenium import webdriver

path = '/Users/.../chromedriver' #path of your driver file
driver = webdriver.Chrome(path)
driver.get("http://www.example.com")

driver.find_element_by_xpath('//*[@id="accept-button"]').click() #click on a button
elements = driver.find_elements_by_class_name('class-example')
data = [element.text for element in elements]

driver.quit()
```

FIGURE 2.2 A snippet code of scraping data.

Python. The request library allows easy HTTP connections with its mighty functions. Beautiful Soup employs this library to reach the website.

The snippet in Fig. 2.3, represents a web scraping with Beautiful Soup [13]. The very next step after loading the required libraries (requests and Beautiful Soup from bs4 package) is to navigate to the website.

An object of the library is created defining the HTML parsing for the specified website. There are various in-built methods to perform the operations on the data. In this section, an implementation with Beautiful Soup is worked out for the readers. Hence, a detailed step-by-step guide will help the readers to scrape data from any HTML website from scratch. Beautiful Soup has a challenge in mining large data from any complex website and it cannot stand with any JavaScript supported website. This is a best fit method to collect data from small scale static web pages.

The readers after comparing the multiple tools to scrape web data can note few inferences. The choice of the scraping tool has different aspects like size of the data, design platform of website (HTML or JavaScript), complexity of the site, dependency to other libraries, etc. The best thing of all these three tools is that all are open sources, so the cost factor nullifies.

2.4.2   Importing required libraries

In this model walk through, all the codes are built on Jupyter environment with Python language [14].

To begin the task, the necessary libraries are to be imported in the Notebook.

import requests
from bs4 import BeautifulSoup

As discussed in 2.4.1.3, in this implementation the library Beautiful Soup will be used. And to use that, the import of requests library is also a prerequisite.

In this implementation, a real estate website is used to scrape the data from.

2.4.3   Accessing website data

On the website, there is a list of properties representing their individual price, features (age, architecture types, etc.) and property descriptions. The task is to capture all these

```
import requests
from bs4 import BeautifulSoup

website = requests.get('https://example.com').text
soup = BeautifulSoup(website, 'html.parser')

headlines = soup.find_all('span', class_='class-example')
data = [headline.text for headline in headlines]
```

FIGURE 2.3 A code snippet with Beautiful Soup.

information in a single json file or csv file format, which further becomes a potential dataset for any prediction of the properties. There can be many facets of possible cognitive analysis and prediction once the data is available. So, the reader will see how to collect the data from the website.

The request library has a function called get () which redirects to the website. In this code sample, a header is included as a parameter to this get (). There are certain occasions to extract data from the website which needs headers to give access to their data. Once the request object is created, request.text helps in getting the text data present in the entire website.

> *url="http://www.pythonhow.com/real-estate/rock-springs-wy/*
> *LCWYROCKSPRINGS/"*
> *req=requests.get (url, headers= {'User-Agent': 'XYZ/3.0'})*
> *content=req.text*
> *print(req)*

From the above code, the req returns the response type after communicating with the website. The response code 200 represents a successful response type from the site. Hence, the expectation is that content will return the entire HTML tagging structure of the website.

2.4.3.1 Output

```
<Response [200]>
'<!DOCTYPE html>\n<! - saved from url= (0110)http://web.archive.org/web/20160127020422/
http://www.century21.com/real-estate/rock-springs-wy/LCWYROCKSPRINGS->\n<html lang="en"
style="margin: 0px;overflow:hidden"><script async="" src="./LCWYROCKSPRINGS1_files/beacon.js">
</script>................ ............ .......<script src="http://web.archive.org/web/20151215220806/
http://chat.xtdirect.com/Chat/MasterServer/Public/Identify.php?packets=5:indentify:location,
http*XOMOMO!web%L0archive%L0org%MOweb%MO20160127020422%MOhttp*XOMOMO!www%L0century21%
L0com%MOreal%K0estate%MOrock%K0springs%K0wy%MOLCWYROCKSPRINGS."type="text/javascript"
id="PCJSF_CommScriptId"></script></div></body><div></div></html>\n'
```

Here the site's HTML content is captured.

But it is complicated to analyze something so raw as well as tedious to look through the entire tags to find out the wanted section of the site data.

2.4.4 Web crawling

Web crawling helps in finding the HTML tag in charge of holding the content of the page. In BeautifulSoup library, the function find_all () traverses to that specific tag. Before using the library, an object needs to be created for the soup. And while creating the object, the function takes care of parsing the entire content of the page.

> *#Creating object for BeautifulSoup*
> *soup=BeautifulSoup(content,'html.parser')*

#Finding the HTML div type tag with class id "propertyRow" to access the list of properties #data
all=soup.find_all("div",{"class":"propertyRow"})

This brings the complete content of the desired section of the website where the filter is only the div class "propertyRow". So, it will extract only the data included in the property row division.

2.4.5 Data extraction

Data extraction is all about purifying the primitive data to some meaningful information. In the discussed use case, this can be done by creating dictionaries of the respective features of each property. The data segregation happens here to send data into different categorical sections like property address, property price, locality, number of beds, area sqft., etc.

#Finding address
data["Address"]=item.find_all("span",{"class","propAddressCollapse"})[0].text
#Finding price
data["Price"]=item.find("h4",{"class","propPrice"}).text.replace("\n","").replace(" ","")
#Finding the count of bedrooms
data["Beds"]=item.find("span",{"class","infoBed"}).find("b").text
#Finding the sqft of the area
data["Area"]=item.find("span",{"class","infoSqFt"}).find("b").text
#Finding the number of full bathrooms
data["Full Baths"]=item.find("span",{"class","infoValueFullBath"}).find("b").text
#Finding the number of half bathrooms
data["Half Baths"]=item.find("span",{"class","infoValueHalfBath"}).find("b").text
#Printing the data
print(data)

2.4.5.1 Output
[{'Address': '0 Gateway',
      'Locality': 'Rock Springs, WY 82901',
      'Price': '$725,000',
      'Beds': None,
      'Area': None,
      'Full Baths': None,
      'Half Baths': None},
   {'Address': '1003 Winchester Blvd.',
      'Locality': 'Rock Springs, WY 82901',
      'Price': '$452,900',
      'Beds': '4',

'Area': None,
'Full Baths': '4',
'Half Baths': None,
'Lot Size': '0.21 Acres'},

.

.

.

{'Address': '19350 E SAGUARO Drive',
    'Locality': 'Black Canyon City, AZ 85324',
    'Price': '$28,995',
    'Beds': None,
    'Area': None,
    'Full Baths': None,
    'Half Baths': None,
    'Lot Size': '0.73 Acres'},
{'Address': '20650 E Amethyst Place',
    'Locality': 'Black Canyon City, AZ 85324',
    'Price': '$15,000',
    'Beds': None,
    'Area': None,
    'Full Baths': None,
    'Half Baths': None,
'Lot Size': '0.31 Acres'}]

The above code appended the entire information in a dictionary which is later exported as a json file. The output shows a list of entries of the property data. Each entry represents each property with its own details like *Address, Locality, Price,* etc. The json data obtained is in a ready structured format. But there are other formats of data to export as well (Fig. 2.4).

2.4.6 Data framing

In data framing, one prepares a well formatted dataset. In Python, there is a library called Pandas, which has a functionality called Data Frame [15]. In the below snippet of code, it is shown how a pandas Data Frame gives a perfect representation of the json data in a tabular format. Again, to use this library it needs to be imported in the notebook.

```
#Importing pandas library
import pandas.
#Converting the json data into a DataFrame
df = pandas.DataFrame(data)
#Printing the dataframe
print(df)
```

	Address	Locality	Price	Beds	Area	Full Baths	Half Baths	Lot Size
0	0 Gateway	Rock Springs, WY 82901	$725,000	None	None	None	None	NaN
1	1003 Winchester Blvd.	Rock Springs, WY 82901	$452,900	4	None	4	None	0.21 Acres
2	600 Talladega	Rock Springs, WY 82901	$396,900	5	3,154	3	None	NaN
3	3239 Spearhead Way	Rock Springs, WY 82901	$389,900	4	3,076	3	1	Under 1/2 Acre,
4	522 Emerald Street	Rock Springs, WY 82901	$254,000	3	1,172	3	None	Under 1/2 Acre,
5	1302 Veteran's Drive	Rock Springs, WY 82901	$252,900	4	1,932	2	None	0.27 Acres
6	1021 Cypress Cir	Rock Springs, WY 82901	$210,000	4	1,676	3	None	Under 1/2 Acre,
7	913 Madison Dr	Rock Springs, WY 82901	$209,000	3	1,344	2	None	Under 1/2 Acre,
8	1344 Teton Street	Rock Springs, WY 82901	$199,900	3	1,920	2	None	Under 1/2 Acre,
9	4 Minnies Lane	Rock Springs, WY 82901	$196,900	3	1,664	2	None	2.02 Acres
10	9339 Sd 26900	Rocksprings, TX 78880	$1,700,000	None	2,560	None	None	NaN
11	RR674P13 Hwy 377	Rocksprings, TX 78880	$1,100,000	None	2,000	None	None	NaN
12	0 Hwy 41	Rocksprings, TX 78880	$1,080,000	None	None	None	None	NaN
13	9339 Sd 26900	Rocksprings, TX 78880	$908,350	None	2,560	None	None	NaN
14	CR450 Hwy 377	Rocksprings, TX 78880	$905,000	None	None	None	None	NaN

FIGURE 2.4 Output- scraped data in a data frame format.

2.4.6.1   Output

This above data is the first 15 entries of property on the website with their respective details. Such formatting makes a comprehensive understanding of the data for further analysis. The next step is to perform an exploration of the dataset.

The pandas library also supports to export the data into different formats like csv, xml, or json. Here in the use case, the data is exported into a csv file name "PropertyDetails."

> *#Exporting Dataframe to CSV*
> **df.to_csv('PropertyDetails.csv', index=False)**

2.4.7   Stages of data transformation

During the architectural discussion of web scraping in Fig. 2.1, it was noted that the data captured from the websites go through different stages of transformation. Initially it is obtained in a raw structure which is hard to analyze. So, the second stage is to imbibe the required portion of the data from the entire site's information. In the third stage, the data gets its final shape and an appropriate structure.

Before concluding with the task, it is noteworthy to understand that while scraping data the end goal is to present it in a clean structured format. This enhances the readability of the data for future predictions on it.

2.5 Web scraping in reality

In this recent competitive era of technologies, the market has the highest demand of bringing the cutting-edge automations to reach to the common people. This makes life much convenient. To add value in this process, data plays one of the key roles.

In this section a very shallow walk will be done to show the readers the various fields where scraped data is getting utilized [16]. But to mention, these applications are just for some broader picture understanding. The real utilization of data cannot be confined into one section. The applications can be enlisted as:

Price Comparisons: This is one of the biggest target zones for the utilization of web scraping. The data collected from the e-commerce sites can be compared among the competitive platforms and can deliver into some capable analysis.

Customer Sentiment Analysis: A massive usage of scraping takes place in extracting customer feedback data and analyzing their sentiments for betterment of customer service.

Newspaper data: The world is busy to spare time in reading through the pages of newspaper. To automate the process of collecting useful data from the newspaper headlines and articles improves the finance and investment workflows.

Data Science: In real-time analytics like fraudulent credit card data, customer relationship management or financial decisions are the areas to inculcate the practice of automation using data extracted from the websites.

Some utilization is always worthy to note in predictive analysis like forecasting, risk managements and business-related decision makings.

Strategic Marketing: Data is the primary ingredient to reach target customers in business. There are marketing types which has a 100% dependency on business-related statistics. The data-driven approach of marketing and the content marketing are one of these highly flourishing strategies governing the business sectors.

2.6 Conclusion

To summarize on the complete idea of web scraping, the few key takeaways are hard to miss. At first, there comes the objective of scraping data from the website. The nature of the data, the prime motive to mine and the authorization to access it are included in the objective of the web scraping. The second functionality is to discover the pieces to combine to achieve the job. The discussed architectural diagram exposes how to crawl into the websites. HTML design fetches the appropriate data out from it. A parsing concept aids the process of understanding the web data before any further operations on it. Also, there are couple of frameworks that equips significant functionalities to extract, filter and represent data. Beautiful Soup being one of the simplest libraries to scrape data at initial level from static websites. Its function in Beautiful Soup like find_all () is powerful enough to refine the data from its entire bulk. Finally, the raw form of data is barely credible to perform any analysis. Consequently, the job is to structure the data. In Python, a library called Pandas is capable of decorating the data in an orderly format.

References

[1] J. Clement, Global Digital Population, 2020 [online] Available : https://www.statista.com/statistics/617136/digital-population-worldwide/.

[2] A. Namoun, A. Alshanqiti, E. Chamudi, M.A. Rahmon, Web design scraping: enabling factors, opportunities and research directions, in: 2020 12th International Conference on Information Technology and Electrical Engineering (ICITEE), 2020, pp. 104–109, https://doi.org/10.1109/ICITEE49829.2020.9271770.

[3] O. Castrillo-Fernández, W. Scraping, Applications and Tools", European Public Sector Information Platform Topic Report No. 2015/10, December, 2015.

[4] R. Diouf, E.N. Sarr, O. Sall, B. Birregah, M. Bousso, S.N. Mbaye, Web Scraping: State-of-the-Art and Areas of Application, 2019 IEEE International Conference on Big Data (Big Data), 2019, pp. 6040–6042, https://doi.org/10.1109/BigData47090.2019.9005594.

[5] V. Singrodia, A. Mitra, S. Paul, A review on web scrapping and its applications, in: 2019 International Conference on Computer Communication and Informatics (ICCCI), 2019, pp. 1–6, https://doi.org/10.1109/ICCCI.2019.8821809.

[6] The 5 Best Programming Languages for Web Scraping. [online] Available: https://prowebscraper.com/blog/best-programming-language-for-web-scraping/.

[7] K. Turk, S. Pastrana, B. Collier, A tight scrape: methodological approaches to cybercrime research data collection in adversarial environments, in: 2020 IEEE European Symposium on Security and Privacy Workshops (EuroS&PW), 2020, pp. 428–437, https://doi.org/10.1109/EuroSPW51379.2020.00064.

[8] James D. Ethics in Web Scraping. [online] Available: https://towardsdatascience.com/ethics-in-web-scraping-b96b18136f01.

[9] H. Wu, F. Liu, L. Zhao, Y. Shao, Data analysis and crawler application implementation based on Python, in: 2020 International Conference on Computer Network, Electronic and Automation (ICCNEA), 2020, pp. 389–393, https://doi.org/10.1109/ICCNEA50255.2020.00086.

[10] H. Yang, Design and implementation of data acquisition system based on scrapy technology, in: 2019 2nd International Conference on Safety Produce Informatization (IICSPI), 2019, pp. 417–420. https://doi.org/10.1109/IICSPI48186.2019.9096044.

[11] K. Varshney. Web Scraping Using Selenium. [online] Available: https://medium.com/@kapilvarshney/web-scraping-using-selenium-836de8677ae5.

[12] F. Andrade. Web Scraping with Beautiful Soup, Selenium or Scrapy?. [online] Available: https://gist.github.com/ifrankandrade/189e0f903773dff419f6cd39e130c906#file-web-scraping-libraries-py.

[13] T.S. Deshpande, S. Varghese, P.D. Kale, M.P. Atre, Incident classification, prediction of location and casualties, in: 2021 8th International Conference on Computing for Sustainable Global Development (INDIACom), 2021, pp. 778–781. https://doi.org/10.1109/INDIACom51348.2021.00139.

[14] A.P. Lorandi Medina, G.M. Ortigoza Capetillo, G.H. Saba, M.A.H. Pérez, P.J. García Ramírez, A simple way to bring Python to the classrooms, in: 2020 IEEE International Conference on Engineering Veracruz (ICEV), 2020, pp. 1–6. https://doi.org/10.1109/ICEV50249.2020.9289692.

[15] K. Willems. Pandas Tutorial: DataFrames in Python. [online] Available: https://www.datacamp.com/community/tutorials/pandas-tutorial-dataframe-python.

[16] H. Patel. How Web Scraping is Transforming the World with its Applications. [online] Available: https://towardsdatascience.com/https-medium-com-hiren787-patel-web-scraping-applications-a6f370d316f4.

Analysis of Covid-19 using machine learning techniques

Dyna Marneni[1,2], Sridhar Vemula[1,3]

[1]DEPARTMENT OF COMPUTER SCIENCE AND ENGINEERING, MATURI VENKATA SUBBA RAO
ENGINEERING COLLEGE, HYDERABAD, TELANGANA, INDIA; [2]RESEARCH SCHOLAR, JNTUH,
HYDERABAD, INDIA; [3]RESEARCH SCHOLAR, OSMANIA UNIVERSITY, HYDERABAD, INDIA

3.1 Introduction

SARS-CoV-2 is a severe acute respiratory syndrome that causes coronavirus infection [1] (Covid-19), and was initially diagnosed in December 2019 in the city of Wuhan, China, and later affected many states in China. The first Covid-19 positive case in India was reported in the state of Kerala on January 30th, 2020. Subsequently, there has been a significant rise [2] in the number of Covid-19 positive cases. Covid-19 is caused by coronavirus. It affects the respiratory tract and has caused major disease outbreaks worldwide. So far, there is no specific medicine known to treat Covid-2019. Humans do not have resistance to this virus and are susceptible to get affected easily through contact with a Covid-2019 infected [1] person. The diseases caused by these kind of viruses could result in serious health issues globally. Persistent health issues later could result in causing death and have an undesirable and devastating effect on social and economic sectors. The disruption of such viruses spread across many regions and would consequently affect the Indian economy [2]. This would compel researchers to handle the pandemic by including statistical modeling and its associated analysis in their study.

The most common characteristics of Covid-2019 virus are dry cough, high fever, sore throat, loss of smell or taste, diarrhea, fatigue, difficulty in breathing, and if the symptoms worsen with respiratory trouble, it is life threatening. As the characteristics of the Covid-2019 virus are related to any other flu symptoms, a person could misinterpret their symptoms and be infected [1] with the Covid-19 virus. In order to identify if a person has been infected with the virus and to further avoid serious complications a thorough analysis of the most significant symptoms of the disease needs to be done. It could be used as an early warning indicator. The key objective of this chapter is to identify which symptoms together influence the occurrence of Covid-19 [3] by using the multiple linear regression model. Also the effect of each symptom is evaluated using a correlation coefficient. We make use of a sample data set [4] for this purpose. For this to be of any importance the sample data must be a true reflection of the population data on which the analysis is to be made. How can we be certain that this is actually the case? Further if the correlation between the symptoms is just by a chance? Such questions are answered in this paper by hypothesis testing. To determine which symptoms are significant to detect the presence of Covid-19, feature selection is used. Measures used to evaluate the model in feature selection is Residual Sum of Squares (RSS) and coefficient of determination (R2 score).

3.2 Literature survey

In this survey we study about the observations and analysis made by authors from different journals related to Covid-2019. Scientists and other contributors have made valuable insights about Covid-2019 and its effect on health and socioeconomic factors globally. From the paper [5], the authors used the different machine learning algorithms like Long Short Term Memory, Decision Tree, and Convolution Neural Network algorithms to estimate the number of confirmed cases, number of deceased, and those who have recovered from Covid-2019. The paper presents an analysis on different data sets based on measures like Mean-squared error and coefficient of determination, R2. The analysis and observations on the said algorithms showed that when CNN algorithm was used the model's performance was best with R2 score of 0.99. Also if the same model is used for pandemic the results could be used to save several lives. According to the authors when the same numerical analysis has been performed in India it was found that about 90 cases would be prevented if the isolation for the outbreak would have been implemented 10 days before rather than on March 25, 2020. The paper concluded by saying that there would be a markdown of 4% in rise of Covid-2019 positive cases if the model is used.

From the paper [6] published in the journal of Emerging trends in Engineering research made a study on Covid-19 [3] data sets to analyze which age group is mostly affected with the disease. Performances are computed here using machine learning algorithms like random forest regressor, Decision Tree Classifier, Gaussian Classifier, and Logistic Regression. The results showed that the people in the category of age 20–30, 30–40, and 40–50 predominantly tested positive for Covid-2019. In this paper, Correlation coefficient is used to understand how the various characteristics of the data set are related to each other and also their influence on the output. The results also showed that Random Forest Classifier and Random Forest Regressor performed well compared to other algorithms in terms of R2 score and Accuracy.

In this paper [7], the authors proposed a Naïve Bayes' classification strategy based on feature correlation to detect new cases of Covid-19 [3] with high speed. There are four stages in the proposed strategy. They are: Selection of Features stage, Clustering of Features stage, Feature Weighting stage, and Correlated Features Naïve Bayes stage (FCNB). FCNB uses feature selection method to choose the most predominantly occurring features of Covid-2019 by using a genetic algorithm to group them into clusters. Each cluster consists of a set of features that are interrelated to each other. In the master feature weighting phase, each Master Feature is assigned a weight based on the importance of characteristics and its correlation. In the FCNB stage patients are labeled by making use of the Naïve Bayes weighted algorithm. In Classification stage, weighted Navies' has been implemented on weights of master features for rapid and accurate interpretations. Ninety-nine percent detection accuracy is achieved through the results of feature-correlated Naïve Bayes stage.

3.3 Study of algorithms

In this section we are using the following machine learning algorithms to analyze the significance of characteristics of Covid-19 [3] and also performance measures like R2 score to check the accuracy of the models.

3.3.1 Multiple regression

Multiple regression is an extended technique in machine learning that makes use of various dependent variables and a constant term to predict [1] the output of a target variable. It is an extension of simple regression model that can be applied when there are more than two explanatory variables. The aim of multiple regression is to find out the association between the various independent variables and the target variable. In this paper we show how the various Covid-19 symptoms influence the output variable, presence of Covid-19. We also build a regression model using Python.

$$Y = \beta_0 + \beta_1 X_1 + \ldots \ldots \ldots + \beta_n X_n + \varepsilon \tag{3.1}$$

Mathematically, in Eq. (3.1) Y is the response variable that we need to estimate. All the X terms are the respective input variables. β_0 is called the intercept value which is equal to the value of Y when all other input variables, $X_1, X_2 \ldots X_n$ terms are 0. Using the regression model, we need to find out the values of these constants (β) by reducing the error function and identify the line that best fits the data. This can be attained by minimizing the Residual Sum of Squares (RSS), which is equal to the sum of squares of the difference between actual output and the observed output as shown in Eq. (3.2).

$$RSS = \sum_{i}^{n} (y_i - \hat{y}_i)^2 \tag{3.2}$$

3.3.1.1 Ordinary least squares (OLS)

In linear regression, we need to find a boundary that minimizes the RSS and fits the data well. This can be achieved through the method of Least Squares. In this method, we draw a line covering all the data points and find out the distance of each point from the line and square the distance and finally sum all the distances. The best fit line is the one which includes all the data points provided the distance of each point from the line is minimum. OLS procedure tries to reduce the RSS value so that the error is also minimum and we can find best fit line to the data.

3.3.1.2   Multiple regression analysis has two main uses
- An Association between the independent variables can be identified through multiple linear regression by using correlation coefficient. In this paper we check the correlation of different symptoms of Covid-19 by finding coefficient matrix.
- An observation of how the independent variables effect the dependent variable can be achieved through multiple linear regression.

3.3.1.3   Collinearity
When more than one input variables are linearly related the association is referred to as collinearity. We have to validate which predictor features correlate to target. We can perform this by calculating the correlation coefficients for each feature against the predictor. We use correlation matrix heat maps as a guide on which features to include. The correlation coefficient values range from -1 to $+1$. If the value of the coefficient is greater than zero then it indicates a positive relationship. Conversely, if the value of the coefficient is less than zero then it indicates a negative relationship and if the value of the coefficient is zero then it indicates that there is no association between the two variables.

3.3.1.4   Hypothesis test
One of the foremost questions that should be resolved while performing Multiple Regression is that, if at least one of the dependent variables is needed in estimating the output then we need to execute a Hypothesis Test to answer this question and confirm our assumptions. It all starts by forming a **Null Hypothesis H_0,** which states that all the coefficients are equal to zero, and there's no effect or association among the explanatory variables and the output variable. This means that none of the independent variables are useful in determining the output variable. The null hypothesis is represented in Eq. (3.3).

$$H_0: \beta_1 = \beta_2 = \beta_3 = 0 \tag{3.3}$$

We also need to define an **Alternative Hypothesis, H_A** which states that at least one of the coefficients is not zero, and there is an effect or significance of the explanatory variables in estimating the target variable. This means that at least one of the independent variables is useful in explaining the target variable. The alternative hypothesis is represented as follows:

$$H_A: \text{at least one } \beta_i \text{ is non} - \text{zero.}$$

If we want to reject the Null Hypothesis, H_0 and be certain that our regression model best fits the data then we need to find a strong statistical evidence. To do this we perform a hypothesis test with the help of **F-statistic** method.

The formula for this method contains the sum of squared estimate of errors and the Sum of Squares. The Statsmodels [8] package in python makes use of OLS approach to find such statistics. F-test can check the significance of multiple coefficients simultaneously. From the test, if the observed value is equal to or approximately one then we cannot reject the Null Hypothesis. If the value of F-statistic is much higher than the null hypothesis will be rejected and we can be certain that at least one or more coefficients are not equal to zero.

3.3.1.5 Feature selection

Running a Multiple Regression model with more than one variable including the insignificant ones will lead to an undesirable complex model. Which of the variables actually effect the target variable? Is there any significance of using those various variables to our model? In order to understand that, we need to execute a method called **Feature Selection**. There are two approaches for feature selection, one is forward selection and the other is backward feature selection. In this paper we use Forward Selection.

Forward feature selection: Here the independent variables are added one at a time beginning with the one with the highest correlation with the target variable. We use RSS and R2 score as measures to check which predictors are to be included in the model. The feature with the smallest value of RSS is the best performer. OLS module in stats package helps in calculating the value of RSS. We then add one more input variable to it and find out the RSS value for the combined variables. The process is repeated until we find out the lowest value of RSS for the combined variables and those variables are incorporated to the model sequentially.

3.3.1.6 Coefficient of determination

R^2 is the measure which is used to understand the variance in data caused by its association between the dependent and the independent variables. This correlation gives information about the goodness of a fit of the model. Fitness of a model measures the distance between all the data points to the regression line. The closest distance from the line to the data points is considered as best fitted. The model is said to be fit by a measure of the value of R^2.

Mathematically, it is calculated as the square of the correlation between the actual output and the estimated output. A value of R^2 as represented in Eq. (3.4) is approximately equal to one indicates that the model is suitable for the data set and a value of R^2 approximately equal to zero shows that the model is not suitable for the data set.

$$R^2 = Corr(y, \hat{y})^2 \tag{3.4}$$

By using these measures, we can check the credibility of association of symptoms that cause Covid-19.

3.3.2 Logistic regression

Logistic Regression is a simple machine learning technique to implement binary classification. Logistic regression [9] describes and estimates the relationship between one

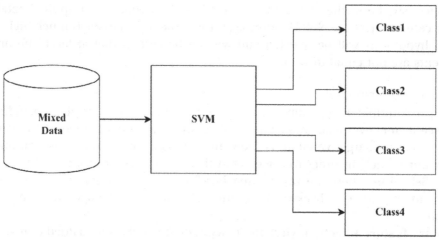

FIGURE 3.1 SVM classification.

dependent binary variable and independent variables. In this paper we make an analysis to determine the effect of symptoms of Covid-19 on the target variable.

3.3.3 Support vector machine

Support vector machine (SVM) is a supervised learning algorithm which is used for classification and regression problems. It is an effective classifier that can be used to solve linear problems. SVM also supports kernel methods to handle nonlinearity. Given a training data, the idea of SVM is that the algorithm creates a line or a hyper plane which segregates the data points into classes. Diagrammatic representation of SVM classification is shown in Fig. 3.1.

From Fig. 3.1, we can find that SVM accepts mixed data, which consists of both linear and nonlinear data and then separates the data into class1, class2, class3, and class4.

Thus SVM creates a decision boundary which ensures that the separation between the classes is maximum. It tries to maximize the separation of the data set and the decision surface.

3.3.4 Building a decision tree

A decision tree follows a tree structure where an internal node represents a feature, the branch represents a condition check by applying a rule, and an output is produced at the end of every child node. The uppermost node in a decision tree is termed as the root node. The idea is to construct a tree for the entire data set and interpret the outcome at every leaf node. It learns to split depending on the feature value. It splits the tree in a recursive pattern call recursive partitioning.

The steps involved behind any decision tree algorithm is as follows:

1. Start with a feature in the data set and make a split based on that feature using Gini index or gain ratio.
2. Make that feature a decision node and split the data set into smaller subsets.
3. Continue splitting based on the features by repeating this process recursively for each leaf node until one of the condition match:
    (a) Leaf nodes are pure such that only one class remains
    (b) A maximum depth is reached such that there are no more leaf nodes.
    (c) A performance metric is reached.

Diagrammatic representation for Decision tree generation is shown in Fig. 3.2.

3.3.5   Random forest regressor

A Random Forest is technique that is used for classification and regression tasks. Random forest as the name suggests, is a model that combines multiple decision trees to find out the output. It makes use of a technique called Bootstrap and Aggregation, commonly known as bagging. The idea is to create several trees and the final prediction is obtained by finding the average of the individual outcomes at each tree or by taking the maximum vote-cast method. Mean of all the outcomes is applied in a regression problem. Maximum of all the votes is applied in a classification problem.

In this paper, we make use of symptoms of Covid-19 as the data set [10] and create a Machine learning predictor or Regressor by fitting the Random Forest Regressor to the data set. We train the model with the test data and perform prediction. Also performance is checked using score method in python.

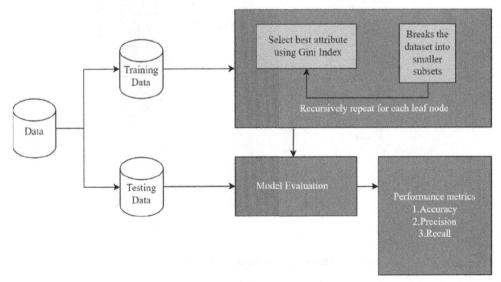

FIGURE 3.2 Decision tree generation.

3.4 Experimental analysis and results

In this paper, we have used SVM, Random Forest, Decision Tree, and Linear Regression models from Python's sklearn library. The Covid-19 symptoms data set is run through all the models said. Performance assessment [11] is also made for all the models using the data set.

3.4.1 Analysis for multiple linear regression [9]

Fig. 3.3 shows correlation matrix for the features of the Covid-19 symptoms dataset. From the figure, we can see that highest correlation is with the High fever symptom and with Difficulty in breathing symptom with the target variable, Infected with Covid-19 (Fig. 3.4). The same analysis is represented through Table 3.1 also.

3.4.1.1 Select the input variable and output variable
Next, we split the features into two categories: output variable or the target variable "y" and independent or input variables "x."

3.4.1.2 Splitting the data set
In this phase we divide the data set into a training set and a test set. By splitting the given data into two separate sets, we can apply training one set and testing on the other set. In this way the model performance on unseen data is performed.

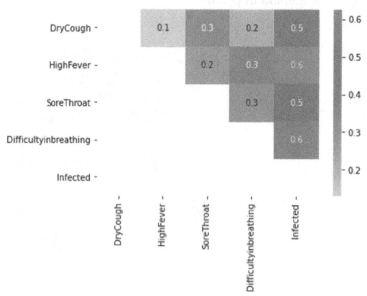

FIGURE 3.3 Correlation heat map.

Table 3.1 Correlation matrix representing correlation coefficient of each symptom against the target variable.

	Dry cough	High fever	Sore throat	Difficulty in breathing	Infected
Dry cough	1.000000	0.129988	0.304869	0.193098	0.539926
High fever	0.129988	1.000000	0.249360	0.329848	0.625718
Sore throat	0.304869	0.249360	1.000000	0.278639	0.490831
Difficulty in breathing	0.193098	0.329848	0.278639	1.000000	0.590464
Infected	0.539926	0.625718	0.490831	0.590464	1.000000

3.4.1.3 Build and train the predictor

Running your predictor to the training data serves as the training part. After the model is trained, it can be used to make predictions on unseen data. The unseen data is the test data set. Test data set is independent of the training data set which will help you have a better view of its ability to generalize.

3.4.1.4 Evaluate performance

The quality of a model is assessed by how well the estimated values match the actual output values. The results are as follows:

Mean Absolute Error: 0.2517655018291291.

Mean-Squared Error: 0.07928848787143285.

Root Mean-Squared Error: 0.2815821156810795.

R Squared Score is: 0.6814301826594216.

From the observations, $R^2 = 0.68$, which is marginally less. Higher value of R^2 indicates that the model is more suitable for the data set. We can improve the regression model by some ways which will be discussed in the subsequent sections.

3.4.1.5 Results of the hypothesis test

We perform a hypothesis test with the help of **F-Statistic** method. This statistic method makes use of RSS and is implemented through the Statsmodels package. The OLS model consists of the short description of all the statistics which is shown Fig. 3.5.

3.4.1.6 Understanding the outputs of the model from the above OLS regression results: is this statistically significant?

3.4.1.6.1 What is the significance of p-value?

P-value is a statistical method that is used to check if the null hypothesis is true or false. It helps to understand if our data has occurred by chance or not. The statistical significance is represented by the value of *P* which lies between 0 and 1. The smaller the *P*-value is, the better is the chance of rejecting the null hypothesis. If the value of *P* is less than .05 then it means that there is only 5% probability of the null hypothesis being correct. So if the value of *P* is less than .05 then it is considered as significant to the model.

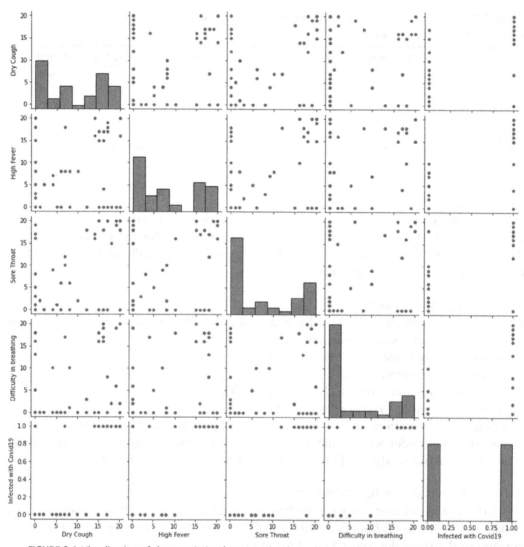

FIGURE 3.4 Visualization of the association between the dependent variable and independent variables.

From the OLS Regression results shown in figure 3.4.1.6, the four features of the data set representing the input variables have some P-values indicating their association with the target variable. We observe that the symptom, Sore throat has a P-value of .261 and can be considered as statistically insignificant. So we can eliminate that feature. This way we can check the significance of each input variable on the response variable with the help of P-value and further consider only those variables of significance for the prediction.

Dep. Variable:	y	R-squared:	0.765
Model:	OLS	Adj. R-squared:	0.731
Method:	Least Squares	F-statistic:	22.78
Date:	Wed, 25 Aug 2021	Prob (F-statistic):	1.84e-08
Time:	14:03:24	Log-Likelihood:	-0.045065
No. Observations:	33	AIC:	10.09
Df Residuals:	28	BIC:	17.57
Df Model:	4		
Covariance Type:	nonrobust		

	coef	std err	t	P>|t|	[0.025	0.975]
const	-0.1129	0.083	-1.361	0.184	-0.283	0.057
Dry Cough	0.0216	0.006	3.435	0.002	0.009	0.035
High Fever	0.0285	0.007	3.917	0.001	0.014	0.043
Sore Throat	0.0079	0.007	1.148	0.261	-0.006	0.022
Difficulty in breathing	0.0209	0.008	2.763	0.010	0.005	0.036

Omnibus:	0.504	Durbin-Watson:	1.566
Prob(Omnibus):	0.777	Jarque-Bera (JB):	0.574
Skew:	0.259	Prob(JB):	0.751
Kurtosis:	2.614	Cond. No.	34.2

FIGURE 3.5 Ols regression results.

3.4.1.6.2   R squared and adjusted R squared?

R squared is a metric that is used to show the variance which is explained by the model. As the number of input variables increase, the value of R square increases and the higher R squared value indicates that data fits well to the model. Adjusted R squared considers only statistically significant variables. So the value of adjusted R squared value is the result that is obtained after the addition of significant variables to the model. So it is recommended to consider adjusted squared value for good model. From the OLS Regression results, 73.1% of the variance can be explained by my model, which is considered as a good value.

3.4.1.6.3 Checking for errors

While building models, studying them and deciding which one is better is an essential step. We need to perform several tests and then analyze the summaries. Sometimes we need to eliminate some variables, sum or multiply them and perform retests. After completing the series of trials and, we need to check the *P*-values, errors that arise from the experiments and performance measures like the value of R square. The model is considered as best fit when the *P*-values are smaller than .05, a higher R2 score and comparatively small errors.

3.4.1.7 *Forward selection*

The predictor with highest correlation is chosen. Then the one by one predictors are added to check the RSS and R2 score. We use RSS and R2 score as measures to check which predictors are to be included in the model. The feature with the smallest value of RSS is the best performer. We check R2 score and RSS for each symptom with the target variable, the results are as shown in Table 3.2.

From Table 3.2 we understand that the value of RSS is least and the value of R^2 is the highest for the variable High Fever. Hence we select High Fever as the first predictor to move forward.

3.4.1.8 *Visualization plots for each symptom against the target variable*

3.4.1.8.1 Difficulty in breathing symptom versus infected

From the heatmap diagram as shown in Fig. 3.6 we can see that the infected is indicated by a value of one and not infected by a value of 0. Also we can visualize from Fig. 3.6 the count of people who have difficulty in breathing for the number of days. For example we can understand that about two persons suffer from difficulty in breathing for about 20 days.

3.4.1.8.2 High fever symptom versus infected

From the heatmap diagram as shown in Fig. 3.7 we can see that the infected is indicated by a value of 1 and not infected by a value of 0. Also we can visualize from Fig. 3.7 the count of people who have High fever for the number of days. For example we can understand that about two persons suffer from High fever for about 10 days.

Table 3.2 R2 score and RSS for every symptom.

Symptom	R2 score	RSS
Dry cough	0.291	8.501
High fever	0.391	7.301
Sore throat	0.24	9.109
Difficulty in breathing	0.348	7.816

3.4.1.8.3   Dry Cough versus infected

From the heatmap diagram as shown in Fig. 3.8 we can see that the infected is indicated by a value of 1 and not infected by a value of 0. Also we can visualize from Fig. 3.8 the count of people who have DryCough for the number of days. For example we can understand that about four persons suffer from DryCough for about 20 days.

3.4.1.8.4   Sore throat versus infected

From the heatmap diagram as shown in Fig. 3.9 we can see that the infected is indicated by a value of 1 and not infected by a value of 0. Also we can visualize from Fig. 3.9 the count of people who have Sore throat for the number of days. For example we can understand that about four persons suffer from Sore throat for about 20 days.

Next we add two symptoms to check the R2 score and RSS score. We are going to use R2 and RSS are used as metrics to evaluate the model after the addition of each symptom. The results are given in Table 3.3.

We observe that for Dry cough and High fever combination, RSS is least and R2 score is highest, which stands as a best case. Next we check for three symptoms by adding one by one which is shown in Table 3.4.

We observe that for the combination of symptoms, Dry Cough, High Fever and Difficulty in breathing RSS is least and R2 score highest. So these symptoms are chosen as predictors for the occurrence of the disease Covid-19 infected.

3.4.2   Evaluation of machine learning algorithms

A relative study of the SVM, Decision Tree, Logistic Regression, Random Forest and Linear Regression is made by R2 score which is calculated by importing r2_score from sklearn.metrics in Python. The results are as shown in Table 3.5.

From the above scores, it is evident that Logistic Regression, SVM and Decision Tree has outperformed Random Forest and Linear Regression models.

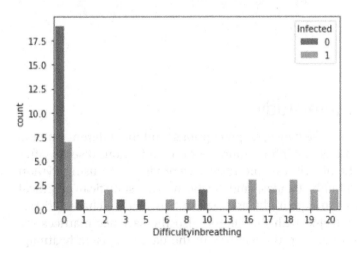

FIGURE 3.6 Difficulty in breathing symptom versus infected.

FIGURE 3.7 High fever symptom versus infected.

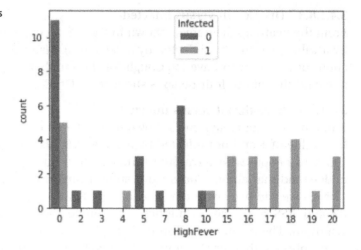

FIGURE 3.8 Dry cough versus infected.

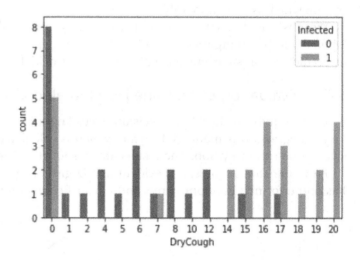

3.5 Conclusion and future study

In this paper, we strived to contrive the concepts, perceptions, and the different variants from the data sets of Covid-19 and global information about the infectious disease. After the execution of the data on the algorithms, the results were visualized using Python libraries such as Numpy, Matplotlib, Pandas, and Seaborn. The statistical graphical representations such as pair plots to explain the association between the characteristics of Covid-2019 and the person effected with the illness. Also the variance matrices are built to interpret the association between the features of the data sets using heatmap.

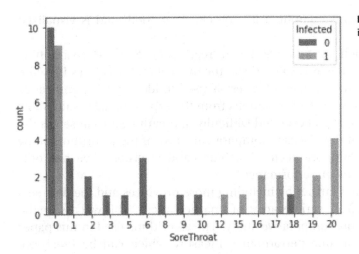

FIGURE 3.9 Sore throat versus infected.

Table 3.3   R2 score and RSS for two symptoms.

Symptom	R2 score	RSS
Dry cough + high fever	0.605	4.734
High fever + sore throat	0.511	5.867
High fever + difficulty in breathing	0.557	5.315

Table 3.4   R2 score and RSS for three symptoms.

Symptom	R2 score	RSS
Dry cough + high fever + difficulty in breathing	0.718	3.372
Dry cough + high fever + sore throat	0.655	4.134
High fever + sore throat + difficulty in breathing	0.627	4.475

Table 3.5   R2 score.

	Model	Score
1	Logistic regression	100.000000
0	Support vector machines	93.333333
4	Decision tree	93.333333
2	Random forest	87.695339
3	Linear regression	68.143018

The significance of the features is computed for the classifiers built and is visualized using barplots.

The prediction models such as Multiple linear regression, SVM, Decision tree, Random Forest, Logistic Regression are evaluated on the basis of metrics like as R2 score and Mean-Squared Error. Forward Selection Process is used to identify the significance of predictor to the models applied. The observations from the experimental results show that the features like Dry Cough, High Fever and Difficulty in breathing are chosen as the predictors for the occurrence of the disease. Comparative study of the algorithms show that the Logistic Regression, Support Vector Machines and Decision Tree has out-performed other models in terms of performance.

The research can be extended further by including more provinces and the data sets of positive cases of the disease from various areas of a particular city. This will assure the precision of the evaluation of this comparative study. The algorithms used in this paper can be evaluated on more than one performance measure which can be Precision, Recall, Area under curve, variance, etc. In future, more ML classifiers and regressors such as GBM, Catboost, XGBoost, and ARIMA can be evaluated on the evolving Covid-19 to broaden our comparative study.

References

[1] L.J. Muhammad, E.A. Algehyne, S.S. Usman, A. Ahmad, C. Chakraborty, I.A. Mohammed, Supervised machine learning models for prediction of COVID-19 infection using epidemiology dataset, SN Comput. Sci. 2 (1) (2021) 1−13.

[2] S.U. Kumar, et al., The rise and impact of Covid-19 in India, Front. Med. 7 (2020) 250.

[3] M. Jamshidi, A. Lalbakhsh, J. Talla, Z. Peroutka, F. Hadjilooei, P. Lalbakhsh, W. Mohyuddin, Artificial intelligence and COVID-19: deep learning approaches for diagnosis and treatment, IEEE Access 8 (2020) 109581−109595.

[4] Covid-19 datasets. https://www.kaggle.com/prakharsrivastava01/Covid19-symptoms-dataset.

[5] A. Kunjir, D. Joshi, R. Chadha, T. Wadiwala, V. Trikha, A comparative study of predictive machine learning algorithms for COVID-19 trends and analysis, in: 2020 IEEE International Conference on Systems, Man, and Cybernetics (SMC), IEEE, October 2020, pp. 3407−3412.

[6] K.B. Prakash, S.S. Imambi, M. Ismail, T.P. Kumar, Y.N. Pawan, Analysis, prediction and evaluation of covid-19 datasets using machine learning algorithms, Int. J. 8 (5) (2020).

[7] N.A. Mansour, A.I. Saleh, M. Badawy, H.A. Ali, Accurate detection of covid-19 patients based on feature correlated Naïve Bayes (FCNB) classification strategy, J. Ambient Intell. Humaniz. Comput. (2021) 1−33.

[8] https://www.statsmodels.org/Statsmodels python package.

[9] E. Gambhir, R. Jain, A. Gupta, U. Tomer, Regression analysis of COVID-19 using machine learning algorithms, in: 2020 International Conference on Smart Electronics and Communication (ICOSEC), IEEE, September 2020, pp. 65−71.

[10] F. Rustam, A.A. Reshi, A. Mehmood, S. Ullah, B.W. On, W. Aslam, G.S. Choi, COVID-19 future forecasting using supervised machine learning models, IEEE Access 8 (2020) 101489−101499.

[11] E. Casiraghi, D. Malchiodi, G. Trucco, M. Frasca, L. Cappelletti, T. Fontana, G. Valentini, Explainable machine learning for early assessment of COVID-19 risk prediction in emergency departments, IEEE Access 8 (2020) 196299–196325.

Further reading

[1] Coronavirus disease (Covid-19): How is it transmitted? https://www.who.int/news-room/q-a-detail/coronavirus-disease-Covid-19-how-is-it-transmitted. December 13, 2020 | Q&A.

[2] Y. Shen, D. Guo, F. Long, L.A. Mateos, H. Ding, Z. Xiu, H. Tan, Robots under COVID-19 pandemic: a comprehensive survey, IEEE Access 9 (2020) 1590–1615.

[3] M.U. Rehman, A. Shafique, S. Khalid, M. Driss, S. Rubaiee, Future forecasting of COVID-19: a supervised learning approach, Sensors 21 (10) (2021) 3322.

[4] M. Satu, K.C. Howlader, M. Mahmud, M.S. Kaiser, S.M. Shariful Islam, J.M. Quinn, M.A. Moni, Short-term prediction of COVID-19 cases using machine learning models, Appl. Sci. 11 (9) (2021) 4266.

4

Discriminative dictionary learning based on statistical methods

G. Madhuri, Atul Negi

OCR LAB, SCIS, UNIVERSITY OF HYDERABAD, HYDERABAD, TELANGANA, INDIA

4.1 Introduction

Due to immense increase in social media, digital business practices, etc., data created, captured, copied, or consumed went from 1.2 trillion GB to 59 trillion GB (2010–2020) (Source: Forbes.com, "54 Predictions About the State of Data in 2021," Gil Press- Forbes). Hence there is a great requirement for faster and efficient methods to categorize or classify data for search or retrieval. In an abstract sense, these methods are well known in literature and are called pattern classification methods. Pattern classification involves efficient representation of data as d—dimensional feature vectors, designing a discriminant function with classification error as a criterion to decide the class membership of a new data vector. Statistical decision theory has been used historically to define the decision boundaries of pattern classes.

4.1.1 Regularization and dimension reduction

When the sample size is small compared to the number of variables, any model trained on such data could be overfitted, i.e., the classification rule learns parameters, noise in the data, and hence cannot classify new samples correctly. Regularization is a method to reduce the complexity of a model by decreasing the importance of some variables to zero. Retaining relevant features which have variance in the data and dropping features with high correlation or low variance results in reduced dimensionality. Principal Component Analysis (PCA) and Independent Component Analysis (ICA) are applied to attain reduced dimensionality. Nonnegative Matrix Factorization (NMF) has been used for dimension reduction in Ref. [1]. According to Ref. [2], to improve the robustness of a classifier in case of few training samples of high dimensionality, features for discrimination and other design parameters such as window size used in Parzen windows approach, number of features used in decision rule, number of neighbors in k-NN method, etc. have to be carefully selected. With the rise in online and mobile applications, a mathematically sound model which replaces handcrafted feature extraction and is capable of working with limited computational resources and few training samples is of interest.

4.1.2   Sparse representation (SR)

SR has its roots in compressed sensing. Olshausen and Field [3] proposed that the sparse representation model is similar to the receptive field properties of sensory cells in the mammalian visual cortex. Field [4] has applied log-Gabor filters on images and the histograms of the resultant output distributions have high kurtosis indicating sparse structure. Field proposed "a high kurtosis signifies that a large proportion of the sensory cells is inactive (low variance) with a small proportion of the cells describing the contents of the image (high variance) being active." These works support the idea of sparse representation of natural images.

With rigorous proofs and with proven error bounds, sparse representation is a viable model for constrained resource-based applications. SR model finds a low dimensional subspace to embed the given high-dimensional signals. This embedding is performed against a fixed basis matrix called *dictionary*. If the dictionary is perfect for the given set of signals, then the input signal or image can be represented with very few columns of the dictionary, with corresponding very few coefficients.

Section 4.2 describes the notation used throughout the article, Section 4.3 gives an account of sparse coding methods based on l_0, l_1 optimizations and statistical modeling based sparse coding methods. Section 4.4 describes the differences between orthogonal, undercomplete and nonorthogonal, overcomplete dictionaries. The similarities and differences between dictionary learning (DL) and other subspace learning methods are also discussed in the same section. Section 4.5 gives a review of statistical methods used in the design of discriminative dictionaries in a variety of applications like MRI data classification, surgeon classification, and level of skill identification based on surgical trial data, histogram feature-based supervised DL for face recognition, etc. Section 4.7 reports usage of CNN-based DL for content and style separation in images and generation of a new set of images using sparse coding-based convolutional neural network (CNN) and convolutional dictionary learning. Results of using a hybrid dictionary learning method to classify high-dimensional data using a simple multilayer perceptron which is a nonparametric statistical approach are also discussed here. Section 4.8 concludes. The categorization among various sparse coding algorithms and DL algorithms are depicted in Fig. 4.1.

4.2 Notation

In this section we introduce our notation. Throughout the article, A denotes a matrix, a denotes a vector, $\|A\|_F = \sqrt{\sum_{i=1}^{m}\sum_{j=1}^{n}|a_{ij}|^2}$ denotes the Frobenius norm of matrix A. A*

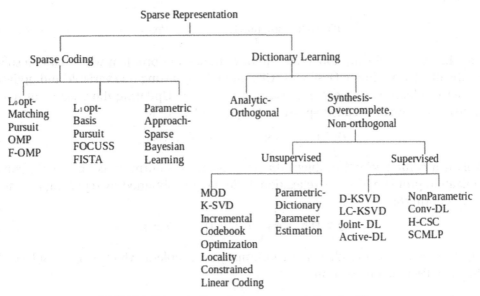

FIGURE 4.1 Categorization of sparse representation algorithms.

denotes complex conjugate transpose (Hermitian Conjugate) of A and $A^\dagger$ denotes Moore–Penrose pseudoinverse of A. For a given set of N training patterns, $Y \in R^{d \times n}$ with dimensionality d, the SR model finds a representation of $Y, Y \approx DX$ subject to $\|x_i\|_0 \leq \mathbf{T}$ $\forall \mathbf{i}$, where $X \in R^{K \times N}$ is the Coefficient Matrix for input signals. $D \in R^{d \times K}$ has K << N columns $\{d_k\}_{k=1}^k$, called atoms and the matrix is called dictionary. To get nontrivial solutions to this problem, the dictionary atoms are constrained to have $\|d_k\|_2 \leq 1, \forall k. \|.\|_0$ is the pseudo l_0— norm which denotes the number of nonzero components of a vector.

l_p — norm of a vector $\mathbf{v}$ is defined as $\|v\|_p = \left(\sum_i |v_i|^p\right)^{(1/p)}$. $Y \approx DX$ is

$$[y1 \quad y2 \quad \cdots yn] \approx [d_1 \quad d_2 \quad \cdots d_K] \begin{bmatrix} x_{11} & x_{12} & \cdots x_{1N} \\ x_{21} & x_{22} & \cdots x_{2N} \\ \vdots & \vdots & \vdots \\ x_{K1} & x_{K2} & \cdots x_{KN} \end{bmatrix}$$

where $y_N \approx x_{1N}d_1 + x_{2N}d_2 + \ldots + x_{KN}d_K$.

Each x_{kN} is the weightage given to the dictionary atom d_k in the representation of Nth training pattern. The problem is formulated as an optimization problem in Eq. (4.1)

$$Q(Y, D, X) = \arg\min_{D,X} \|Y - DX\|_F^2 \tag{4.1}$$

subject to $\|x_i\|_0 \leq T$, $\forall i$. Using the Lagrange multiplier method, Eq. (4.1) becomes Eq. (4.2).

$$Q(Y, D, X) = \arg \min_{D,X} \|Y - DX\|_F^2 + \lambda \|x_i\|_0 \tag{4.2}$$

Eq. (4.2) is a nonlinear, nonconvex, joint optimization problem which can be solved using the Block Coordinate Descent (BCD) method [5]. Fixing one variable and updating the other results in two linear optimization problems. Updating the coefficient matrix w.r.t. fixed dictionary is called sparse coding given by Eq. (4.3)

$$Q(Y, D, X) = \arg \min_{X} \|Y - DX\|_F^2 + \lambda \|x_i\|_0 \tag{4.3}$$

This is a combinatorial problem due to pseudo l_0 − norm, making it a nonconvex optimization problem. Convex relaxation of Eq. (4.3) is obtained by replacing l_0 − norm with l_1 − norm [6].

$$Q(Y, D, X) = \arg \min_{X} \|Y - DX\|_F^2 + \lambda \|x_i\|_1 \tag{4.4}$$

Eq. (4.4) is a nonsmooth convex optimization problem which can be solved. DL problem is discussed in Section 4.4.

4.3 Sparse coding methods

In this section, we give a broad overview of the classification of sparse coding algorithms, based on the norm used for regularization. Sparse coding algorithms based on l_0—norm regularization are easy to implement and thus most popular. Matching Pursuit (MP) [7], Orthogonal Matching Pursuit (OMP) [8], Fast OMP [9], etc., find the sparse coefficient matrix in Eq. (4.3), using a greedy approach. The coefficient of dictionary atom d_k which is highly similar to the input is updated first and the residual after subtracting the contribution of d_k multiplied with the coefficient, is again matched with the dictionary atoms. Though these methods work well, they give suboptimal sparsity levels and sometimes local minima as solutions.

Basis Pursuit (BP) [10], Generalized Lasso [11], Focal Underdetermined System Solver (FOCUSS) [12] are some of the important methods of sparse coding using l_1—norm optimization in Eq. (4.4). Such piecewise linear approximations provide a guarantee of maximally sparse unique solution to the sparse coding problem. A probabilistic model for representing an observed pattern in a lower dimensional space with respect to (w.r.t.) an optimum dictionary and with a prior on the coefficient vector is given in sparse Bayesian learning [13]. Sparsity-inducing prior acts as a means of regularization. Each pattern $\mathbf{y}$ is represented as $\mathbf{y} = D\mathbf{x} + \mathbf{n}$ where $\mathbf{n}$ is additive Gaussian noise with variance σ^2. Now, the likelihood function to be maximized is Eq. (4.5).

$$p(y|D) \propto \int p(x)p(y|D, x)dx \tag{4.5}$$

Several approximations to Eq. (4.5) have been proposed in Refs. [3,14−16] to obtain approximate maximum likelihood (AML) estimates of coefficient vector **x** which maximizes the log-likelihood function. A collection of training patterns $\{y_i\}_{i=1}^N$ are assumed to be independent and different assumptions or approximations about the coefficient vectors $\{x_i\}_{i=1}^N$ result in different estimates. For example, in Refs. [14,16], components of each coefficient vector are assumed to be independently identically distributed (i.i.d.) with a Laplacian prior to promote sparsity i.e., $p(x_i) \propto \exp(-|x_i|/\alpha)$ where α denotes the parameter diversity.

4.3.1 Importance of statistical concepts in sparse coding methods

Assumptions about data and the sampling method used determine the performance of parametric methods. Some of the Bayesian sampling techniques used in pattern recognition are rejection sampling, ratio of uniforms, importance sampling, Markov Chain Monte Carlo (MCMC) methods, and slice sampling [17,18].

In Ref. [19], sparse coefficients of time series data have been estimated using Gibbs sampling and importance sampling methods. Gibbs sampler cannot explore the entire posterior distribution but takes samples from just a single mode of the posterior distribution. It is difficult for Gibbs sampler to escape local maxima [20]. However, Gibbs sampler combined with annealing techniques can help in faster convergence.

Partially Collapsed Gibbs (PCG) sampler replaces some conditional distributions with marginal distributions to overcome the limitations of the standard Gibbs sampler as described by Van Dyk and Park in Refs. [21,22]. In Ref. [20], Bayesian inference on the unknown parameters corresponding to each sparse coefficient is conducted using samples generated by PCG sampler. These samples asymptotically follow the joint posterior distribution of the unknown model parameters and their hyperparameters. Such samples can closely approximate the joint maximum a posteriori estimate of the coefficients and the dictionary.

Importance sampler is not good for finding sparse approximations as it depends on the proposal distribution used. Importance sampler samples from a distribution (proposal) and finds the expectations w.r.t. the target distribution.

Priors used in sparse approximations. Generally, the class conditional probability densities (assume features are continuous), $p(\mathbf{y}|C_i)$ are unknown. If the form of the $p(\mathbf{y}|C_i)$ is known, but its parameters like mean and variance are unknown, these unknown parameters are estimated if some prior information is known about these parameters and then the Bayes' decision rule is applied. Bayesian framework for estimation of parameters starts with specifying a probabilistic model from which marginal and posterior distributions can be evaluated. When we have large number of training patterns, the general prior applied is Gaussian prior. Though Gaussian prior works very well, sparsity-inducing Laplacian or Cauchy priors act as a way of regularization and allow working with fewer variables than Gaussian case. Jeffrey's prior is invariant w.r.t change of coordinates and hence works well as a prior for scale parameters. In Ref. [23], the author has described several priors on the coefficient vector which induce sparsity. The generalized Gaussian prior is given by Eq. (4.6).

$$p(\mathbf{x}|\alpha,\beta) = \prod_{j=1}^{n} \mathscr{GG}(x_j|\alpha,\beta) \tag{4.6}$$

where

$$\mathscr{GG}(x_j|\alpha,\beta) = \frac{\alpha\beta}{2\Gamma(1/\beta)}e^{-\alpha|x_j|^{\beta}}$$

The shape parameter value $\beta = 2$ gives Gaussian prior which corresponds to l_2-norm regularization in Eq. (4.1). $\beta = 1$ gives Laplacian prior which is equivalent to l_1-norm regularization. The scale parameter α squeezes or stretches and along with location and shape parameters, determines the shape of a distribution. When compared to Gaussian prior, Laplacian prior and those with $0 < \beta < 1$ are good sparsity-inducing priors. When the application is compression-based, a higher level of sparsity is desired.

If the prior and the posterior are from the same family of probability distributions, then the prior is a conjugate prior for the likelihood function. For example, in Ref. [20], additive Gaussian noise has variance σ^2, with Inverse Gaussian prior. Such conjugate priors help in arriving at a closed-form posterior, avoiding numerical integration. The authors [20] have used PCG sampler to generate samples of the joint probability distribution of model parameters and hyperparameters, where the prior on coefficient vector $\mathbf{x}$ is Bernoulli-Gaussian (BG) distribution with parameters λ_k and component variance b_k^2 for each x_k. The hyperprior on λ_k is beta distribution.

A generic method for sparse coding using Bayesian approach is given in Algorithm 1. For simplicity, one-dimensional signals $\mathbf{y} = (y_1, ..., y_N)$ are considered and corresponding errors in $\in = (\in_1, ..., \in_N)$, coefficients of $\mathbf{y}$, $\mathbf{x} = (x_1, ..., x_K)$ have to be determined using dictionary $D \in R^{1 \times K}$. In Ref. [24], the authors have used a flexible prior based on original data. But, highly sparse priors like Jeffrey's prior result in multimodal posteriors and hence the problem of local optima arises.

In Ref. [25], the coefficients are assumed to follow Cauchy distribution which is a heavy-tailed distribution and is a member of the Levy-alpha-stable family of distributions. The Cauchy proximal operator has been defined and the Cauchy Convolutional Sparse Coding algorithm has been proposed to learn sparse coefficients to minimize the representation loss.

For example, Sparse Bayesian Learning (SBL) is a Bayesian approach to find sparse coefficient vectors of given observations. Multiple snapshot SBL (M-SBL) is used for a data set of N observations constituting input data Y. The corresponding sparse coefficient matrix $X \sim N(\mu, \Sigma)$.

In [26], the authors have assumed Gaussian hyperprior and achieved results comparable to the state-of-the-art. Though Gaussian hyperprior does not induce a high level of sparsity, the SBL algorithm which achieves a maximally sparse solution even with a random dictionary [13], is used in DoA estimation. The sparsity level in each coefficient vector is automatically determined at the point of convergence [27].

Another approach to sparse coding is to generate samples from $p(x|y, \theta)$ and $p(\theta|y, x)$ for MCMC sampling methods. To estimate $p(x,\theta|y)$, Gibbs sampler gives $X \sim p(x|y, \theta)$. These samples approximate **x**. Now, samples $\theta \sim p(\theta|y, \widehat{x})$, are used to approximate $\widehat{\theta}$. A technical review of Bayesian approaches to sparse coding methods has been given in Ref. [23].

4.4 Dictionary learning

The origins of research into dictionary learning are in ICA. ICA minimizes the dependence among vector components by imposing independence up to second-order [28] i.e., the variables are linear combinations of unknown latent variables which are also assumed to be independent. For a random vector **v** with finite covariance C_v, ICA finds a pair of matrices {M,N}, N being diagonal whose entries are sorted in descending order, such that $C_v = MN^2M^*$. Similar to DL, the directions here are also orthogonal, with unit norm constraint on columns of M. The dictionary entries are real numbers and the largest modulus in each column of M is a positive real number. Sparse representation is closely related to ICA with these conditions and hence can be used as a preprocessing tool, just like ICA, before applying Bayesian detection and classification [28].

Algorithm 1. Bayesian sparse coding procedure.

Procedure BayesianSparseCoding(y, D, threshold)

$$y = Dx + \epsilon$$

$$posterior = \frac{likelihood \times prior}{evidence}$$

$p(x|y, \theta) \propto p(y|x, \theta_1) \cdot p(x|\theta_2)$ where $\theta = (\theta_1, \theta_2)$ are hyperparameters.

To estimate $\widehat{x}$, first estimate $\widehat{\theta}$. Assign a prior to θ with fixed hyper-hyperparameters, θ_0.

$$p(x, \theta|y, \theta_0) = \frac{p(y|x, \theta_1)p(x|\theta_2)p(\theta|\theta_0)}{p(y|\theta_0)}$$

Full Bayesian Approach: Using *Joint MAP* to estimate (x, θ)

$$\left(\widehat{x}, \widehat{\theta}\right) = \arg \max_{x,\theta}\{p(x,\theta|y,\theta_0)\}$$

OR

Using *Evidence Maximization:* Integrate out parameters to estimate hyperparameters $\widehat{\theta}$ using $p(\theta|y,\theta_0) = \int p(x,\theta|y,\theta_0)dx$. Now, $\widehat{\theta} = \arg \max_{\theta}\{p(\theta|y,\theta_0)\}$ using *MAP or Expectation Maximization (EM) or Maximum Likelihood Estimation.*

Bayesian EM: Consider y as incomplete data, x as hidden random variable. (y, x) is complete data. ln p(y, x|θ) is complete data log-likelihood.

E-step: $Q(\theta, \theta^{(k)}) = E_{p(x|y,\theta^{(k)})}[\ln p(y, x|\theta) + \ln p(\theta)]$

M-step: $\theta^{(k)} \arg \max_{\theta} Q(\theta, \theta^{(k-1)})$

Repeat E-step, M-step until

$$Q(\theta, \theta^{(k)}) - Q(\theta, \theta^{(k-1)}) \leq \text{threshold}$$

$$\hat{\theta} = \theta^{(k)}$$

Output: $\hat{x} = \arg \max_{x} \left\{ p\left(x|y, \hat{\theta}\right) \right\}$ using MAP estimation.

end procedure

Fixing X obtained from Eq. (4.3), the joint optimization of D, X in Eq. (4.2) is reduced to a linear optimization problem using block coordinate descent method [29]. Updating dictionary D w.r.t a fixed coefficient matrix X results in Eq. (4.7) and this learning phase to update D is called dictionary learning (DL).

$$Q(Y, D, X) = \arg \min_{D} \|Y - DX\|_F^2 + \lambda \|x_i\|_0 \qquad (4.7)$$

where $\|d_k\| \leq 1, \quad k = 1, 2, ..., K, \quad$ K is the size of dictionary D.

4.4.1 Orthogonal dictionary learning

Initially, mathematical transforms were applied on original data columns to get orthonormal dictionaries called analytic dictionaries. Such Wavelet dictionaries, Fourier dictionaries have incoherent atoms, orthogonal to each other, hence opted for compression-based applications. The level of sparsity achieved is very good with orthogonal dictionaries. Though PCA is capable of capturing major variance in data, minor details which are crucial for discrimination, are not captured. Moreover, the number of significant eigenvalues is specified by the user. These limitations of PCA could be overcome by a synthesis dictionary comprising original data as atoms and then iteratively updated such that the representation error is minimal, with better representations and faster convergence [30].

Though Nonnegative Matrix Factorization (NMF) works well for compressed representations of data, in the case of natural images, NMF does not perform well when compared to overcomplete sparse representations [31].

4.4.2 Overcomplete dictionary learning

Representation of natural images is rich when the redundancy in data is utilized in the form of overcomplete dictionaries. Atoms of an overcomplete dictionary are selected such that their number K is small compared to the data size N, but larger than the input dimensionality i.e., $K >> d$. Unlike under complete, orthogonal dictionaries, these overcomplete dictionaries are used in reconstruction-based applications like image denoising, inpainting where missing or corrupted part of an image is reconstructed.

An overcomplete dictionary works in contrast with vector quantization (VQ), in which each sample is mapped to exactly one prototype. DL algorithms could be used to update prototype vectors as in Ref. [32], where dictionary learning helps in better quantization of ECG patterns.

4.4.3 Structured dictionary learning

When the training set comprised of features along with their class labels, structured dictionaries could be generated. Subdictionaries of all classes are grouped to form a global shared dictionary which represents features shared by all classes. Subdictionaries have atoms used to represent features, particularly of a specific class. Minimal reconstruction error in Eq. (4.7) w.r.t. subdictionaries decides the label of test pattern. When the number of classes increases, the computation of structured subdictionaries becomes expensive. A single shared dictionary whose atoms have features of each class, as well as common features shared by all classes, saves memory and time. In Ref. [33], to learn a discriminative structured dictionary, a reconstruction error term is designed such that a given class of data is represented best by the global dictionary and the corresponding class dictionary but not by other class dictionaries. Fisher criterion, i.e., minimal intraclass scatter and maximal interclass scatter, is imposed on sparse coefficient vectors, making both the coefficients and the dictionary atoms discriminative, leading to better classification results. Within-class-scatter is given by $S_\omega(X) = \sum_{i=1}^{C} \sum_{x_k \in X_i} (x_k - \text{mean}_i)(x_k - \text{mean}_i)^T$ and Between-class-scatter is given by $S_B(X) = \sum_{i=1}^{C} (\text{mean}_i - \text{mean})(\text{mean}_i - \text{mean})^T$, where mean_i is the mean sparse coefficient vector of class i and mean is the mean sparse coefficient vector of all the data. Fisher Discrimination Dictionary Learning (FDDL) uses alternating optimization method with Fisher discrimination-based sparse coding in Eq. (4.8).

$$L(X_i) = \arg \min_{X_i} \{\gamma(Y_i, D, X_i) + c_1 \|X_i\|_1 + c_2 g(X_i)\}, \tag{4.8}$$

where $g(X_i) = \left\{ \text{tr}(S_B(X)) - \text{tr}(S_\omega(X_i)) + \eta \|X\|_F^2 \right\}$ could be computed by finding the eigen values of the scatter matrices $S_B(X)$ and $S_w(X_i)$. Thus, simple statistical concepts used in FDDL, help in learning a discriminative dictionary. Unsupervised and supervised methods of learning such structured dictionaries are given in Section 4.4.4 and Section 4.4.5.

4.4.4   Unsupervised DL algorithms

Unsupervised algorithms for dictionary learning result in generative or representative dictionaries, usually applied in image denoising, deblurring, and inpainting. The missing pixels of an image can be reconstructed with the help of generative dictionaries. In each iteration, Method of Optimal Directions (MOD) [34] updates the dictionary by computing pseudo-inverse of the coefficient matrix, which causes slow convergence. Another unsupervised algorithm, K-SVD [35] is a generalization of the k-means algorithm, which converges faster due to simultaneous update of both coefficient vectors and dictionary atoms. Only the elements corresponding to nonzero components of the coefficient vector are considered to compute residual signal, E and Singular Value Decomposition (SVD) is applied to diagonalize the residual, $E = E\Delta V^T$. The first column of U gives an updated atom and the product of the first diagonal element and the first row of V gives an updated coefficient vector. Retaining only the major part of the signal in the form of a few large singular values effectively reduces noise and gives a better representation of the signals. Incremental Codebook Optimization [36] and Locality Constrained Linear Coding [37] are other unsupervised DL algorithms.

4.4.5   Supervised DL algorithms

Supervised DL gives discriminative dictionaries for the classification of patterns using labels of patterns in the formulation of the objective function. Face recognition in the presence of obstructions and different moods and postures is an important application where discriminative dictionaries are used. Discriminative-KSVD (DKSVD) learns a discriminative dictionary by incorporating label information into the objective function of K-SVD.

$$\begin{cases} \langle D, W, X \rangle = \underset{D,W,X}{\arg\min} \|Y - DX\|_F + \alpha\|H - WX\|_F + \beta\|W\|_F, \\ \text{sub.} \quad \text{to} \quad \|X\|_0 \leq \tau. \end{cases} \tag{4.9}$$

The matrix $\begin{pmatrix} D \\ \sqrt{\alpha}W \end{pmatrix}$ is always normalized columnwise, so the regularization penalty $\|W\|_2$ can be dropped to get

$$\begin{cases} < D, W, X > = \underset{D,W,X}{\arg\min} \left\| \begin{pmatrix} Y \\ \sqrt{\alpha}H \end{pmatrix} - \begin{pmatrix} D \\ \sqrt{\alpha}W \end{pmatrix} X \right\|_F, \\ \text{sub.} \quad \text{to} \quad \|X\|_0 \leq \tau. \end{cases} \tag{4.10}$$

The label matrix H is approximated by a classifier matrix W and the coefficient matrix X, using alternating optimization (BCD), given by Eq. (4.9). With α as regularization parameter, *K*-SVD algorithm is applied to optimize Eq. (4.10). A similar approach to learning a discriminative dictionary is Label Consistent KSVD (LCKSVD) [38]. If dictionary atoms are coherent, then there is a multiple-representation problem. So, a compact dictionary is preferred with which similar signals(from the same class) can be described by roughly the same set of atoms with almost similar coefficients. Application of statistical methods in feature extraction as well as determining the size of the dictionary and the dictionary columns results in better discriminative dictionaries.

4.5 Statistical concepts in dictionary learning

The problem of identifying a dictionary relies on the assumptions of statistical independence and non-Gaussian distribution set as prior [39]. The ratio of majority and minority class cardinality could be high leading to high misclassification costs. A probabilistic model for sparse representation-based classification has been given in Ref. [40], to address the problem of class imbalance in the data set. A cost-sensitive classification rule based on a Bayesian framework with sparse coefficients as features has not only improved accuracy but also reduced misclassification cost.

4.5.1 Histogram of oriented gradients (HoG)

In the case of high dimensionality, feature descriptors are used to avoid unnecessary computations involved in classification. Histogram of oriented gradients (HoG) is a feature descriptor used to define an image by the pixel intensities and intensities of gradients of pixels. Gradients define the edges of an image, so extraction of the HoG feature descriptor is the same as extracting edges.

Histogram of Oriented Gradients generates gradients at each point of the image providing invariance to occlusions, illumination, and expression changes. In Ref. [41], group sparse coding with HoG feature descriptors is used to achieve good results on face recognition.

4.5.2 Use of correlation analysis in dictionary learning

Correlation is the value of association between two independent or one independent and other dependent variables, determined by measuring the correlation coefficient (Pearson, Kendall, Spearman) and also the direction of their relationship i.e., positive correlation or negative correlation. Quantification of this association involves computing correlation coefficient ranging between $[-1,1]$. In Ref. [42], Pearson product-moment correlation coefficient is combined with the sparse reconstruction error of samples for face recognition. While reconstruction error tries to reduce the error between the test sample and same class samples, the Pearson correlation coefficient maximizes the error between the test sample and other class samples, for improved classification results.

Canonical correlation analysis (CCA) is an extension of bivariate to multivariate analysis. When several factors influence a single outcome, it is multivariate data and the corresponding correlation analysis is called CCA. In Ref. [43], the unknown block structure of the dictionary is explored using the correlation among dictionary atoms. This method gives control over the size of blocks. The maximum correlation quotient between the test sample and training samples and the reconstruction residual are weighted in the decision function to determine the label of the test signal.

4.6 Parametric approaches to estimation of dictionary parameters

Parametric approaches make some assumptions about the population distribution from which the training data originated. The central limit theorem is crucial to these assumptions. The theorem states that if a sufficiently large number of random samples are drawn (with replacement) from any population with mean μ and variance σ^2, then the distribution of sample means will be approximately Gaussian. Whenever there is uncertainty about the probability model of data, a Gaussian probability model is assumed, and the population parameters are derived from the model.

The parametric approach to DL assumes a known distribution from which the columns of the dictionary are drawn and tries to estimate the parameters of the distribution, such as size K and the atoms themselves, by using maximum likelihood maximization to derive mean and covariance of dictionary column distribution. Full posterior estimates are provided using a Bayesian framework, which takes care of uncertainty and unseen data generally observed in biomedical applications. For representing an observed pattern in a lower dimensional space w.r.t. a coefficient vector and with a prior on the dictionary parameters $\theta = \left(\{d_K\}_{k=1}^{K}, K \right)$, the likelihood function to be maximized is Eq. (4.11).

$$p(y|x)\alpha \int p(y|\theta, x)p(\theta)d\theta. \tag{4.11}$$

Approximate Maximum Likelihood estimation of an unknown but deterministic dictionary using Eq. (4.11) is equivalent to Method of Optimal Directions (MOD) when the noise is assumed to be Gaussian [44]. In Ref. [45], an algorithm to find a joint maximum a posteriori probability (MAP) estimate of an unknown random initial dictionary and the corresponding coefficient matrix, is given.

In Ref. [46], a Bayesian approach has been employed to estimate dictionary atoms and dictionary size K along with the sparse coefficient vector hyperparameters. Additive noise is assumed whose variance is modeled from a gamma distribution with unknown parameters. Each dictionary atom or column has been assumed to be randomly drawn from a uniform distribution with components from [0,1]. Such uniform prior is non-informative, so this assumption is equivalent to taking a random initial dictionary, whose columns have unit norm. The coefficient vectors have been modeled as a zero-mean Gaussian where the covariance matrix is determined by hyperparameters which are assumed to be independently gamma-distributed.

The dictionary atom parameters, hyperparameters on coefficient vectors, noise variance are determined by approximating to a MAP estimate, obtained by iteratively maximizing the log-posterior density w.r.t. each of them, keeping the others fixed. This approach is equivalent to the block coordinate descent technique employed to optimize Eq. (4.2).

A closed-form solution to maximizing likelihood function in Eq. (4.11) is intractable, but Monte Carlo methods like Gibbs sampler and Metropolis–Hastings are used to approximate closed-form posteriors of dictionary variables [47]. A Markov chain (MC) is said to be ergodic or irreducible if it is eventually possible to reach every state from each state with positive probability. In Ref. [47], uniform ergodicity properties of high-dimensional MC, which imply convergence to a stationary distribution independent of the initial states, have been discussed.

To approximate posteriors of the dictionary, Groupwise sampling and aggregation have been used to identify groupwise similar functional brain networks of different persons in Ref. [48]. Signal sampling and sparse coding on task fMRI data for learning a shared dictionary within a group of persons have helped in identifying and examining common cortical functional networks at an individual level and population level. The authors have used no sampling, random sampling, uniform random sampling, two-ring, and four-ring sampling methods, and the corresponding statistical significance tests have been conducted.

Data-driven overcomplete dictionaries enable flexible representations of data and the quality of an overcomplete dictionary could be determined using diversity measures like the distance between atoms, reconstruction error, coherence among atoms. The Babel measures and entropy from information theory measure the randomness in a system. A high value of entropy denotes spread of atoms in a dictionary [49].

Active DL updates dictionary atoms from the information in training data, using different strategies. Selecting the most useful sample by uncertainty sampling and by generalization error are classical strategies. The sample whose label cannot be decided is called uncertainty sample and can be decided using posterior probability, margin sampling, and entropy-based methods [50].

When the samples are complexly structured like trees and sequences, entropy-based queries retrieve informative samples for dictionary building. The uncertainty sample based on entropy is given by

$$y^* = \arg \max_{y} \left(- \sum_{i} P_\theta(\text{labels}_i | y) \log P_\theta(\text{labels}_i | y) \right),$$

where θ is the set of dictionary parameters.

When the training set contains both labeled and unlabeled samples, the informativeness of samples could be decided by the probability distribution of class-specific reconstruction error, which determines how well the current dictionary can discriminate the sample.

In Ref. [51], the authors have used both reconstruction error of a sample w.r.t. shared dictionary and its entropy on the probability distribution over class-specific reconstruction error, to determine the dictionary. Here, the level of discrimination of the dictionary is given by the entropy on the probability distribution of error of labeled samples, and the level of representation is given by the distribution of the error of unlabeled samples.

4.6.1   Hidden Markov model (HMM): discriminative dictionary learning

With hidden Markov model (HMM), it is possible to describe the sparsity profile as each hidden state represents a set of non-zero coefficients. In Ref. [52], the problem of sparse representation has been modeled as an HMM. The approach in this paper has combined filtering based on HMM and manifold-based dictionary learning for estimating both the nonzero coefficients and the dictionary. An equivalence relation, partitioning the set of dictionaries into equivalence classes, has been introduced. A direct search for the equivalence class which contains the true dictionary has been used. The observations are decoupled using a new technique called change-of-measure, so that the observations are all uniformly, identically distributed.

Expectation-maximization has been used to recursively update state in the MC i.e., the coefficient matrix X with Gaussian prior, transition matrix of MC, and the dictionary.

Sparse HMM has been used in Ref. [53], to model surgical gestures, where the dictionary is a set of basic surgical motions. The algorithm to learn a dictionary for all gestures and an HMM grammar describing the transitions among gestures has been proposed here. New motion data is classified based on these dictionaries and grammar. Viterbi algorithm is used for surgeme classification.

Given a surgery trial $\{y_t \in R^d\}_{t=1}^T$, assign a surgeme label $v_t \in \{1, 2, ..., V\}$ to each frame y_t. Skill-level from $I \in \{1, 2, ..., L\}$ is assigned to the trial $\{y_t\}_{t=1}^T$. The surgeme label is a hidden (unobserved) state modeled as a Markov process with transition probability $q_{v'v} = p(v_t = v | v_{t-1} = v')$. Thus, an observation at time t, y_t, depends on hidden state v_t through the emission probability density $p(y_t|v_t)$, which is generally assumed to be Gaussian or a mixture of Gaussians.

Also, y_t is expressed as a superposition of atoms from a dictionary corresponding to gestures. Hence, y_t depends on another hidden variable x_t, i.e.,

$$y_t = D_{v_t} x_t + noise.$$

For each hidden state v_t, a Laplacian prior is imposed on x_t, to get a sparse latent variable, given in Eq. (4.12).

$$p(x_t | v_t = v) = \left(\frac{\lambda_v}{2}\right)^K e^{-\lambda_v \|x_t\|_1}, \tag{4.12}$$

where $\lambda > 0$ is parameter and K is the size of dictionary D_{v_t} corresponding to v_t. Now,

$$p(y_t | \upsilon_t = \upsilon, x_t = x) \sim \mathcal{N}(D_\upsilon x, \sigma_\upsilon^2 I),$$

where D_υ is an overcomplete dictionary corresponding to surgeme v.

Bayesian expectation maximization is applied to learn all the transition probabilities $\{q_{s,s'}\}$ and the parameters of each surgeme model $\Theta_V = (D_V, \sigma_V^2, \lambda_V)$, for each $\upsilon \in \{1, 2,..., V\}$.

To get the surgeme labels $\{\upsilon_t\}_{t=1}^T$ of a given trial $\{y_t\}_{t=1}^T$, a dynamic programming approach has been given. If the number of states is finite, then the algorithm converges.

For skill-level classification, three Sparse HMM models are learned for expert, intermediate, and novice levels. The level of skill is determined by using the Viterbi algorithm [53].

4.7 Nonparametric approaches to discriminative DL

Unsupervised and supervised DL algorithms discussed in Section 4.4.4 and Section 4.4.5 are nonparametric approaches to DL [47]. Parametric dictionaries consider uncertainty in data and avoid local optima. This property of parametric dictionaries improves generalization of sparse representation model. In supervised DL algorithms [33,38,54,55], sparse codes consistent with class labels are generated for both generative and discriminative models.

In Ref. [56], the objective function is formulated combining classification error and the representation error of both labeled and unlabeled data, with a constraint on the number of coefficients. All these algorithms are tersely mathematically formulated, tested on data sets for face recognition like Extended YaleB, AR data set, and handwritten numerals data sets MNIST and USPS.

If the form of $p(y|C_i)$ is unknown, there are nonparametric approaches like Parzen windows, K-nearest neighbor rule, MLP with backpropagation, to estimate $p(y|C_i)$ from the observed data. To improve generalization, data-based methods require huge data. According to Ref. [57], a simple neural network with one hidden layer is capable of solving any problem. MLP does not depend on any assumptions about the data and hence is a nonparametric method used to decide the boundaries based on the observed data [58].

With the increase in input dimensionality, the number of hidden neurons increases exponentially. CNNs and deep belief networks with several hidden layers are being used in computer vision and pattern recognition, to achieve the best classification results. In deep neural networks, where data paucity could affect generalization, auto-encoder is applied for dimensionality reduction. When the training samples are limited and feature extraction is carried out by several hidden layers, there could be problems like vanishing gradient and overfit. The learning time increases as the gradient vanishes in back-propagation [59].

If the feature extraction step of MLP could be replaced with sparse representation, the classifying capability of MLP could be used to classify data with high dimensionality and fewer samples.

In Ref. [60], a one-to-one correspondence between the sparse coding step, and deep CNNs has been proposed, representing images using wavelet analysis, sparse coding, and DL. Dense signal gives the scale, while SR that selects a few dictionary atoms, gives the detail. Hierarchical convolutional sparse coding (H-CSC) and convolutional DL have been used alternatingly, to generate a different set of images combining the content of one set of images with the style of another set of images [61].

To overcome the limitations of both DL and deep learning, a hybrid method has been proposed, selecting optimal weights and picking the best performing compact architecture empirically, in Ref. [62]. Sparse coefficients of samples of the same class are similar and those of different classes are quite different when computed using a single shared dictionary [38]. Here, the authors have used this property of sparse coefficients and Discriminative K-SVD to learn a dictionary to classify data sets that have a large number of classes and a huge class imbalance ratio.

For example, Telugu OCR data set UHTelPCC [63] has high class imbalance as shown in Fig. 4.5. Telugu script characters have structural complexity which makes their image feature extraction complex. Also, there is confusing pairs problem as given in Fig. 4.2, very commonly found in Dravidian scripts.

A hybrid method that makes use of the sparse codes as input features, avoids tedious feature extraction overhead in deep networks as shown in Fig. 4.3, leading to a compact MLP architecture.

Initializing $W^{(0)} = (X^TX + \lambda I)^{-1}X^TH$, sparse codes generated using Eq. (4.10) are given as input to a simple MLP with two hidden layers as shown in Fig. 4.4. The MLP architecture has a dense layer (with ReLU activation), a batch normalization layer, a dropout layer, another dense layer (with ReLU activation). The output layer (with softmax activation) corresponds to categorical labels of the data set. The addition of the batch

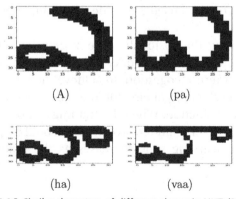

(A)   (pa)

(ha)   (vaa)

FIGURE 4.2 Similar characters of different classes in UHTelPCC [62].

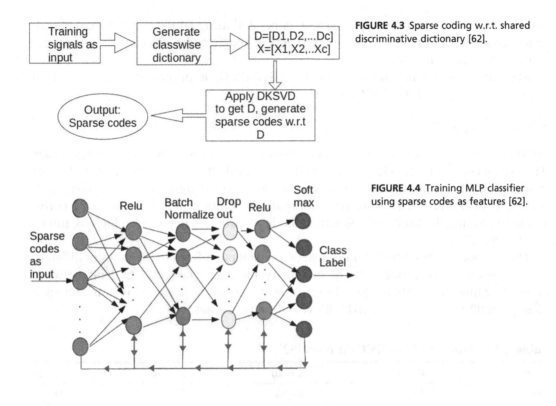

FIGURE 4.3 Sparse coding w.r.t. shared discriminative dictionary [62].

FIGURE 4.4 Training MLP classifier using sparse codes as features [62].

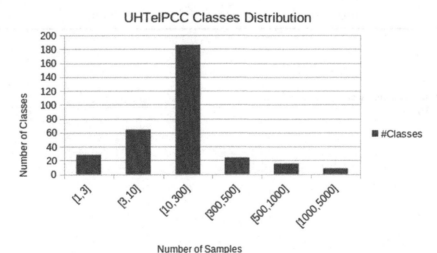

FIGURE 4.5 Number of classes on Y-axis containing number of samples ranging over X-axis [62].

normalization layer between hidden layers maps the nonlinear features to the linear part of the activation function. Dropout layer has been applied to eliminate the problem of overfit. The MLP is trained on sparse codes generated using DKSVD, and evaluated with sparse codes of test images. Train and test sets of sparse codes are generated w.r.t. the same shared dictionary [62].

Algorithm 2 [62], has been tested on UHTelPCC, a printed Telugu connected component data set and MNIST data set.

4.7.1   UHTelPCC

UHTelPCC is a Telugu data set, contains 70,000 binary connected components of size 32×32 pixels from 325 classes. (UHTelPCC is available at http://scis.uohyd.ac.in/chakcs/UHTelPCC.zip.) These 70,000 samples are divided into training (50,000), validation (10,000) and test (10,000) sets. Computation times reported in Table 4.1 correspond to training the MLP and validating. Model accuracy is depicted in Fig. 4.6, model loss in Fig. 4.7.

The method has been tested on sparse codes generated from dictionaries of different sizes. The choice of proper size of dictionary for each class is a tradeoff between computing time and accuracy [62]. From Table 4.1, dictionary size of 20 atoms for each class gives 98.7% accuracy for UHTelPCC with dimensionality 1024.

Table 4.1   Results on UHTelPCC data set [62].

#Atoms	K = 16	K = 20	K = 24	K = 26
Time	19s	24s	27s	32s
Accuracy	97.9	98.7	98.73	98.91
F1-score	0.9856	0.9963	0.9991	0.9998

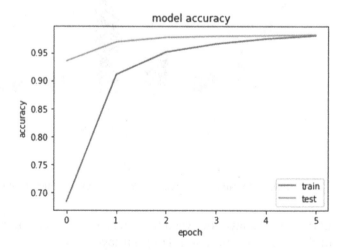

FIGURE 4.6 Model accuracy on UHTelPCC [62].

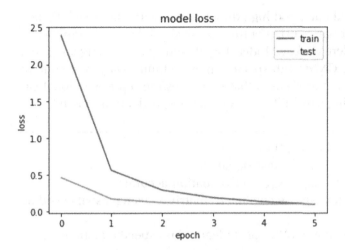

FIGURE 4.7 Model loss on UHTelPCC [62].

Table 4.2   Results on MNIST data set [62].

#Atoms	K = 14	K = 16	K = 18	K = 23
Time	18s	21s	22s	32s
Accuracy	95.34	95.12	96.32	96.4

4.7.2 MNIST

MNIST [65] is a handwritten numerals dataset of 60,000 samples for training and 10,000 for testing. The dimensionality is 784 and a dictionary size of 18 atoms gives 96.3% accuracy.

Reduced training and testing times for both UHTelPCC and MNIST, from Tables 4.1 and 4.2, suggest the low computational complexity of the model. This nonparametric method of learning classifier weights supports the idea of using statistical concepts in sparse coding as well as DL.

4.8 Conclusion

The transformation of DL from orthogonal transforms to overcomplete analytic transforms to overcomplete synthesis dictionaries is followed by parametric DL. In this review article, we present an overview of using probabilistic models, with different priors and hyperpriors on variables, parametric and nonparametric approaches to parameter estimation, used in sparse representation algorithms. Sampling techniques used in sparse representation to overcome problems like multimodal data, class imbalance in data, unlabeled data mixed with labeled data, and high dimensionality are discussed. Design of structured, overcomplete dictionaries using entropy analysis of data and examples of research articles presenting Hidden Markov Models for DL and sparse coding are given. Research articles which combine CNNs with sparse representation to separate content and style in images as well as a hybrid method that combines the representational capabilities of DL with the classifying capabilities of a neural network are discussed.

Algorithm 2. Sparse code-trained MLP (SCMLP).

    input: Training signals, $Y_c, c = 1, 2, ..., C$, test signals $\mathbf{X_i}$

    output: Sparse Coefficient matrix X_{train}, X_{test} and validation accuracy.

1. For each class of signals Y_c, obtain sparse coefficient matrix X_c using a sparse coding algorithm.
2. After sparse coding, use ApproximateKSVD [64] to learn class-specific dictionary, D_c, $c = 1, 2,..., C$.
3. Concatenate class-wise dictionaries and input signals i.e.,
   $$D = [D_1, ... D_c], Y = [Y_1, Y_2, ..., Y_c]$$
4. Apply Discriminative-KSVD by adding label matrix H in the objective function and obtain D from Eq. (4.9).
5. Extract first d rows of D and normalize to get shared dictionary D^s.
6. For the training set, find the sparse coefficient matrix X_{train} w.r.t. D^s using OMP. Store X_{train}.
7. Feed the sparse codes X_{train} of training signals to the MLP in Fig. 4.4.
8. Find the sparse codes X_{test} of test signals w.r.t D^s using OMP and feed them to the trained model to evaluate the performance of the model.

References

[1] S. Tsuge, M. Shishibori, S. Kuroiwa, K. Kita, Dimensionality reduction using non-negative matrix factorization for information retrieval, in: 2001 IEEE International Conference on Systems, Man and Cybernetics. e- Systems and e-Man for Cybernetics in Cyberspace (Cat. No. 01CH37236), vol. 2, IEEE, 2001, pp. 960–965.

[2] S.J. Raudys, A.K. Jain, et al., Small sample size effects in statistical pattern recognition: recommendations for practitioners, IEEE Trans. Pattern Anal. Mach. Intell. 13 (3) (1991) 252–264.

[3] B.A. Olshausen, D.J. Field, Sparse coding with an overcomplete basis set: a strategy employed by v1? Vis. Res. 37 (23) (1997) 3311–3325.

[4] D.J. Field, What is the goal of sensory coding? Neural Comput. 6 (4) (1994) 559–601.

[5] A. Beck, L. Tetruashvili, On the convergence of block coordinate descent type methods, SIAM J. Optim. 23 (4) (2013) 2037–2060.

[6] S. Schuler, C. Ebenbauer, F. Allgower, l0-system gain and l1-optimal control, IFAC Proc. Vol. 44 (1) (2011) 9230–9235.

[7] S. Mallat, Z. Zhang, Matching pursuits with time-frequency dictionaries, IEEE Trans. Signal Process. 41 (12) (1993) 3397–3415, https://doi.org/10.1109/78.258082.

[8] Y.C. Pati, R. Rezaiifar, P.S. Krishnaprasad, Orthogonal matching pursuit: recursive function approximation with applications to wavelet decomposition, in: Proceedings of the 27th Annual Asilomar Conference on Signals, Systems, and Computers, 1993, pp. 40–44.

[9] S.-H. Hsieh, C.-S. Lu, S.-C. Pei, Fast omp: reformulating omp via it- eratively refining l < inf > 2 </inf >-norm solutions, in: 2012 IEEE Statistical Signal Processing Workshop (SSP), 2012, pp. 189–192, https://doi.org/10.1109/SSP.2012.6319656.

[10] S. Chen, D. Donoho, Basis pursuit, in: Proceedings of 1994 28th Asilomar Conference on Signals, Systems and Computers, vol. 1, IEEE, 1994, pp. 41–44.

[11] N. Morioka, S. Satoh, Generalized lasso based approximation of sparse coding for visual recognition, Adv. Neural Inf. Process. Syst. 24 (2011) 181–189.

[12] I.F. Gorodnitsky, B.D. Rao, Sparse signal reconstruction from limited data using focuss: a reweighted minimum norm algorithm, IEEE Trans. Signal Process. 45 (3) (1997) 600–616.

[13] M.E. Tipping, Sparse bayesian learning and the relevance vector machine, J. Mach. Learn. Res. 1 (June 2001) 211–244.

[14] M.S. Lewicki, B.A. Olshausen, Probabilistic framework for the adaptation and comparison of image codes, JOSA A 16 (7) (1999) 1587–1601.

[15] T.-W. Lee, M.S. Lewicki, M. Girolami, T.J. Sejnowski, Blind source separation of more sources than mixtures using overcomplete representations, IEEE Signal Process. Lett. 6 (4) (1999) 87–90.

[16] M. Lewicki, T. Sejnowski, Learning overcomplete representations, Neural Comput. 12 (2000) 337–365. https://doi.org/10.1162/089976600300015826.

[17] Sampling Methods, 2021. https://ermongroup.github.io/cs228-notes/inference/sampling/. (Accessed 3 August 2021).

[18] R.M. Neal, Bayesian methods for machine learning, NIPS Tutorial 13 (2004).

[19] T. Blumensath, M.E. Davies, Monte Carlo methods for adaptive sparse approximations of time-series, IEEE Trans. Signal Process. 55 (9) (2007) 4474–4486.

[20] N. Dobigeon, J.-Y. Tourneret, Bayesian orthogonal component analysis for sparse representation, IEEE Trans. Signal Process. 58 (5) (2010) 2675–2685.

[21] D.A. Van Dyk, T. Park, Partially collapsed gibbs samplers: theory and methods, J. Am. Stat. Assoc. 103 (482) (2008) 790–796.

[22] T. Park, D.A. Van Dyk, Partially collapsed gibbs samplers: illustrations and applications, J. Comput. Graph Stat. 18 (2) (2009) 283–305.

[23] A. Mohammad-Djafari, Bayesian approach with prior models which enforce sparsity in signal and image processing, EURASIP J. Adv. Signal Proc. 2012 1 (2012) 1–19.

[24] D.P. Wipf, B.D. Rao, An empirical bayesian strategy for solving the simultaneous sparse approximation problem, IEEE Trans. Signal Process. 55 (7) (2007) 3704–3716.

[25] P. Mayo, O. Karakus, R. Holmes, A. Achim, Representation learning via cauchy convolutional sparse coding, IEEE Access 9 (2021) 100447–100459.

[26] P. Gerstoft, C.F. Mecklenbrauker, A. Xenaki, S. Nannuru, Multisnapshot sparse Bayesian learning for DOA, IEEE Signal Process. Lett. 23 (10) (2016) 1469–1473.

[27] O. Williams, A. Blake, R. Cipolla, Sparse bayesian learning for efficient visual tracking, IEEE Trans. Pattern Anal. Mach. Intell. 27 (8) (2005) 1292–1304.

[28] P. Comon, Independent component analysis, a new concept? Signal Process. 36 (3) (1994) 287–314. https://doi.org/10.1016/0165-1684(94)90029-9.

[29] Y. Xu, W. Yin, A block coordinate descent method for regularized multi- convex optimization with applications to nonnegative tensor factorization and completion, URL, SIAM J. Imag. Sci. 6 (3) (2013) 1758–1789. https://doi.org/10.1137/120887795.

[30] Z. Zhang, Y. Xu, J. Yang, X. Li, D. Zhang, A survey of sparse representation: algorithms and applications, IEEE Access 3 (2015) 490–530. https://doi.org/10.1109/ACCESS.2015.2430359.

[31] M. Shokrollahi, S. Krishnan, Non-negative matrix factorization and sparse representation for sleep signal classification, in: 2013 35th Annual International Conference of the IEEE Engineering in Medicine and Biology Society (EMBC), 2013, pp. 4318–4321. https://doi.org/10.1109/EMBC.2013. 6610501.

[32] T. Liu, Y. Si, D. Wen, M. Zang, L. Lang, Dictionary learning for VQ feature extraction in ECG beats classification, Expert Syst. Appl. 53 (C) (2016) 129–137. https://doi.org/10.1016/j.eswa.2016.01.031.

[33] M. Yang, L. Zhang, X. Feng, D. Zhang, Fisher discrimination dictionary learning for sparse representation, in: 2011 International Conference on Computer Vision, 2011, pp. 543–550.

[34] K. Engan, S.O. Aase, J.H. Husoy, Method of optimal directions for frame design, in: 1999 IEEE International Conference on Acoustics, Speech, and Signal Processing. Proceedings. ICASSP99 (Cat. No.99CH36258), Vol. 5, vol. 5, 1999, pp. 2443–2446. https://doi.org/10.1109/ICASSP.1999.760624.

[35] M. Aharon, M. Elad, A. Bruckstein, K.-S.V.D. rm, An algorithm for designing overcomplete dictionaries for sparse representation, IEEE Trans. Signal Process. 54 (11) (2006) 4311–4322. https://doi. org/10.1109/TSP.2006.881199.

[36] J. Mairal, F. Bach, J. Ponce, G. Sapiro, Online learning for matrix factorization and sparse coding, J. Mach. Learn. Res. 11 (1) (2010).

[37] J. Wang, J. Yang, K. Yu, F. Lv, T. Huang, Y. Gong, Locality-constrained linear coding for image classification, in: 2010 IEEE Computer Society Conference on Computer Vision and Pattern Recognition, IEEE, 2010, p. 33603367.

[38] Z. Jiang, Z. Lin, L.S. Davis, Label consistent k-svd: learning a discriminative dictionary for recognition, IEEE Trans. Pattern Anal. Mach. Intell. 35 (11) (2013) 2651–2664. https://doi.org/10.1109/ TPAMI.2013.88.

[39] R. Gribonval, K. Schnass, Dictionary identification: sparse matrix- factorization via l1-minimization, IEEE Trans. Inf. Theor. 56 (7) (2010) 3523–3539. https://doi.org/10.1109/TIT.2010.2048466.

[40] Z. Liu, C. Gao, H. Yang, Q. He, A cost-sensitive sparse representation based classification for class-imbalance problem, Sci. Program. 2016 (2016).

[41] Y. Li, C. Qi, Face recognition using hog feature and group sparse coding, in: 2013 IEEE International Conference on Image Processing, 2013, pp. 3350–3353. https://doi.org/10.1109/ICIP.2013.6738690.

[42] Y. Xu, J. Cheng, Face recognition algorithm based on correlation coefficient and ensemble-augmented sparsity, IEEE Access 8 (2020) 183972–183982. https://doi.org/10.1109/ACCESS.2020. 3028905.

[43] N. Kumar, R. Sinha, Improved structured dictionary learning via correlation and class based block formation, IEEE Trans. Signal Process. 66 (19) (2018) 5082–5095. https://doi.org/10.1109/TSP.2018. 2865442.

[44] K. Engan, K. Skretting, J.H. Hus0y, Family of iterative ls-based dictionary learning algorithms, ils-dla, for sparse signal representation, Digit. Signal Process. 17 (1) (2007) 32–49.

[45] K. Kreutz-Delgado, B. Rao, K. Engan, T. Lee, T. Sejnowski, Learning Overcomplete Dictionaries and Sparse Representations, Preparation for Submission to Neural Computation, 2000.

[46] T.L. Hansen, M.A. Badiu, B.H. Fleury, B.D. Rao, A sparse bayesian learning algorithm with dictionary parameter estimation, in: 2014 IEEE 8th Sensor Array and Multichannel Signal Processing Workshop (SAM), IEEE, 2014, pp. 385–388.

[47] T. Chaspari, A. Tsiartas, P. Tsilifis, S.S. Narayanan, Markov chain Monte Carlo inference of para-metric dictionaries for sparse bayesian approximations, IEEE Trans. Signal Process. 64 (12) (2016) 3077–3092.

[48] B. Ge, X. Li, X. Jiang, Y. Sun, T. Liu, A dictionary learning approach for signal sampling in task-based fmri for reduction of big data, Front. Neuroinf. 12 (2018) 17. https://doi.org/10.3389/fninf.2018.00017. URL, https://www.frontiersin.org/article/10.3389/fninf.2018.00017.

[49] P. Honeine, Entropy of overcomplete kernel dictionaries, Bulletin of Mathematical Sciences and Applications 16 (11 2014). https://doi.org/10.18052/www.scipress.com/BMSA.16.1.

[50] J. Xu, H. He, H. Man, Active Dictionary Learning in Sparse Representation Based Classification, arXiv Preprint arXiv:1409.5763, 2014.

[51] C. Zheng, F. Zhang, H. Hou, C. Bi, M. Zhang, B. Zhang, Active discriminative dictionary learning for weather recognition, Math. Probl Eng. 2016 (2016).

[52] L. Li, A. Scaglione, Learning hidden markov sparse models, in: 2013 Information Theory and Applications Workshop (ITA), IEEE, 2013, pp. 1–10.

[53] L. Tao, E. Elhamifar, S. Khudanpur, G.D. Hager, R. Vidal, Sparse hidden markov models for surgical gesture classification and skill evaluation, in: International Conference on Information Processing in Computer-Assisted Interventions, Springer, 2012, pp. 167–177.

[54] J. Mairal, F. Bach, J. Ponce, G. Sapiro, A. Zisserman, Supervised dictionary learning, in: Proceedings of the 21st International Conference on Neural Information Processing Systems, NIPS'08, Curran Associates Inc., USA, 2008, pp. 1033–1040. URL, http://dl.acm.org/citation.cfm?id=2981780.2981909.

[55] Q. Zhang, B. Li, Discriminative k-svd for dictionary learning in face recognition, in: 2010 IEEE Computer Society Conference on Computer Vision and Pattern Recognition, 2010, pp. 2691–2698. https://doi.org/10.1109/CVPR.2010.5539989.

[56] D. Pham, S. Venkatesh, Joint learning and dictionary construction for pattern recognition, in: 2008 IEEE Conference on Computer Vision and Pattern Recognition, 2008, pp. 1–8. https://doi.org/10.1109/CVPR.2008.4587408.

[57] G. Cybenko, Approximation by superpositions of a sigmoidal function, Math Control Signals Syst. 2 (4) (1989) 303–314.

[58] A.K. Jain, R.P.W. Duin, J. Mao, Statistical pattern recognition: a review, IEEE Trans. Pattern Anal. Mach. Intell. 22 (1) (2000) 4–37.

[59] S. Hochreiter, The vanishing gradient problem during learning recurrent neural nets and problem solutions, Int. J. Uncertain. Fuzziness Knowledge-Based Syst. 6 (02) (1998) 107–116.

[60] J. Zazo, B. Tolooshams, D. Ba, H.J.A. Paulson, Convolutional dictionary learning in hierarchical networks, in: 2019 IEEE 8th International Workshop on Computational Advances in Multi-Sensor Adaptive Processing (CAMSAP), IEEE, 2019, pp. 131–135.

[61] H.-J. Seo, Dictionary Learning for Image Style Transfer, Ph.D. Thesis, Harvard College, 2020.

[62] G. Madhuri, M.N. Kashyap, A. Negi, Telugu OCR using dictionary learning and multi-layer per-ceptrons, in: 2019 International Conference on Computing, Power and Communication Technologies (GUCON), IEEE, 2019, pp. 904–909.

[63] R. Kummari, C. Bhagvati, UHTelPCC: a dataset for Telugu printed character recognition, in: Recent Trends on Image Processing and Pattern Recognition 862, 2018, pp. 1–13.

[64] R. Rubinstein, M. Zibulevsky, M. Elad, Efficient implementation of the k- svd algorithm using batch orthogonal matching pursuit, CS Technion 40 (January 2008).

[65] D. Ciregan, U. Meier, J. Schmidhuber, Multi-column deep neural networks for image classification, in: 2012 IEEE Conference on Computer Vision and Pattern Recognition, IEEE, 2012, pp. 3642–3649.

5

Artificial intelligence—based uncertainty quantification technique for external flow computational fluid dynamic (CFD) simulations

Srinivas Soumitri Miriyala[3], Pramod D. Jadhav[2], Raja Banerjee[2], Kishalay Mitra[1]

[1]DEPARTMENT OF CHEMICAL ENGINEERING, INDIAN INSTITUTE OF TECHNOLOGY, HYDERABAD, TELANGANA, INDIA; [2]DEPARTMENT OF MECHANICAL AND AEROSPACE ENGINEERING, INDIAN INSTITUTE OF TECHNOLOGY HYDERABAD, HYDERABAD, TELANGANA, INDIA; [3]GLOBAL OPTIMIZATION AND KNOWLEDGE UNEARTHING LABORATORY, DEPARTMENT OF CHEMICAL ENGINEERING, INDIAN INSTITUTE OF TECHNOLOGY HYDERABAD, HYDERABAD, TELANGANA, INDIA

5.1 Introduction

In the quest to capture the exact physics involved in a complex engineering processes, there is a surge in adopting detailed methodologies while building the corresponding mathematical models. As a result, these first principles based models often turn to be computationally intensive in nature. This trend of moving toward building computationally expensive models is well supported by the rapid evolution of high performance computing resources and parallel computing frameworks. However, the iterative methods such as, sensitivity analysis, optimization and control, where the repeated execution of these models is a necessity, remain unattainable in real time, due to the time-expensive proposition of this approach [1]. An example could be the study of uncertainty propagation, when the considered model comprises of computationally intensive solvers based on computational fluid dynamics (CFD) [2]. One of the ways of realizing these iterative activities in realistic time frame is to emulate such time-expensive first principles—based models (FPMs) accurately and replace them with the data-driven surrogate models [3]. The idea is to generate limited number of input-output data using high-fidelity model and then use surrogates to create a close replica of the original time-expensive model. Once such well validated surrogate models are in place, they can be used extensively in iterative studies e.g., optimization, uncertainty quantification, etc. [4—11].

For example, analysis of variance (ANOVA)-based uncertainty propagation using surrogate models for complex vehicular design was proposed in Ref. [4], where the authors developed a novel adaptive sampling strategy aimed at minimizing the training sample points. Zhang et al. [5] used neural networks, fuzzy inference systems, adaptive neuro-fuzzy inference systems and support vector machines to predict ten-day ahead inflow conditions in a reservoir in China and performed the uncertainty analysis through these methods using ANOVA. This analysis considered the uncertainties due to the surrogate models and showed in this work that these are more significant than the uncertainties arising due to the physiological inputs. Studies like these show the importance of building highly accurate surrogate models. Authors in Ref. [6] focused on design of plastic injection modeling where initially they determined the significant operating conditions from a superset of features and then built an artificial neural network (ANN) in the reduced dimensional space for performing evolutionary surrogate based optimization. In Ref. [7], surrogate assisted uncertainty quantification was performed to estimate the sensitivity of parameters in Spalart-Allmaras (SA) model, used for emulating the compressor and aerodynamic stall. This uncertainty analysis provided statistically significant heuristics for implementing the turbulence models in simulations involving CFD.

Machine learning (ML) based uncertainty estimation is now becoming popular among researchers as these ML models prove to be time-cheap substitutes for first principles based models if they are trained accurately and tested thoroughly [8−10]. At the same time, it is also necessary to check the uncertainties arising due to inaccuracies in the predictions of ML models, as illustrated in the work by Begoli et al. [11]. Among various ML models, ANNs are known to be the best through the numerous recent works published in journals of international repute. Though ANNs have demonstrated the ability to handle complex nonlinearities in the data, ANNs have several avenues through which inaccuracies can crop up. These avenues are listed in the following points:

- Design of ANN is heuristics based.
- Activation functions are chosen based on users' experience.
- Training data size is chosen based on some given set of heuristic rules.
- ANNs often suffer from getting overfitted.

Hence, a unique ANN constructing framework which can simultaneous estimate all the aforementioned hyperparameters is essential for successful implementation of these networks as surrogate models for the FPMs.

Recent works aimed at intelligent estimation of these hyperparameters are gaining immense popularity [12−21]. These works are now shaping up a major area of research called automated machine learning (autoML), where focus is on eliminating the heuristics involved in building the ML models [22]. Neural architecture search (NAS) is a subtopic under autoML which deals with hyperparameter estimation in ANNs [23]. Research in NAS is mainly performed using (a) reinforcement learning [24], (b) Bayesian statistics [25], and (c) multicriteria decision-making. Some notable works in NAS include the design of ANN using a mixed integer nonlinear programming problem (MINLP) [26], multiobjective optimization problem based on prediction accuracies of different outputs [27] and ANN design based on minimization of a combination of training and testing loss [28]. Instead of combining the conflicting objectives, more suitable criteria like the Akaike Information Criteria (AIC) [29], can be used as the single objective. Further, the idea of combining the conflicting objectives is also criticized by many in literature. It is often suggested to solve those formulations as multiobjective optimization problems (MOOPs) [30]. NAS strategies only focus on architecture design, whereas there are other hyperparameters that affect the neural networks significantly. Other prominent works such as [31] focused on optimal training sample size estimation and used an ensemble of surrogate models. But, none of the works discuss about simultaneous estimation of optimal training sample size and activation function along with the architecture. The following is a brief list of challenges in modeling ANNs and constructing them as efficient surrogates:

- Models tend to overfit the data and lose generalization ability.
- Consideration of large number of inputs and problem of optimal training size is not addressed.
- Simultaneous determination of sample size and NAS considering overfitting is not addressed.

In this chapter, the aim is, therefore, to present a methodology that can perform surrogate building using ANNs along with the design of their hyperparameters in a holistic sense. The multiobjective framework helps in balancing the conflict between the parsimony and generalization ability of ANNs. Overfitting of multilayered perceptron (MLP) networks due to large number of training samples is also prevented by implementing a novel optimal sample size estimation algorithm. Since the proposed algorithm contains more than one conflicting objectives, it results in a set of Pareto optimal solutions demonstrating the trade-off among all the objectives. AIC measure is used to select a single ANN model from this list. It is a suitable model evaluation criterion which is well-known for penalizing the complexity in the models, thereby preferring the models with less overfitting [29]. The success of this approach has been demonstrated while performing surrogate assisted uncertainty analysis in design of a cruciform missile system. First, the optimal ANN model is built to establish the map between the following three inputs: (1) flux, (2) discretization order, and (3) models for turbulence and three outputs; (1) lift coefficient (C_L), (2) drag coefficient (C_D), and (3) coefficient of the rolling moment (C_M), whose data is obtained from ANSYS fluent software, while simulating the supersonic-flow over the missile using CFD. Then the sensitivity of these inputs are studied using the ANOVA approach resulting in first-of-its-kind quantification of uncertainties arising due to design parameters while studying the cruciform missile system. This task is accomplished in two steps: first, optimally trained surrogate models are built, which closely mimic the time-expensive CFD model using minimum possible high-fidelity data points obtained through the proposed algorithm; and second, these optimal surrogates are used to generate the outputs in the design of experiments required for uncertainty analysis, subsequently leading to the application and results of ANOVA.

The rest of the chapter contains a section on formulation which first briefly describes the model for emulating the compressible flow over the cruciform missile and then presents the proposed algorithm in detail for optimal construction of ANNs using the evolutionary NAS strategy, followed by the section on results and discussions which present the analysis conducted in this study and finally a brief summary of work as presented in the conclusions section.

5.2 Formulation

5.2.1 Governing equations and model for compressible flow over missile

The supersonic-flow over the missile body is governed by two dominant forces: lift and drag which are generated as a result of differences in pressure and friction, respectively. The corresponding pressure force $\vec{F}_P$ and the viscous force $\vec{F}_V$ are defined in Eqs. (5.1) and (5.2), respectively, where pressure and shear stress are denoted by P_F and τ_w. The surface area vector in Eqs. (5.1) and (5.2) is denoted by $d\vec{A}$.

$$\vec{F}_P = \int -P_F d\vec{A} \tag{5.1}$$

$$\vec{F}_V = \int \tau_w d\vec{A} \tag{5.2}$$

These effect of these forces on the missile body can be quantified using the coefficients of lift and drag as defined in Eqs. (5.3) and (5.4), respectively, where fluid density is given by ρ, cross-sectional area is given by A, and velocity is given by u. Further, the interactions between vortex and wings results in generation of immense amount of turbulence which leads to the rolling moment $\vec{M}$ which can be determined using $\vec{F}_P$ and $\vec{F}_V$. The effect of rolling moment on the missile body is quantified using the parameter coefficient of rolling moment as defined in Eq. (5.5) where L is the wing span.

$$C_L = \frac{\left|\vec{F}_P\right|}{\frac{1}{2}\rho A u^2} \tag{5.3}$$

$$C_D = \frac{\left|\vec{F}_V\right|}{\frac{1}{2}\rho A u^2} \tag{5.4}$$

$$C_M = \frac{|\vec{M}|}{\frac{1}{2}\rho u^2 L} \qquad (5.5)$$

In this study, we considered the control volume V where the governing equations for one component fluid are defined on a differential area segment dA to model the aforementioned properties in integral Cartesian form. Flux vector splitting (FVS) is a popular type of discretization that is currently applied as a result of improvement in computational facilities. Another well-known technique for discretization is the flux difference splitting (FDS). While FVS methods are known to be more robust than FDS methods, particularly in shear layers, they are considered less accurate. In such regions, on the contrary, variants of FDS such as Roe's method are known to be more efficient. Thus, advection upstream splitting method (AUSM) is developed as a trade-off between the FDS and FVS methods and is currently in implementation for modeling the shear layer regions. In this work, the impact of different flux schemes on C_D, C_L, and C_M is studied. This is the first objective of the uncertainty analysis conducted in this work. In general, another commonly used parameter setting in ANSYS FLUENT for solving the CFD simulations is determination of properties at centers of the cells. In order to evaluate the face values which are utilized in the Naiver-Stokes equation, the values at cell centers are considered for interpolating the desired values at face centers. This work focuses on uncertainty arising due to upwind schemes used for interpolation in ANSYS FLUENT on C_D, C_L, and C_M as the second objective of uncertainty analysis. Lastly, from several turbulence models, SA model was selected, by performing sensitivity analysis, for solving the transport equation involving the turbulent kinematic viscosity term. This is to enable the modeling of compressible flow over the missile body with severe pressure gradients. The SA Model in ANSYS FLUENT is parametrized by several constants and the variation in the value of one particular constant C_{b1} was observed to effect the outputs significantly. Thus, the effect of uncertainty arising due to SA model by varying C_{b1} is set as the third objective of surrogate assisted uncertainty analysis.

Fig. 5.1 shows the geometry of the typical cruciform tactical missile system considered in the present study. These details are obtained from the literature [32].

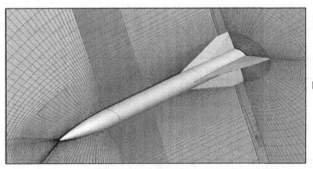

❑ Geometrical description
  ➢ Fuselage
  ➢ 4 Wings
  ➢ 4 Fins (or tails)
  ➢ Diameter = 35 mm
❑ Boundary Conditions
  ➢ Roll angle = 22.5°
  ➢ Angle of incidence = 21.7°
  ➢ Mach number = 2
  ➢ Total pressure = 1.8×10^6 Pa
  ➢ Total temperature = 300 K

FIGURE 5.1 Missile geometry along with meshing of the domain [32].

A computational domain of hemispherical inlet with a radius of 80D, (where D is missile diameter) and cylindrical far-field is used to enclose the missile system. The origin of the computational domain was set at the tip of the missile nose. A mesh with a total of 53 lakh elements was created to run the simulations and generate the outputs. Several such function calls were needed to generate the training and testing data. The ANN models were then trained using the proposed algorithm and subsequently used in the surrogate assisted uncertainty analysis.

5.2.2 Evolutionary neural architecture Search strategy

A hidden layer in the neural network can be geometrically realized as a hyperplane capable of modeling linearly separable data. With the nature of data not known a priori, it is better to have more number of such layers in the ANN instead of assuming that one layer might be sufficient to model the data, especially when its nature is nonlinear. If it is checked in a flexible manner with an option open for zero nodes to a certain fixed number of nodes appearing in several hidden layers during the NAS process, the result might show actually whether one layer is sufficient or more than one layer is needed for mapping the input-output relationship. On the other hand, the increase in the number of layers might lead to the case of overfitting due to the increase in number of parameters. We can observe the first trade-off here which sets the aim to go for higher accuracy with a smaller number of parameters to avoid overfitting. The second trade-off appears as a result of overfitting due to overtraining. Keeping the architecture fixed in an ANN, a greater number of samples give flexibility to fit the data well. However, more data points for training can lead to the case of overfitting clearly showing us the second trade-off between number of data points considered for training and predictability. This sets the platform for a three objective optimization formulation: predictability maximization with minimization of total number of parameters in ANN and minimization of training data (see Table 5.1).

As mentioned before, a generic NAS algorithm has been formulated, where several number of layers can be assumed a priori and the number of nodes in those layers can be found by an optimization algorithm. ANNs with maximum of three hidden layers were adopted in this work. However, the structure of the algorithm is flexible to adopt that

Table 5.1 Nonlinear programming problem formulation for evolutionary NAS strategy.

Nonlinear objectives	Integer decision variables
Minimize $-R^2$ Minimize N Minimize P	Number of nodes (N_i) in the three hidden layers and choice of activation defined by binary variable N_{AF} (1 for log-sigmoid and 2 for tan-sigmoid) are the decision variables. The lower and upper bounds for N_1, N_2 and N_3 are 1 and 8, 0 and 7, and 0 and 7, respectively.

change if accuracy obtained is not up to the mark using the three intermediate layers. Therefore, the number of nodes in each of the hidden layers and choice for activation function are considered as the decision variables for the MOOP. Though the first layer is having 1 as the lower bound for the number of nodes, it is 0 as lower bound for the second and third hidden layers. This ensures 0 to appear for second and third layer as the number of nodes, if necessary, showing the emergence of a single layered ANN if found suitable to represent the data. Any other number than 0 appearing in the second and third layer can signify the need for an ANN with higher number of layers. Nondominated sorting genetic algorithm, NSGA-II, which is one of the well-established MOOP algorithms, has been used to solve this problem [30]. Each population member will represent an architecture. Once the architecture is given by NSGA-II, ANN is trained using the high-fidelity data (whose optimal number will be determined) and test accuracy is evaluated in terms of R^2 (Eq. 5.6), where y and $\widehat{y}$ are the original and predicted values, respectively.

$$R^2 = \left(\frac{\text{cov}(y, \widehat{y})}{\sqrt{\text{var}(y)\text{var}(\widehat{y})}} \right)^2 \tag{5.6}$$

$$\text{cov}(y, \widehat{y}) = n \sum_{i=0}^{n} y^{(i)}\widehat{y}^{(i)} - \sum_{i=0}^{n} \widehat{y}^{(i)} \sum_{i=0}^{n} y^{(i)} \quad \text{and} \quad \text{var}(y) = n \sum_{i=0}^{n} y^{(i)2} - \left(\sum_{i=0}^{n} y^{(i)} \right)^2$$

This completes the evaluation of all objectives (R^2, P and N) for the given population. The algorithm's flowsheet is shown in Fig. 5.2.

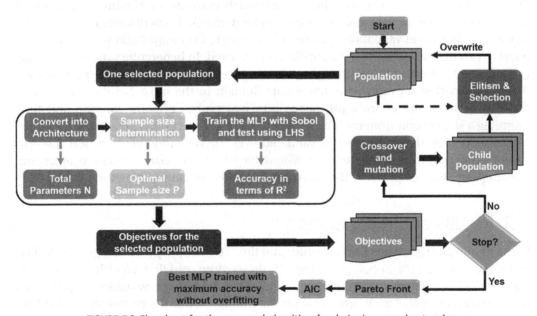

FIGURE 5.2 Flowsheet for the proposed algorithm for designing neural networks.

5.2.3   Determination of sample size for training the ANN

The sample size determination (SSD) algorithm is inspired by the K-fold Cross Validation technique of machine learning. This is often used to identify the most generalized model among several alternatives available. The methodology is explained here:

(a) Once an MLP architecture [3-N_1-N_2-N_3-1] is chosen, a sample size is selected which is popularly used as K times the number of inputs (here it is 3), where K is number of folds. These points are created using Sobol sampling scheme, one of the methods for creating samples most uniformly in the input space. The computationally expensive FPM is used to generate the corresponding outputs.

(b) Next, the generated set of input-outputs is divided into K folds and data in K-1 folds are used for training a model and data in the left over fold is used for calculating the validation error. Since this can be performed in K random ways, the mean of all these errors is considered as the cross validation (CV) error.

(c) The number of points are then incremented and the corresponding CV error is evaluated as described in previous steps. Two consequent CV values and their corresponding sample sizes are used for evaluating the slope of the CV error vs. sample size curve which is then compared with a tolerance value.

(d) These steps are continued till a termination criterion is satisfied. A possible criterion can be defined as no further change in validation error (usually expressed by a tolerance) with increase in sampling size.

It has been shown in the literature that the K-fold based approach prevents overfitting. However, one can realize that the approach consumes a significant amount of computational time as the validation needs to be performed repetitively. To prevent the repeated training and validation, and thereby reducing the computational complexity, a novel hypercube sampling based algorithm is presented. In hypercube sampling, a single validation set is formed by sampling the input space in the most uniform manner such that the validation set represents the inputs domain in the most holistic way. This is done by dividing the entire input space into hypercubes of same volume and then sampling a single point from each hypercube. The rest of the points are used for training. Model trained on the training set is validated over the validation set to report the validation error. The remaining steps of K-fold based sample size estimation algorithm remain same. Fig. 5.3 presents the working of hypercube sampling pictorially (Fig. 5.4).

5.3 Results and discussions

In the input space generated by C_{b1}, order and flux, 100 sample points are generated first using low-discrepancy Sobol [33] plan. At these input variable combinations, high-fidelity CFD simulations are run to generate the output space variables (C_D, C_L, and C_M). Keeping 20 points aside from this set for testing, rest of the 80 points are used for training and validation of the ANN surrogate model. For the sake of brevity, results

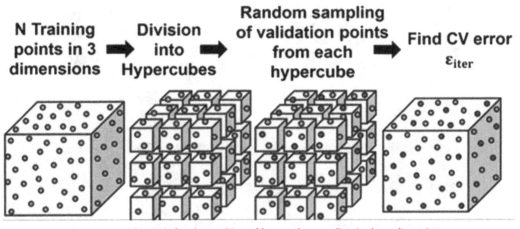

FIGURE 5.3 Schematic for the working of hypercube sampling in three dimensions.

corresponding to only one output (i.e., C_D) are shown. As the proposed Neural Architectural Search algorithm is multiobjective in nature, it resulted into a three dimensional Pareto front (Fig. 5.4), where each solution in the set represents a different ANN architecture keeping a balance among accuracy, size and model parsimoniousness. Pareto solutions are obtained using an established multiobjective evolutionary algorithm called NSGA-II. The choice of NSGA-II is purely related to the legacy of this group of researchers and this could have been replaced by any other well-established multi-objective evolutionary algorithm. NSGA-II has been run several times with different initial populations to avoid aberrations in results due to stochasticity involved in the algorithm. It has been also tested for sufficiently large number of generations in order to ensure that the Pareto points obtained are all rank one and the Pareto solutions got enough opportunity to converge. These points are presented in Table 5.2 along with their AIC values, which represent the extent of overfitting each model corresponding to every Pareto solution possesses. The model with minimum AIC value has been chosen and shown as highlighted in Table 5.2. Similarly, Pareto solutions corresponding to other two outputs (C_L, and C_M) are also generated separately and the solution with the least AIC value from each set is chosen as shown in Table 5.3. The results show the presence of multiple hidden layers in the networks justifying the necessity of studying a NAS algo-rithm. In fact, the competitive Pareto solutions as shown in Fig. 5.4 speak about the importance of the proposed NAS study instead of setting the surrogate hyperparameters based on individual experience. It has been observed that all 80 simulations are utilized for building these highly accurate surrogate models (R^2 values being close to 1) in spite of aiming for minimum size determination as an objective. So, the proposed approach is found to build highly accurate surrogates.

For studying the uncertainties due to the single factors (C_{b1}, order, and flux) and the interactions between them on C_D, C_L, and C_M, the optimally trained ANN surrogate

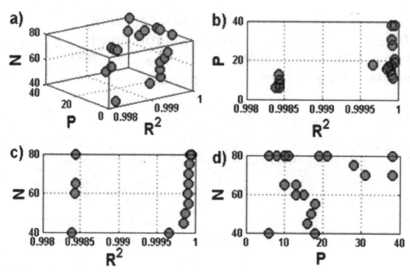

FIGURE 5.4 (A) Three dimensional Pareto front as the solution of NSGA-II. Subfigures (B–D) show the representation of Pareto front in two dimensions.

Table 5.2   List of solutions corresponding to the Pareto front shown in Fig. 5.4.

N_1	N_2	N_3	N_{AF}	R^2	P	N	AIC
1	1	0	2	0.998434	8	80	−684.871
1	0	0	1	0.998385	6	40	−283.561
1	0	0	2	0.998434	6	80	−688.868
1	2	1	2	0.998425	13	60	−500.495
1	1	0	1	0.998434	8	80	−573.975
1	1	1	1	0.998436	10	80	−570.074
1	1	1	2	0.998431	10	65	−550.869
2	0	0	2	0.999937	11	80	−936.072
2	1	0	2	0.999912	13	65	−730.999
2	1	1	2	0.999906	15	60	−663.406
2	2	0	2	0.999871	17	50	−528.579
2	5	2	1	0.999954	38	80	−800.452
2	2	1	2	**0.999951**	**19**	**80**	**−946.119**
2	1	3	1	0.999953	21	80	−833.376
3	0	0	1	0.999854	16	45	−408.081
3	1	0	2	0.999669	18	40	−376.664
3	3	0	1	0.999924	28	75	−729.482
3	5	0	2	0.999918	38	70	−745.239
3	1	0	1	0.999914	18	55	−531.792
6	0	0	1	0.999917	31	70	−664.857

Table 5.3 Selected architectures for emulating the nonlinear CFD maps corresponding to C_D, C_M, and C_L.

Feature	N_1	N_2	N_3	N_{AF}	R^2	P	N
C_D	2	2	1	2	0.999951	19	80
C_M	6	0	0	2	0.991923	31	80
C_L	2	0	0	2	0.999999	11	80

models are used next. These ANN models are simulated to generate the outputs for the inputs obtained according to the DoE for ANOVA presented in Table 5.4. Subsequently, ANOVA analysis was performed on these outputs and the corresponding results are presented in Table 5.5. In one-way ANOVA, a linear model is assumed between a single categorical input and the variable of interest. The null hypothesis (H_0) is defined as: all the groups/levels follow a same distribution and therefore have same group means. The alternative hypothesis (H_1) states that at least one group follows a different distribution and thus effects the variable of interest differently than the other groups. Utilizing the DoE in Table 5.4, the test metric is evaluated and compared with the critical value obtained from the F distribution and a decision is taken based on the comparison between test metric and critical value. In case of ANOVA, if test metric ($F_{calculated}$) < critical value ($F_{critical}$), the Null hypothesis is accepted, otherwise the alternate hypothesis is accepted. In n-way ANOVA when $n \geq 2$, the effect of interactions is studied in addition to the main effects.

The Null hypothesis is considered a the absence of interaction effects on the outputs.

- On the outputs C_D and C_L, Flux variations had statistically significant main effect but no effect has been shown on C_M.
- On the one hand, all outputs are affected by the variation in order of discretization, but none of them is affected by C_{b1}, when altered alone.

Table 5.4 Plan for Design of Experiments to carry out the three-way ANOVA analysis to evaluate the main effect and interaction effect of three inputs on the desired output Y (in this case C_{b1}). It is assumed that there exist n number of points in each level and j, q, and k indicate the number of levels for turbulence model, flux and order, respectively.

	C_{b1j}		
$Flux_q$	1	Y_{1jqk}	$Order_k$
	2	Y_{2jqk}	
	...	...	
	i	Y_{ijqk}	
	n	Y_{njqk}	

Table 5.5 Three-factor ANOVA analysis to study the statistical significance of main and interaction effects of considered design feature parameters (F_1: Flux, F_2: Order of discretization, and F_3: SA model constant C_{b1} on the variables of interest for confidence level alpha $= 0.01$).

Input feature	Variable of interest	Test metric ($F_{calculated}$)	Critical value ($F_{critical}$)	Result
F_1	C_D	13.68	6.90	Accept H_1
F_1	C_L	17.90	6.90	Accept H_1
F_1	C_M	02.38	6.90	Accept H_0
F_2	C_D	13.53×10^1	4.83	Accept H_1
F_2	C_L	11.03×10^1	4.83	Accept H_1
F_2	C_M	49.9×10^1	4.83	Accept H_1
F_3	C_D	0.08	4.83	Accept H_0
F_3	C_L	0.08	4.83	Accept H_0
F_3	C_M	0.12	4.83	Accept H_0
F_1 and F_2	C_D	64.29×10^3	4.84	Accept H_1
F_1 and F_2	C_L	14.26×10^4	4.84	Accept H_1
F_1 and F_2	C_M	11.76×10^1	4.84	Accept H_1
F_1 and F_3	C_D	0.02	4.84	Accept H_0
F_1 and F_3	C_L	0.02	4.84	Accept H_0
F_1 and F_3	C_M	0.17	4.84	Accept H_0
F_2 and F_3	C_D	0.10	3.53	Accept H_0
F_2 and F_3	C_L	0.10	3.53	Accept H_0
F_2 and F_3	C_M	0.10	3.53	Accept H_0
F_1, F_2 and F_3	C_D	1.71	3.56	Accept H_0
F_1, F_2 and F_3	C_L	1.19	3.56	Accept H_0
F_1, F_2 and F_3	C_M	1.94	3.56	Accept H_0

- Flux and order combined affects C_D and C_L significantly higher compared to C_M.
- C_{b1} and Flux combined and C_{b1} and order combined show no effects on outputs.
- Further, no three-factor interactions were found to be statistically significant on any of the outputs C_D or C_L or C_M.

5.4 Conclusions

In this chapter, the idea of ANN surrogate assisted uncertainty quantification in design study of tactical missile system is presented. First, the shortcomings of ANNs are identified and a novel evolutionary strategy for automated hyperparameter estimation is presented. This algorithm based on the trade-off between complexity and accuracy of the network is formulated as a multiobjective optimization problem whose decision variables are the hyperparameters of the network: architecture, activation function and optimal sample size. Next, flux type, order, and turbulence model parameter are identified as inputs, and the effect of uncertainty in these variables is studied on the lift coefficient, the drag coefficient, and the coefficient of the rolling moment. The data is generated using the high-fidelity simulations, and optimal ANNs are trained between these inputs and outputs with an accuracy of around 99%. Subsequently, the uncertainty quantification is performed using the predictions obtained from the optimal ANN model and ANOVA analysis. This method quantified for the first time, the uncertainties associated with design aspects while modeling the compressible flow over tactical missiles.

Acknowledgments

Authors acknowledge the support provided by the Department of Science and Technology, Government of India through the National Supercomputing Mission project [IISC/CHE/F089/2020–21/G354] for this work.

References

[1] R.G. Regis, July). A survey of surrogate approaches for expensive constrained black-box optimization, in: World Congress on Global Optimization, Springer, Cham, 2019, pp. 37–47.

[2] A. Mogilicharla, T. Chugh, S. Majumdar, K. Mitra, Multi-objective optimization of bulk vinyl acetate polymerization with branching, Mater. Manuf. Process. 29 (2) (2014) 210–217.

[3] Y. Jin, Surrogate-assisted evolutionary computation: recent advances and future challenges, Swarm Evol. Comput. 1 (2) (2011) 61–70.

[4] K. Kwon, N. Ryu, M. Seo, S. Kim, T.H. Lee, S. Min, Efficient uncertainty quantification for integrated performance of complex vehicle system, Mech. Syst. Signal Process. 139 (2020) 106601.

[5] X. Zhang, H. Wang, A. Peng, W. Wang, B. Li, X. Huang, Quantifying the uncertainties in data-driven models for reservoir inflow prediction, Water Resour. Manag. 34 (4) (2020) 1479–1493.

[6] Q. Feng, L. Liu, X. Zhou, Automated multi-objective optimization for thin-walled plastic products using Taguchi, ANOVA, and hybrid ANN-MOGA, Int. J. Adv. Manuf. Technol. 106 (1) (2020) 559–575.

[7] X. He, F. Zhao, M. Vahdati, Uncertainty quantification of spalart–allmaras turbulence model coefficients for simplified compressor flow features, J. Fluids Eng. 142 (9) (2020) 091501.

[8] B. Chen, D.R. Harp, Y. Lin, E.H. Keating, R.J. Pawar, Geologic CO_2 sequestration monitoring design: a machine learning and uncertainty quantification based approach, Appl. Energy 225 (2018) 332–345.

[9] S. Chan, A.H. Elsheikh, A machine learning approach for efficient uncertainty quantification using multiscale methods, J. Comput. Phys. 354 (2018) 493–511.

[10] Y. Zhou, S. Zheng, Uncertainty study on thermal and energy performances of a deterministic parameters based optimal aerogel glazing system using machine-learning method, Energy 193 (2020) 116718.

[11] E. Begoli, T. Bhattacharya, D. Kusnezov, The need for uncertainty quantification in machine-assisted medical decision making, Nat. Mach. Intell. 1 (1) (2019) 20–23.

[12] P. Nowakowski, K. Szwarc, U. Boryczka, Vehicle route planning in e-waste mobile collection on demand supported by artificial intelligence algorithms, Transp. Res. D Transp. Environ. 63 (2018) 1–22.

[13] F. Ahmad, A. Abbasi, J. Li, D.G. Dobolyi, R.G. Netemeyer, G.D. Clifford, H. Chen, A deep learning architecture for psychometric natural language processing, ACM Trans. Inf. Syst. 38 (1) (2020) 1–29.

[14] E. Moen, D. Bannon, T. Kudo, W. Graf, M. Covert, D. Van Valen, Deep learning for cellular image analysis, Nat. Methods 16 (12) (2019) 1233–1246.

[15] S. Ardabili, A. Mosavi, M. Dehghani, A.R. Várkonyi-Kóczy, Deep learning and machine learning in hydrological processes climate change and earth systems a systematic review, in: International Conference on Global Research and Education, Springer, Cham, September 2019, pp. 52–62.

[16] A. Bauer, A.G. Bostrom, J. Ball, C. Applegate, T. Cheng, S. Laycock, J. Zhou, Combining computer vision and deep learning to enable ultra-scale aerial phenotyping and precision agriculture: a case study of lettuce production, Hortic. Res. 6 (1) (2019) 1–12.

[17] S. Sajeev, A. Maeder, S. Champion, A. Beleigoli, C. Ton, X. Kong, M. Shu, Deep learning to improve heart disease risk prediction, in: Machine Learning and Medical Engineering for Cardiovascular Health and Intravascular Imaging and Computer Assisted Stenting, Springer, Cham, 2019, pp. 96–103.

[18] P. Schneider, W.P. Walters, A.T. Plowright, N. Sieroka, J. Listgarten, R.A. Goodnow, G. Schneider, Rethinking drug design in the artificial intelligence era, Nat. Rev. Drug Discov. 19 (5) (2020) 353–364.

[19] N. Justesen, P. Bontrager, J. Togelius, S. Risi, Deep learning for video game playing, IEEE Trans. Games 12 (1) (2019) 1–20.

[20] N. Naik, B.R. Mohan, Stock price movements classification using machine and deep learning techniques-the case study of indian stock market, in: International Conference on Engineering Applications of Neural Networks, Springer, Cham, May 2019, pp. 445–452.

[21] C.Y. Lu, D. Suhartanto, A.I. Gunawan, B.T. Chen, Customer satisfaction toward online purchasing services: evidence from small & medium restaurants, Int. J. Appl. Bus. Res. 2 (01) (2020) 1–14.

[22] F. Hutter, L. Kotthoff, J. Vanschoren, Automated Machine Learning: Methods, Systems, Challenges, Springer Nature, 2019, p. 219.

[23] T. Elsken, J.H. Metzen, F. Hutter, Neural architecture search: a survey, J. Mach. Learn. Res. 20 (1) (2019) 1997–2017.

[24] Y. Jaafra, J.L. Laurent, A. Deruyver, M.S. Naceur, Reinforcement learning for neural architecture search: a review, Image Vis. Comput. 89 (2019) 57–66.

[25] N. Nayman, A. Noy, T. Ridnik, I. Friedman, R. Jin, L. Zelnik-Manor, Xnas: Neural Architecture Search with Expert Advice, arXiv preprint arXiv:1906.08031, 2019.

[26] V. Dua, A mixed-integer programming approach for optimal configuration of artificial neural networks, Chem. Eng. Res. Des. 88 (1) (2010) 55–60.

[27] F. Boithias, M. El Mankibi, P. Michel, Genetic algorithms based optimization of artificial neural network architecture for buildings' indoor discomfort and energy consumption prediction, in: In Building Simulation 5, June 2012, pp. 95–106 (Tsinghua Press). No. 2.

[28] A.R. Carvalho, F.M. Ramos, A.A. Chaves, Metaheuristics for the feedforward artificial neural network (ANN) architecture optimization problem, Neural Comput. Appl. 20 (8) (2011) 1273–1284.

[29] H. Akaike, Information theory and an extension of the maximum likelihood principle, in: Selected Papers of Hirotugu Akaike, Springer, New York, NY, 1998, pp. 199–213.

[30] K. Deb, Multi-objective optimisation using evolutionary algorithms: an introduction, in: Multi-objective Evolutionary Optimisation for Product Design and Manufacturing, Springer, London, 2011, pp. 3–34.

[31] J. Müller, C.A. Shoemaker, Influence of ensemble surrogate models and sampling strategy on the solution quality of algorithms for computationally expensive black-box global optimization problems, J. Glob. Optim. 60 (2) (2014) 123–144.

[32] M. Khalil, A. Hashish, H.M. Abdalla, A preliminary multidisciplinary design procedure for tactical missiles, Proc. Inst. Mech. Eng. G J. Aerosp. Eng. 233 (9) (2019) 3445–3458.

[33] A. Forrester, A. Sobester, A. Keane, Engineering Design via Surrogate Modelling: A Practical Guide, John Wiley & Sons, 2008.

6

Contrast between simple and complex classification algorithms

Divy Dwivedi[1], Ashutosh Ganguly[2], V.V. Haragopal[3]

[1]MEDIBUDDY, BENGALURU, KARNATAKA, INDIA; [2]EMALPHA, MUMBAI, INDIA; [3]AIZENALGO
PRIVATE LIMITED, KOMPALLY, HYDERABAD, INDIA

6.1 Introduction

Statistical methods form the backbone of many industries that use data as a primary source of information for insights, modeling and predictions. Statistics mainly deals with contrivances to understand data and extract valuable information from the same. A prominent topic in statistical learning is classification. Classification is the process of tagging a particular data point to its corresponding category to group it with other similar data points. Data can be qualitative and quantitative, a data set is said to be quantitative if one can assign numerical values to explanatory variables to make sense of the data. If one can get meaningful numerical answers to questions like "How much?" or "How many?" then quantitative data is in play. On the other hand, qualitative data is nonnumerical data that approximates and characterizes the subjectivity of the variables, for example, the door is black, the sky is blue. For the example of qualitative data, one cannot typically measure the value because the color states the quality of the door. In this chapter, one primarily learns the approaches to quantify and predict the qualitative responses, identify useful variables for classification, and fit algorithms to see their performance on the data set.

In our chapter, the authors have taken up the data set of various music genres. Music taste for each individual may vary largely, therefore music is divided into many different genres which further have their subcategories. The first challenge in this classification is identifying the features based on which one can differentiate a genre of music from another. Music genres are hard to describe systematically and consistently due to their inherently subjective nature.

Therefore, this chapter answers the following questions:

- What features of music can be used to classify them into various genres?
- How accurately can music genres be classified using classical statistical methods and complex algorithms like neural networks?

Statistical Modeling in Machine Learning. https://doi.org/10.1016/B978-0-323-91776-6.00016-6

Here the authors have used the GTZAN data set and have classified them into 10 categories. By the end of this chapter, one will have a clear understanding and appreciation for the various statistical methods as well as the capabilities of a neural network. It makes sense that one compares the results of different methods over the same data set as that broadens the perspective of a reader to get a practical understanding of the process involved and singularly look at the performance of the algorithm. The following section explores the features of music which are going to be used in the coming sections to classify the data set into the respective categories.

Few research papers have outlined in detail the wants of the GTZAN data set [1]. They explore the faults and inefficacies of the data set. It is highly recommended to go through the paper to get a deeper understanding of the data set and more importantly how to deeply analyze a data set.

6.2 Data preprocessing and feature extraction

This section entails the information about the data set used and the features used to classify the musical tracks into their respective genres. The GTZAN data set contains music of 10 different genres, each sample of music being 30 seconds long. The 10 genres in the data set are rock, pop, classical, blues, country, disco, jazz, metal, reggae, and hip-hop. The music files are of wav format and are single-channel recorded at 22000 and 50 (22050) Hz [2]. The total number of musical pieces in the data set is 10,000. Additionally, the authors have used Python 3 as the platform for running and creating our model. Python libraries make it easier for analysis. Here, "librosa" is used to extract the features and analyze the audio files. The authors have used NumPy and Pandas for data preprocessing and feature extraction, as for modeling Scikit Learn is used from which K-nearest neighbor (KNN), linear discriminant analysis (LDA), and quadratic discriminant analysis (QDA) models have been directly imported. TensorFlow and Keras are used for training the neural network, finally, Matplotlib is used for graphical representations of results.

6.2.1 Data preprocessing

After extracting the features mentioned above, let's split the data set into training and testing data sets in a 4:1 ratio. The authors have resolved the categories. One does so using the LabelEncoder and OneHotEncoder library from Python. The authors have extensively used IBM SPSS Statistics to analyze the data set. On extraction, one sees that in total there are 60 features that have been considered in the chapter. Of those, let's drop "length" and "filename," which refer to the length of each audio file, i.e., 3 s long, and the name of the audio file respectively.

6.2.2 Feature study

This section entails the features used to classify the audio files. Feature extraction is an important part of a study. This can make a huge difference in the results and inferences obtained from the data set. Therefore it is highly recommended to make a thorough study of the data set to identify the best features. The following features have been extracted from the audio files to make a reasonable classification.

6.2.2.1 Chroma

The chroma is a set of midlevel features representing the tonal content; it is used since it is not affected by the change in instrumentation, timbre or dynamics. Chroma features are important while doing semantic analysis of the audio file, examples could include recognition of the set of chords being played and/or how similar or dissimilar two audio files are based on their harmonic nature. One first performs short-time Fourier transforms on the audio files, due to its exponential nature and then take a logarithm of it to bring the representation into a linear scale. In simple words, this feature consists of 12 bands that represent musical notes, namely, C, C#, D, D#, E, F, F#, G, G#, A, A#, B [3]. All frequencies for an audio file are associated with one of these bands and hence an audio file has now been converted into a linear and discrete visual representation of it.

6.2.2.2 Root-mean-square (RMS)

RMS is an important feature since it can be used to represent the energy captured by a wave of fixed length. RMS is an indicator of loudness, for instance, louder sounds will contain more energy hence a metal genre sound wave will have a higher RMS than a wave belonging to the blues genre [3]. An interesting fact about the RMS is that it is less sensitive to outliers because for each frame of an audio signal, rather than taking a single value, the sum of the root square of the mean is taken, which makes it a good feature to take into account. It can be easily calculated using the following formula:

$$rms = \sqrt{\frac{x_1^2 + x_2^2 + \ldots + x_n^2}{n}} \tag{6.1}$$

6.2.2.3 Centroid

Centroid is a feature in the spectral set that can be calculated by weighing the frequencies and then taking their mean

$$centroid(t) = \frac{\sum_i S[k_i, t] * freq[k_i]}{\sum_j S[j, t]} \tag{6.2}$$

where i and j range from 0 to n − 1. Here t represents the time frame (1 s), n is frequency values per second, S represents a spectrogram of the sound wave which can be generated using Fourier transformation on the wave and *freq* is the collection of frequencies found in the kth row of the spectrogram. Spectral centroid can be used to understand the brightness of the sound wave. Intuitively a certain collection of higher frequencies would make a wave sound brighter than if it were replaced by the same collection of frequencies in the lower octave.

6.2.2.4   Bandwidth

Bandwidth is another feature in the spectral feature set that can be calculated by subtracting the modulus of the highest and modulus of the lowest frequencies. It can be measured using the following formula:

$$\text{bandwidth}(t) = \sum_i S[k_i, t] * (freq[k_i, t] - \text{centroid}[t]^p)^{1/p} \tag{6.3}$$

Where i takes values from 0 to n − 1. The rest of the symbols have the same meaning as in the spectral centroid. The default value of p is set to 2, which makes it a second-order bandwidth. The unit of bandwidth is Hertz. This can be used to specify the dynamic range of the sound.

6.2.2.5   Zero-crossing rate

This feature gives us a count of how many times the wave crosses the x-axis. It is an important feature to take into account because it gives us a representation of the smoothness of the wave by telling the number of times it becomes from positive to negative and vice versa. For instance, the zero-crossing rate of an audio file of the metal genre will be higher than that of audio belonging to the blues genre. It can be calculated using the following expression:

$$zcr = \frac{1}{2} \sum_i |sgn(s_i) - sgn(s_{i+1})| \tag{6.4}$$

Where i takes values from $t * K$ to $(t + 1) * K - 1$, sgn is the signum function, s_i is the amplitude of the ith frame and K is the number of frames in the audio signal. One of the major applications of zero crossing rate (ZCR) is in distinguishing between high pitched sounds and percussive sounds.

6.2.2.6   Roll-off

The third feature in the set of spectral features, roll-off gives an idea about the shape and size of the sound wave. The roll-off (spectral) for an audio wave is a benchmark frequency decided by the modeler below which a certain constant amount of wave energy (say 50%) lies.

6.2.2.7   Tempo

This is the number of beats a fixed length of our sound wave displays. Here the fixed length is 1 s hence the unit beats per second (bps).

6.2.2.8   Mel-frequency cepstral coefficients

The Mel-frequency cepstral coefficients (MFCCs) are a set of features just like chroma or spectral. MFCCs were developed at MIT during the late 1960s to study the echoes in seismic audio [4]. It also is used to model the characteristics of the human voice. MFCCs are one of the most important characteristics of the sound which will be used in this project. For calculating the cepstrum, first take the discrete Fourier transform of the signal, followed by a logarithm which is finally again taken a Fourier inverse of, hence the spectrum of a spectrum called cepstrum.

6.2.2.9   Harmony

Harmony is the superposition of sounds, which can be analyzed by a human ear. This means when particular sound frequencies with specific amplitude and frequency, pitches(tones or notes) or chords occur simultaneously. It is easier to understand the harmony in musical terms, for instance, in a piano staff, the key A4 has the standard frequency like 440 Hz and the pitch of the same is 69. Now, one knows that the notes repeat in a cycle of 12, therefore, this brings us to a pitch 81, which is A5, having the frequency 880 Hz, an integral multiple of 440 Hz (two times). Thus, if for an audio wave, consider that A4 is a fundamental frequency, then A5 is harmony for the same.

6.3 Data modeling

6.3.1   Fitting linear discriminant analysis

LDA is a classification method similar to logistic regression with a different way to model $P(Y = m|X = x_m)$. Here rather than calculating the probability earlier mentioned, let's try to calculate $P(X = x_m|Y = m)$ and then flip it around using the Bayes theorem [5]. The assumptions involved in this method are:

1. π_k denotes the prior probability that a randomly chosen item from the data set belongs to the *m*th class.
2. Let $f_k(X)P(X = x_m|Y = m)$ denote the density function of X for an observation that comes from the *m*th class. The larger $f_k(X)$ the higher is the probability that a data point (observation) belongs to the *m*th class.
3. For LDA, f(x) can be assumed of the form

$$f(x) = \frac{1}{(2\pi)^{p/2}\sum^{1/2}} exp\left(-\frac{1}{2}(x-\mu)^T \sum^{-1}(x-\mu) \right) \qquad (6.5)$$

A fundamental assumption for deriving the above equation is that observations come from an MGD (multivariate Gaussian distribution). μ_m is the class-specific mean vector and Σ is the common covariance matrix which is common to all M classes.

Now, the function above is substituted into Bayes' theorem, and the posterior probability for each observation is calculated to assign the respective class to it [3]. The function more generally used is a rearrangement of posterior probability which can be obtained by taking a log and using little algebra, the following is obtained

$$\delta_m(x) = x^T \Sigma^{-1} \mu_m - \frac{1}{2}\mu_m^T \Sigma^{-1}\mu_m + log\pi_m \qquad (6.6)$$

The Bayes' classifier assigns an observation $X = x$ to the class for which the above function yields the largest value. The data set is resolved into three components namely 'ld1', 'ld2', and 'ld3', and then the explained variance ratio is calculated in order to observe how much of the total variance is being captured by resolved components. As a rule of thumb one should keep adding components as long as they do not reach an explained variance ratio of >0.7, in this case, the authors got the explained variance ratio matrix to be [0.47109338, 0.21177223, 0.10037413], the components add up to 0.78, hence good to go.

Below images show how a complex and intertwined data set now becomes easily classifiable with distinct borders (Fig. 6.1).

Upon fitting this LDA on the music genre–training data set, the training set accuracy obtained was 68.4%. However, the training set accuracy is of little importance, the true measure of a model is denoted by the test set accuracy, which is evaluated to be 67%. The test set accuracy obtained, though not very high, still is acceptable, considering the training set accuracy as the peak. This accuracy is less compared to the complex neural network which can be used for the same classification problem, but the computation cost of the model will be far higher than a simple LDA which gives 67% accuracy. The confusion matrix of the LDA classification is shown below.

The classes 0 to 9 represent blues, classical, country, disco, hip-hop, jazz, metal, pop, reggae, and rock, respectively. The result can be improved if the cross-validation set is used for the classification. Upon using the cross-validation set, the training set accuracy increases to 77.4% and the cross-validation set accuracy tends to increase to approximately 68%, which is a slight improvement to the normal training test split. The results of the confusion matrix are shown in Fig. 6.2.

The results obtained in the LDA at a glance might not seem very significant and accurate but, if considering the computation costs involved and the results obtained, they tend toward decent results. However, the results obtained from the LDA can still be improved. As mentioned earlier, the LDA is a linear classification algorithm that tends to assume that the covariance matrix is common to all the classes, which may result in a high bias for a model. Therefore, how significantly the model will differ in accuracy if somehow an algorithm is used that tends to reduce the bias by increasing the degree of the classifier. In the coming section, the QDA is discussed which gives surprising results. Fig. 6.2 underlines the performance of the classifier concerning the actual values of the test set. This will give us a visual intuition of the performance of the LDA algorithm.

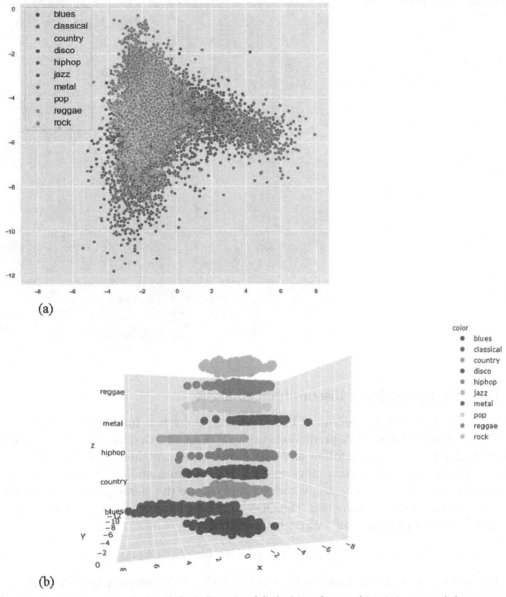

FIGURE 6.1 Two-dimensional (2D) and three-dimensional (3D) plots, after resolving it in two and three components. (A) 2D plot of LD1 versus LD2 in x and y axis respectively. (B) 3D plot of LD1, LD2, LD3 in x-, y-, and z-axes, respectively.

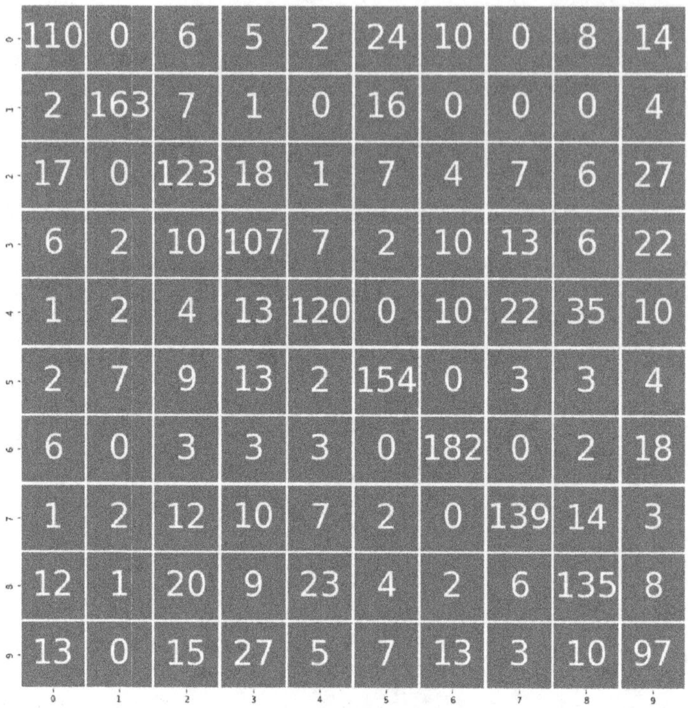

FIGURE 6.2 Results on the test set from LDA.

	0	1	2	3	4	5	6	7	8	9
0	110	0	6	5	2	24	10	0	8	14
1	2	163	7	1	0	16	0	0	0	4
2	17	0	123	18	1	7	4	7	6	27
3	6	2	10	107	7	2	10	13	6	22
4	1	2	4	13	120	0	10	22	35	10
5	2	7	9	13	2	154	0	3	3	4
6	6	0	3	3	3	0	182	0	2	18
7	1	2	12	10	7	2	0	139	14	3
8	12	1	20	9	23	4	2	6	135	8
9	13	0	15	27	5	7	13	3	10	97

6.3.2 Fitting quadratic discriminant analysis

QDA works on a similar logic as of LDA, however, there is a fundamental difference between the two [1]. Like LDA, predictions using QDA assume that the data points which are extracted from each class follow an MGD. However, as explained earlier, unlike LDA, which assumes a common covariance matrix, QDA assumes and calculates covariance matrix for each class, XN (μ_m, Σ_m) where μ_m and Σ_m are the class-specific vectors and covariance matrix for the mth class. The QDA calculates $\delta m(x)$ and classification occurs to that class for which the entity $\delta_m(x)$ is the highest.

$$\delta_m(x) = -\frac{1}{2}(x - \mu_m)^T \Sigma_m^{-1}(x - \mu_m) - \frac{1}{2}log|\Sigma_m| + log\pi_m \qquad (6.7)$$

We performed a similar analysis for QDA as well, the results obtained were as expected, the accuracy of the test set (which matters) and the training set improved drastically. The confusion matrix for QDA is shown in Fig. 6.3. The classes 0 to nine represent the following

0 - blues
1 classical
2 country

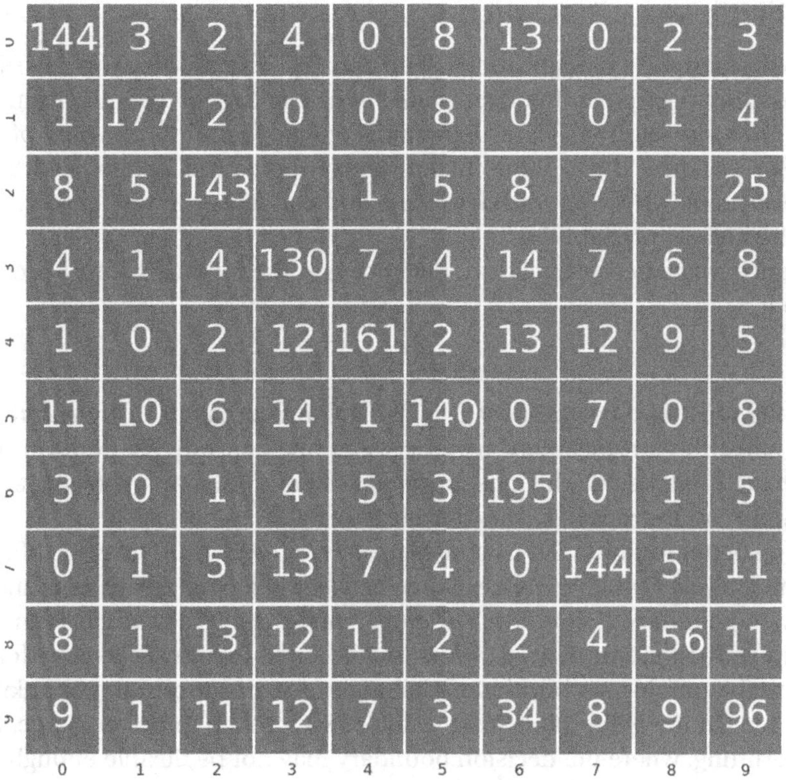

	0	1	2	3	4	5	6	7	8	9
0	144	3	2	4	0	8	13	0	2	3
1	1	177	2	0	0	8	0	0	1	4
2	8	5	143	7	1	5	8	7	1	25
3	4	1	4	130	7	4	14	7	6	8
4	1	0	2	12	161	2	13	12	9	5
5	11	10	6	14	1	140	0	7	0	8
6	3	0	1	4	5	3	195	0	1	5
7	0	1	5	13	7	4	0	144	5	11
8	8	1	13	12	11	2	2	4	156	11
9	9	1	11	12	7	3	34	8	9	96

FIGURE 6.3 Results on the test set from QDA.

3 - disco
4 - hip-hop
5 - jazz
6 - metal
7 pop
8 - reggae
9 - rock

On fitting the data set to the Quadratic Discriminant model, the accuracy of the training set shot to 81% and that of the test set came to 74%. This is a huge difference from the LDA which only gave out an accuracy of 68%. Thus, it is safe to assume that the music genre data set doesn't follow a linear model.

Fig. 6.3 underlines the performance of the classifier for the actual values of the test set. This will give us a visual intuition of the performance of the QDA algorithm.

6.3.3   Fitting k-nearest neighbors

KNN is a nonparametric classification algorithm [6]. The idea of this algorithm is to assign a class to a new incoming data point by measuring its distance from K-nearest points and then deciding its class depending on the plurality of votes of K-nearest members. For instance, if $K = 1$ the incoming data point will be assigned to the class of the nearest point. KNN applies Bayes' rule and classifies the test observation to the class with the highest probability.

The distance calculated is Euclidean distance and the following formula can be used to calculate it.

$$d(a, b) = \sqrt{\Sigma_i (a_i - b_i)^2} \qquad (6.8)$$

What is the most optimal value of K is debatable and has been time and again discussed but since the advancement of processors one can quickly perform a check of accuracy versus K and decide the most optimal K. Below is a graph showing accuracy on the y-axis and K on the x-axis, K varies from 1 to 100.

To develop better intuition, let us consider a few values for K. The value of K has to be chosen in such a way that it forms a balance between the complexity, bias and variance of the decision boundary. A very low value of K might result in a function that is highly flexible and overfits the training set, which might result in an excellent performance of the model on the training set but the model might poorly perform on an unknown data set, the test set. On the other hand, if the value of K is chosen to be very large, this might lead to underfitting, where the decision boundary may not be flexible enough to classify the observations correctly.

Fitting KNN in the data set yielded a whopping accuracy of 94.4% on the test set whereas it produced an accuracy of 87.5% (Fig. 6.4).

The data set chosen in this chapter tends to be one of the most elegant examples to show how beautifully can the KNN perform at low values of K and give us a decision boundary which is close to the Ideal Bayes decision boundary, moreover, performs competitively with the more complex classification algorithms using neural networks.

6.3.4   Feedforward neural networks

Neural networks form the bleeding-edge research domain for the field of artificial intelligence. These are the quintessential models being used to solve a large spectrum of problems from computer vision to natural language processing. Feedforward Neural Networks or Multilayer Perceptrons form the building blocks of the state of art machine learning models like convolutional neural networks (CNN) and recurrent neural networks (RNN) [7]. When trained with appropriate data and hyperparameters, they can perform surprisingly better than any other algorithms known. Deep learning has been used for achieving many ambitious goals like making artificial intelligence (AI) proficient in Chess and Go, improving the performance of self-driving cars, predicting financial

	0	1	2	3	4	5	6	7	8	9
0	165	2	1	1	2	5	1	0	2	0
1	1	182	1	0	0	9	0	0	0	0
2	10	4	174	6	2	2	0	0	7	5
3	3	2	3	164	4	1	0	0	2	6
4	2	0	4	7	187	0	2	5	9	1
5	5	10	4	1	2	170	0	0	1	4
6	0	0	1	6	5	0	201	0	0	4
7	0	2	7	16	3	3	0	154	4	1
8	1	2	4	2	5	0	0	3	202	1
9	4	3	9	10	4	4	4	2	1	149

(a)

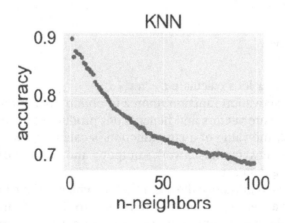

(b)

FIGURE 6.4 (A)Underlines the performance of the classifier to the actual values of the test set. This will give us a visual intuition of the performance of the KNN algorithm. (B) Results on the test set from KNN.

markets. They have been realized using a complex system of neural networks. It is of extreme importance to look into this idea to develop a deeper understanding of these algorithms and their architectures.

Feedforward Neural Networks are the culmination of many functions stacked together, giving them the name networks, they are a kind of mesh of simple nonlinear functions wrapped together to form complex functions [8]. The model is blended with a directed acyclic graph depicting how the functions are formed together, for instance, let's say four functions are chained together in the form $f(x) = f_1 (f_2 (f_3 (f_4 x)))$, these four functions form the most basic structure of a neural network. Here f_1 is called the first layer, f_2 is called the second layer and so on. This multilayer sequential structure gives rise to the name deep learning. The final layer of the structure is called the output layer. The fundamental working of a neural network is to evaluate a function $f(x)$ for a data set such that it is closest to the hypothetical function $f * (x)$ from which the data set is assumed to have originated. Each point in the training data set is used to get approximate $f * (x)$ and all the points are then used to remove noise to evaluate the most concerning ones. Each point x is labeled with a $y \approx f * (x)$. The algorithm must itself decide based on the input x and the output y how to use the layers to get the desired output, but the training data set does not mention what the role of each layer should be, since the training data set does not tell the layers involved in how to implement the functions, they are called the hidden layers.

Let's now delve deeper into the architecture and mathematics of how exactly a neural network works. The following diagram represents the architecture of a basic neuron and has the following properties

$x =$ input
$w =$ weight
$b =$ bias
$f =$ activation function
$y =$ output

With the initial input x let's calculate $z = wx + b$.

We then apply the activation function upon z to obtain final output y, it's important to note that w, x, and b are vectors and hence a dot product of w and x is taken.

Once y is obtained, the value of a cost function is calculated which is basically the sum of the differences between the network's output y and the actual value of y which is taken from the training data set.

The next step then is to minimize the cost function which is done by backpropagation where the partial derivatives of z, w and b with respect to C (cost function) is calculated.

We repeat the whole process until a desired state of accuracy is reached.

With this short description of a neural network, let's dive into the practical implementation of the same for our music genre data set. Along the way, the authors will keep pausing and give necessary theory wherever needed.

6.3.5 Fitting feedforward neural networks

6.3.5.1 Gradient descent

The initial setup for making the model stays the same as the previous sections. First extracting the features from the audio file and then splitting the data set into validation and the training sets. Further, the training set is divided into mini-batches of 128. Using mini-batches is an important step for any neural network. Mini-batch gradient descent is the middle ground of the batch gradient descent and the stochastic gradient descent. Here neither the whole data set is used like in the case of batch gradient descent nor a single data point is used like in the case of stochastic gradient descent, rather the training set is divided into mini-batches of fixed smaller size, generally in the size of exponents of 2. The mini-batch is then fed to the neural network as input and the mean gradient is then evaluated on the mini-batch which is then used to update the weights associated with the neural network. This series of steps is repeated for all the mini-batches for a certain number of epochs to get the optimal result.

6.3.5.2 Activation functions

Choosing the correct activation function for a neural network is a crucial step. It is important to mention that a neural network has to have a nonlinearity introduced in the system. Failure to do so makes a neural network nothing more than a linear regression function. This is fairly intuitive as the composition of many linear functions is nonetheless a linear function.

There are many possible functions to be the nonlinear function required, however, not every function can be used to get the best results. Krizhevsky et al.[11] extensively mentions the finer details of functions like sigmoid, softmax, tanh, ReLU [8,9,11]. The authors have used the ReLU function for the hidden layers and the softmax function for the output layer. ReLU function $f(x)$ is defined as follows

$$f(x) = max(0, x) \tag{6.9}$$

$f(x)$ although seems a simple function, has come out to be the first choice of activation function for most neural networks. It works surprisingly well and has been shown in Krizhevsky et al.[11]. The softmax function is shown as follows

$$\sigma(z) = \frac{e^z}{\sum_{j=1}^{k} e^{zj}} \tag{6.10}$$

The softmax function is a function that turns a vector of K real values into a vector of K real values that sum to 1. The output of the function is always between 0 and 1, which can be used as a probability score. The input can be positive or negative but the output is always a positive value bounded by 0 and 1. This function seems similar to another function that is known very well, the sigmoid function, right? Often this function is referred to as soft argmax function or multi-class logistic function. This is a generalization of the binary classifier-a sigmoid function that is used as an output layer for binary classification problems. The sigmoid function is the special case of the softmax function.

6.3.5.3 Regularization

It is fairly easy to make a neural network or any model in fact which can perform at 100% accuracy on the training set. However, this might be counterproductive to our needs and has an antithetical result. When the accuracy of a model is increased on a training set, it is trained using the training data set only. Given the strong computational power it has, it can evaluate weights in a way to accurately fit the training data set. The statistical condition for a model performing with such high accuracy on the training set is the dilemma of getting a compromise between bias and variance.

First, let's understand the significance of bias and variance in a model and for that, one needs to understand bias and variance. Bias is the difference between the actual value and the predicted value on a training set. The function $f(x)$ which needs to be evaluated, if assumed to be a simple polynomial, would lead to loss of information as the function will not be able to capture the relevant details about the target variable and the input data causing underfitting of the model. On the other hand, if $f(x)$ is assumed to be a very complex polynomial, it will learn the data in a way to fit the function rather than understanding the data which will lead to overfitting of the model. Variance indicates how much the target function for a model would change if the training data set provided is changed. Naturally, our accuracy of the results is directly related to the estimate of the target function $f(x)$. One would want our model to generalize the problem in the best possible way so that when put to test by the real-world data that our model is not trained on, it gives out accurate results. It would defeat the purpose of the model if it drastically changes the estimated target function on changing the training data set. For an ideal model, one should strive to have low bias and low variance.

To avoid overfitting, which turns out to be the more frequent problem, there are many ways to do so. The one more important to us here is the dropout. It is one of the most used methods of regularization in a neural network model [7,10]. Dropout simply is like the name suggests. It means leaving out or ignoring a few of the neurons or activation units during the training process. Fundamentally, during each stage of training, each neuron can be kept with a probability of p which conversely can be stated as that each neuron can be dropped off with a probability of $1 - p$. This brings randomness in the model and in each stage of training a different pair of neurons are activated, compelling the model to make use of all the units and understand the data set rather than learning the data set. Thus dropout forces the model to learn more robust features that help in making more accurate predictions.

6.3.5.4 Model building

After the data processing and selecting the relevant activation functions, comes the task of aligning them in a sequence to get the best possible results. Each layer is followed by a dropout layer with a value of 0.3, and there are five such layers which are ultimately followed by an output layer. The first layer has 1024 neurons, the second layer has 512 neurons, the third layer has 256 neurons, the fourth layer has 128 neurons, and the fifth layer has 64 neurons. The activation function used in all the previously mentioned layers is the ReLU function defined above. The final layer is having 10 neurons, corresponding

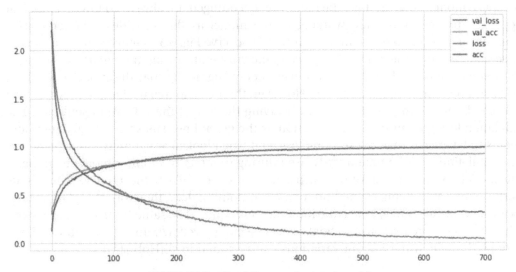

FIGURE 6.5 Results of the neural network model.

94	0	2	1	1	1	1	0	1	1
0	99	0	0	0	2	0	0	0	0
1	0	90	2	0	5	0	1	0	3
0	1	1	93	0	0	0	1	1	5
0	0	3	1	90	1	1	3	3	0
1	3	2	0	0	96	0	0	0	0
1	0	0	1	0	0	98	0	0	2
0	0	1	1	0	0	0	99	0	1
0	0	4	1	0	1	0	1	94	1
0	0	5	1	1	1	0	1	2	91

FIGURE 6.6 Confusion matrix for the test set on Neural Network.

to the 10 genres of music classification the authors are performing. It uses the softmax function. The model runs for 700 epochs and gives a fantastic accuracy of 93% on the test set. The results of the test set have been mentioned in the confusion matrix shown in Fig. 6.6.

It is rather fascinating to see a graphical representation of the results obtained in a complex model such as this. Readers are encouraged to always follow a graphical approach to understand the working of the model as this provides a better under-standing and intuition of the model. Now let's observe Fig. 6.5 to make some reasonable observations about the model. As one sees, the cross-entropy loss(green) associated with the model keeps on decreasing with every epoch signifying that the model is working and learning in the desired direction. However, this does not tell us that the model is well and good. This is only confirmed by observing the loss of the validation set(blue). The validation loss goes in the same direction as the green line. The value of validation loss seems to be greater than the training loss, and when looked at carefully, it makes sense. The validation set is simultaneously used during training for monitoring the learning of the model, as that is the kind of a pseudotest set on which the model evaluates its performance after being trained on a batch of training data, therefore that data is rela-tively new for the model, hence the increased loss value. The same thing can correspond with the accuracy(red) and validation set accuracy(yellow). Implementation of the proposed model with different variations can be found in GitHub [12,13].

6.4 Conclusion

After applying all the three fundamental classification algorithms, the authors have a very conclusive result. The performance of the algorithm can be arranged in the following order LDA < QDA < KNN < neural networks. The neural network gives us the best results of about 93% accuracy, which can be considered very high based on the simplicity of the network.

LDA is closely related to logistic regression. It is a parametric classification algorithm that assumes the decision boundary to be linear, the only difference between the two is in the terms of the number of classes in which both of them are classified. Logistic Regression works best for the cases in which the number of classes to classify is two whereas, LDA can serve for the linear boundary of more than two classes, KNN is a nonparametric algorithm that makes no assumptions about the shape of the decision boundaries. QDA serves as an intermediate classification model between linear LDA and nonparametric KNN. QDA assumes the decision boundary to be quadratic which works in this case because it serves as the equivalent point between the bias and the variance. It tends to settle the conflict between the low bias caused due to LDA and the high bias caused due to KNN.

This conversation, however, changes with the introduction of models as complex as neural networks. Neural networks have proved to be very effective and most importantly accurate at complex tasks that can only be dreamed of using the classical methods at hand. The interesting part is that a neural network is a kind of black box. One can fine-tune it to give very accurate results but the stage of progress cannot be known and it is an area of research. Few methods are used by which one can get a rough idea as to what has been the process of learning till that step but that is still far from being called a definitive method. Nevertheless, neural networks are turning out to be a very effective tool for a variety of problems and they have reached a level where they can outperform humans. An example is image classification. There are networks like ResNet, GoogleNet, and VGGNet that have outperformed humans in the classification task. Amid all the good characteristics of the neural network, there are a few problems associated with them. The biggest one is the computational complexity of neural networks. If not fine-tuned in the most efficient way, they pose a serious challenge in training them. To get an intuition about the same, our fairly simple neural network that on a practical scale, takes a small data set, calculates over 757,000 parameters to train the model and this took a fair amount of time. Therefore, one needs to be very careful with the implementation of a neural network.

From the analysis of a music data set for over more than 1000 tracks, one can conclude that the music tracks have very distinguishing properties which allow them to form a very well defined cluster. This observation is the key reason why the Neural Network classification approach performs so well even though it is one of the simplest classification algorithms in classical statistics. It will be an interesting problem to see if the same approach can be applied to the genres of Indian classical music. The hypothesis that the authors currently have suggests that it can be pretty difficult to classify Indian classical music as they possess a lot of similarities in the way it can be perceived by the listeners. However, this will be an interesting topic to study.

References

[1] B.L. Sturm, The GTZAN dataset: its contents, its faults, their effects on evaluation, and its future use, arXiv Prep. arXiv: 1306.1461 v2 [cs.SD] (2013) 1–29.

[2] J. Yoon, H. Lim, D.-W. Kim, Music genre classification using feature subset search, Int. J. Mach. Learn. Comput. 6 (2) (2016) 134–138.

[3] A.S. Atti, T. Painter, Venkatraman, Audio Signal Processing and Coding", John Wiley & Sons, Hoboken, 2006.

[4] N. Srivastava, et al., Dropout: a simple way to prevent neural networks from overfitting, JMLR 15 (1) (2014) 1929–1958.

[5] https://stats.oarc.ucla.edu/spss/dae/discriminant-function-analysis.

[6] D.R. Flederus, Enhancing Music Genre Classification with Neural Networks by Using Extracted Musical Features, 2020.

[7] B.L. Sturm, On music genre classification via compressive sampling, in: Multimedia and Expo (ICME), 2013 IEEE International Conference on, IEEE, July, 2013, pp. 1–6.

[8] J. Ian, Goodfellow, Yoshua Bengio and Aaron Courville, Deep Learning, MIT Press, 2016.

[9] V. Nair, G.E. Hinton, Rectified linear units improve restricted Boltzmann machines, in: Proc. 27th International Conference on Machine Learning, 2010.

[10] G.E. Hinton, N. Srivastava, A. Krizhevsky, I. Sutskever, R.R. Salakhutdinov, Improving neural networks by preventing co-adaptation of feature detectors, arXiv Prep. arXiv: 1207.0580v1 [cs.NE] (2012) 1–18.

[11] Krizhevsky, et al., ImageNet Classification with Deep Convolutional Neural Networks, 2012.

[12] https://github.com/divydvd.

[13] https://github.com/AshutoshGanguly123/Classification-Algorithms.git.

7

Classification model of machine learning for medical data analysis

Rohini Srivastava, Shailesh Kumar, Basant Kumar

DEPARTMENT OF ELECTRONICS AND COMMUNICATION ENGINEERING, MOTILAL NEHRU NATIONAL INSTITUTE OF TECHNOLOGY ALLAHABAD, PRAYAGRAJ, UTTAR PRDESH, INDIA

7.1 Introduction

The World Health Organization (WHO) states that 55% of the deaths worldwide are caused by the diseases which can be cured completely if diagnosed earlier [1]. Early diagnosis helps in preventing life threatening diseases and can reduce mortality rate by a significant amount. Machine learning is a key to diagnose many diseases at earlier stage such as cancer, glaucoma, strokes, heart disease etc. There are several machine learning techniques available for various disease detection such as neural network (NN), radial basis function (RBF), support vector machine (SVM), deep learning, etc. In medical data analysis using machine learning, classification of diseases can be performed by three learning models: supervised, unsupervised, and reinforcement learning. The machine learning (ML) model is trained with labeled data in the supervised learning and thus the output is predicted using that labeled input data. In unsupervised learning, the input data is unlabeled unlike the supervised learning. In this type of learning, the machines learns itself by clustering and analysis without human interventions. Reinforcement learning differs from supervised learning in a way that supervised learning has both training data and labels for training the models whereas in reinforcement learning, there is no label but the reinforcement agent determines how the job has to be accomplished. It trains itself through experience even if the training dataset is unavailable [2]. It gives sequence of decision for training of ML model. These learning techniques provide the training data for the model. After training, target function is estimated with learning algorithm and input-output mapping, thus classification is performed. ML can also be classified on the basis of outputs. If classification is between only two classes: true or false, then it is called binary classification, whereas if the output of classification is in finite number, then it is called Multiclass classification. Besides of binary and multiclass classification, if the ML output is in continuous values, then it will be known as regression. Nowadays, ML is used in almost every area, however, in medical field it has provided potentials to the medical experts to prognosis and diagnosis of many life threatening disease [2–6].

Statistical Modeling in Machine Learning. https://doi.org/10.1016/B978-0-323-91776-6.00017-8

In medical data analysis, ML has now become a vital process or even it is used as a tool in many medical fields. Diseases like, glaucoma, diabetic retinopathy, cardiovascular abnormalities, breast cancer and even tumor can be diagnosed earlier by using ML. Generalized block diagram for ML based medical data classification is shown in Fig. 7.1.

Fig. 7.1 shows a generalized block diagram of ML used for medical data classification. First we need patient's data which may be structures or unstructured. These data are preprocessed for removing various noise, then necessary features are extracted. After feature extraction, the data is divided into three groups: training, testing and validation: and fed to the L algorithm. ML algorithm processed the data and classify the disease.

There are several literatures available regarding medical data analysis using different machine learning techniques. Evangelia et al. [7] have presented a comparative review on machine learning versus Logistic Regression (LR) for medical data analysis. They have considered 71 studies (case) from Medline for the comparative analysis. In Refs. [5,8], also LR review has been presented for medical data analysis. Shrabani et al. [9] have presented heart disease classification utilizing decision tree (DT) and neural network (NN). They have used University of California at Irvine (UCI) machine learning database for the prediction of heart disease. Wenqian et al. [10] presented K-means clustering and decision tree algorithm for prediction of type 2 diabetes. They have used hybrid model for accurate prediction of diabetes. Zahangir et al. [11] have presented classification of various medical data using random forest (RF). They have used feature ranking technique for classification of heart disease, breast cancer, diabetes, hepatitis etc. Ahmad et al. have presented lymph disease classification using RF. They have used optimization technique (Genetic Algorithm) along with classification model (RF) to increase the accuracy of the classifier. There are several literature available for support vector machine (SVM) and NN based classifier for classification of various medical data [12−21]. Anam et al. have presented SVM classifier for arrhythmia classification. They have proposed a

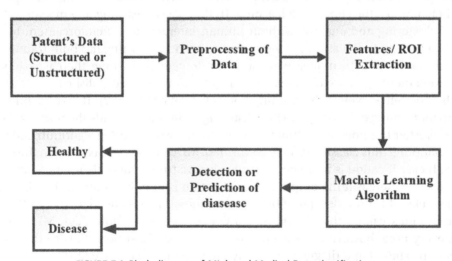

FIGURE 7.1 Block diagram of ML based Medical Data classification.

multiclass classifier to classify electrocardiogram (ECG) into 16 classes. Abdar et al. [22] have presented NN based classifier to detect liver diseases such as bilirubin, total protein, albumin etc. Sridhar et al. [23] have presented NN based classifier for predicting brain disease such as stroke, tumor, brain injuries etc. Machine learning (ML) has made medical diagnosis and prognosis very much easier. Rest of the chapter is organized as follows: ML techniques for various disease classification is presented in Section 7.2. Section 7.3 comprises of disease detected by machine learning. Challenges in ML based classification of medical data and summary have been discussed in Section 7.4 and Section 7.5 respectively.

7.2 Machine learning techniques for diseases detection

Artificial intelligence (AI) is a growing field with a variety of daily life practical applications and currently active research topics. ML has become a vital process in prognosis/ diagnosis of several diseases. It is evident that the machines have become more intelligent over a long period of time. These days, people use capable software to automate daily routine, audio or photo recognition, diagnose decisions in the medical field, and many research fields in science [24,25]. The machines must have certain capabilities that allow an AI based system to extract pattern from raw data. This ability is known as machine learning (ML). ML can be introduced as enabling a computer program for solving real world problems and make appropriate decisions [26]. To solve the objective functions, the complicated concepts of the program could be learned by a computer using a hierarchy of many simple ideas. The graphical representation of these concepts which are assembled by several layers, will be very deep. For this reason, this is known as AI deep learning [27,28]. Some of MI model is explained in details, namely Support vector machine and Radial Basis Function Neural Network.

7.2.1 Logistic regression (LR)

LR is basically used for multivariable methods in which calculation of original data coefficient is more complex. In multivariable methods, the data is divided into two forms: dependent (output) and independent (input). LR is used for predicting the dependent variable using independent variable and the dependent variable is categorical data such as yes/no, pass/fail, true/false etc. It is very much similar to linear regression, however in linear regression, dependent variable is not categorical. LR is a supervised learning algorithm and it uses sigmoid function as activation function. Sigmoid function is simply trying to convert the independent variable into an expression of probability that ranges between 0 and 1 with respect to the dependent variable, thus predictions are converted into probabilities that ranges between 0 and 1. Sigmoid function can be expressed as in Eq. (7.1)

$$y = \frac{1}{1 + e^{-x}} \tag{7.1}$$

Where x is input variable and y is output variable, i.e., categorical variable. The graph for LR is shown in Fig. 7.2.

LR provides the binary classification and tries to map the real value of independent variables in the interval of 0 and 1. A cut-off is assigned, some of the data points will be above cut-off points and vice versa and thus the two classes will be defined. It might be possible that some data points will be precisely positioned with the cut-off, then that data will be unclassified, which is a very rare situation in LR. For accurate classification the data must be free of missing points. There are three types of LR: Binary LR, Polytomous LR, Ordinal LR [4]. Binary LR is same as discussed above whereas Polytomous LR is used for multiclass problem. In this type of situation binary LR is applied to each dichotomous variable. However, this will result in several different analyses for only one categorical response. When the dependent variable is categorized in ordinal form like "health status good/moderate/bad," then binary and polynomial both will not work properly. In this case, proportional odds models are used to classify the independent variable in ordinal form.

If the input and output data are given as (x, y) the lost function is cross-entropy lost function, which is defined as Eq. (7.2)

$$\cos t(h_\theta(x), y) = -y \log(h_\theta(x)) - (1-y)\log(1 - h_\theta(x)) \tag{7.2}$$

And for m data points $(x^{(i)}, y^{(i)})$ total cost function

$$J(\theta) = -\frac{1}{m}\left[\sum_{i=1}^{m} y^{(i)} \log(h_\theta(x^{(i)})) - (1-y^{(i)})\log(1 - h_\theta(x^{(i)}))\right] \tag{7.3}$$

The cross-entropy cost function provides a convex function for sigmoid activation function. Here, gradient descent (GD) algorithm is used for optimization of cost function, which finally provides parameters. The gradient descent algorithm is explained below to calculate parameter θ. The following three steps are repeated to obtain min $J(\theta)$:

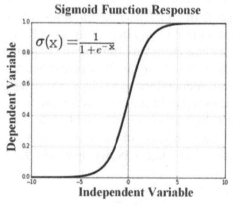

FIGURE 7.2 Graph of LR.

1. Get the data points
2. Calculate the cross-entropy loss
3. Apply gradient descent

$$\{$$

$$\theta_j = \theta_j - \alpha \frac{\partial}{\partial \theta} J(\theta)$$

simulteneosly update all θ_j

$$\}$$

Above algorithm provides all the parameters with optimum value (cost function is minimum).

7.2.2 Decision tree

On certain condition, a graphical demonstration of all the potential explanations to a decision is known as decision tree. It is a type of nonparametric supervised learning. Primarily decision tree can be used for classification such as binary or multiclass. Apart from classification, it can be used for regression as well. The structure of classifier looks like tree, where features of data are represented by the nodes, decision rule by the branches and the output of classifier is represented by each leaf node. Decision consists of three nodes namely, root node (RN), decision node (DN), and leaf node (LN). *RN* depicts the entire population and sample and it gets distributed into two or more homogeneous sets. *DN* is used to take choice whereas *LN* cannot be further divided into more nodes. Decision are made based on feature of input data. The process of dividing the *roots/sub node* into different parts on the basis of some condition is known as *Splitting.*

There are two reasons for using decision tree (1) it shows tree like structure and the logic behind classification is easily understood. (2) It usually try to follows human thinking ability while taking decision. Classification procedure begins with root node which expand later into branches/**Sub tree** (formed by splitting the tree/node) hence it is known as decision tree. Classification and regression tree algorithm (CART) is used to form a tree. A decision tree working flow diagram is shown in Fig. 7.3.

The splitting is decided based on these parameters:

Information gain (IG): On the basis of an attribute, the decrement in the entropy after a data set is split is known as **IG**. Decision node is decided by IG by proper selection of attributes. Construction of decision tree is achieved by the attributes having the highest IG and based on IG, the node is split to form tree.

$$\text{Information gain} = \text{entropy(SS)} - [(\text{weighted average} \times \text{entropy(each feature)})] \qquad (7.4)$$

Where entropy (SS) is the entropy of the sample space.

Gini Index: It is also called Gini impurity which calculates the probability of a attribute which is wrongly predicted when chosen arbitrarily. Lower value of Gini index is

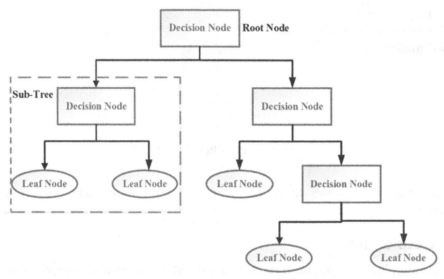

FIGURE 7.3 Decision tree working flow diagram.

preferred on higher value Gini index attribute. Gini index splits the node into binary using (classification and regression trees) CART algorithm.

$$\text{Gini index} = \sum_i p_i^2 \qquad (7.5)$$

Pruning: Pruning is basically removing unwanted branches or nodes from the tree to make the optimal decision tree. If tree has a smaller number of nodes or branches it might not get all the information from the input data and if tree has more number of nodes, the chance of overfitting is high. So the technique to reduce the size of decision tree without compromising accuracy is called pruning. There are two types of pruning such as cost complexity pruning and reduced error pruning.

7.2.2.1 Build our decision tree

Step I: Compute entropy for the data set
Step II: Which node to select as root node based on information gain
Step III: Which node to select further
Step IV: What should I do play: pruning

7.2.2.2 Advantage

- It needs less preprocessing as compared to other algorithms.
- It is easy to understand because it follows same process as human does in real life to make decision.
- It provides the way to reflect about all possible output and useful to solve decision related problem.

7.2.2.3 Disadvantage

- It is complex algorithm.
- It may suffer from overfitting problem.
- For multiclass classification, decision tree would become complex.

7.2.3 Random forest

Random forest (RF) is a type of ensemble techniques. Ensemble techniques can be defined as combination of multiple models. There are two types of ensemble techniques such as bagging and boosting. Bagging is also known as boot strap aggregation. RF is example of bagging technique. RF is used to overcome the overfitting problem of decision tree. Bagging consists of many prediction model with same type. It is used to reduce variance, not bias. Primarily, bagging is used to overcome overfitting issues in a prediction model. All the predictions get equal weight in bagging technique. The number of weights should be odd in bagging because it depends upon voting system of predictions. Training is done with randomly drawn data from data set with replacement for all the model in bagging technique.

Random forest classifier uses bagging techniques where decision tree classifier is used as base learner. Random forest consists of many trees, and each tree predicts his own classification and the final decision makes by model based on maximum votes of trees (Fig. 7.4). There is very simple and powerful concept behind RF—the wisdom of crowd. A variety of uncorrelated trees acts as a group will overtake the result of any

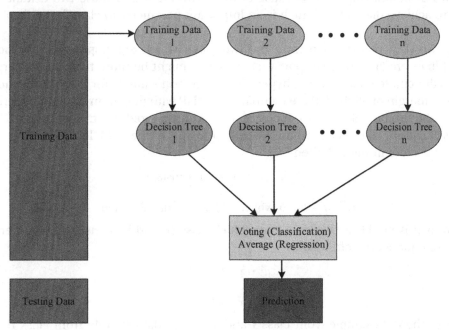

FIGURE 7.4 RF classifier working flow diagram.

separate constitute model. The less correlation among models is the key of success of RF. The main motive for this outcome is that the trees defend each other from their separate errors. If some trees are wrong and many others are right then group of trees make decision in correct direction. Since the RF combines a variety of trees for classification of input data. There are assumptions for RF classifier: there should be less correlation between the estimation of individual tree and second, feature variable of data should have actual value based on that classifier predicts correct outcome rather than estimated outcome. The RF is very useful because training procedure is faster than other algorithms and RF classification accuracy is high. It works efficiently with large data set and even huge amount of data is disappeared. The RF is used mostly in banking, medicine, land use, and marketing.

7.2.4 Support vector machine

It lies in the supervised learning and used for classification and regression. It performs binary and multi classification and for the classification a decision boundary is always drawn to discriminate the two classes which is known as hyperplane (HP). This HP decides that the new data belongs to either class I or class II. When some of the data points of the considerable class is nearer to the HP of the opponent class, then sometimes it is harder to classify that data. Therefore, for accurate classification a line will be drawn very much closure to that particular data point and parallel to the decision boundary. Similar process is repeated for the data points in the opponent class. Now two distances will occur because of these two lines, and these two distances are added to achieve the margin which is an important parameter to classify the data using SVM.

The two data points (each from the two class) which are responsible for the two parallel lines are known as support vectors. There might be more than one hyperplane and it is difficult to choose which hyperplane is the best suitable for the classification. In that case, margin of all the HP are compared and HP with maximum margin hyperplane (MMH) should be selected to increase the accuracy, reduce the error and lesser misclassification. To understand the working principal of the SVM, let us consider u and w as two vectors in the HP, then

$$\vec{w} \cdot \vec{u} \geq c, \text{ where c is constant} \tag{7.6}$$

$$\vec{w} \cdot \vec{u} + b \geq 0 \text{ which is known as decision rule} \tag{7.7}$$

Now as it is not clear that data from which class should be consider here, therefore the above equation is reframed as

$$\vec{w} \cdot \vec{x}_+ + b \geq 1 \tag{7.8}$$

$$\vec{w} \cdot \vec{x}_- + b \leq 1 \tag{7.9}$$

Where x_+ the data sample from class I and x_- is the data sample from class II. Now

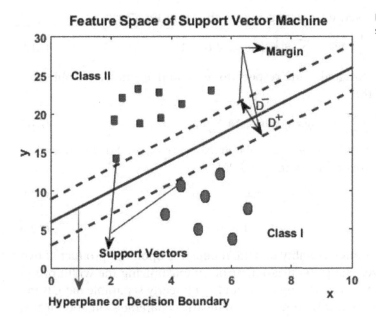

FIGURE 7.5 Feature space of support vector machine.

consider y_i such that

$$y_i = +1 \rightarrow \text{class I} \tag{7.10}$$

$$y_i = -1 \rightarrow \text{class II} \tag{7.11}$$

The maximum margin is shown in above Fig. 7.5. The point x_i on the margin shown in Eq. (7.12)

$$y_i\left(\overrightarrow{w} \cdot x_i + b\right) - 1 = 0 \tag{7.12}$$

Width of the margin for point x_i in the margin shown in Eqs. (7.13) and (7.14)

$$\text{width} = (x_+ - x_-) \cdot \frac{\overrightarrow{w}}{\| w \|} \tag{7.13}$$

$$\text{width} = \frac{2}{\| w \|} \tag{7.14}$$

Width of the margin has to be maximum for better classification. Next equation shows the objective function is an example of constrained optimization.

$$\max \frac{2}{\| w \|} \tag{7.15}$$

$$\min \frac{1}{2} \| w \|^2 \text{ subject to } y_i\left(\overrightarrow{w} \cdot x_i + b\right) - 1 = 0 \tag{7.16}$$

Lagrange optimization is used for convergence as shown in Eq. (7.17)

$$L = \frac{1}{2} \| \vec{w} \|^2 - \sum \alpha_i \left[y_i \left(\vec{w} \cdot x_i + b \right) - 1 \right]$$
(7.17)

Now expression is differentiated with respect to $\vec{w}, \alpha$ and b, and new objective function is shown in Eq. (7.18)

$$\max L = \sum_{i=1}^{N} \alpha_i - \frac{1}{2} \sum_{i=1}^{N} \sum_{j=1}^{N} \alpha_i \alpha_j y_i y_j x_i x_j \text{ subject to } \sum \alpha_i y_i = 0 \text{ and } 0 \leq \alpha \leq c$$
(7.18)

Quadratic programming gives α and w is calculated using Eq. (7.19). Furthermore, b is calculated using support vectors as show in Eq. (7.20).

$$\vec{w} = \sum \alpha_i y_i x_i$$
(7.19)

$$y_i \left(\vec{w} \cdot x_i + b \right) - 1 = 0$$
(7.20)

SVM does not depend on dimensionality of data. It depends upon dot product of two input vectors. SVM minimizes the generalization error by maximizing the width of hyperplane. Data are of two types: linearly separable and nonlinearly separable data. If the data is not linearly separable then it is known as nonlinearly separable data and for this type of data, the above mentioned method will result in misclassification. For this type of problem, "kernel" functions are introduced, which takes low-dimensional feature space (LDFS) as input and provides the output in high-dimensional feature space (HDFS). A 2D nonseparable feature space will be fed into kernel as input which converts it into 3D high-dimensional feature space as depicted below in Fig. 7.6A and B.

7.2.5   Radial basis function neural network (RBFNN)

RBFNN is a specific class of NN with one hidden layer for application to the problem of supervised learning [29–33]. RBF is a nonparametric regression type problem. There is very little or no prior information about the function that is being predicted, known as a nonparametric function. RBF networks are linear networks, and functional form of $h(x)$ for the linear model is shown in Eq. (7.21). The model h is defined as a set of a linear combination of K functions, which is often called basis function. The typical basis function is Gaussian basis function, described by Eq. (7.22). The standard form is expressed by Eq. (7.23)

$$h(x) = \sum_{n=1}^{N} w_n \varphi_n(x)$$
(7.21)

$$\Phi(x) = \exp\left(-\frac{\| x - x_n \|^2}{\sigma^2} \right)$$
(7.22)

$$h(x) = \sum_{n=1}^{N} w_n \exp\left(-\frac{\| x - x_n \|^2}{\sigma^2} \right)$$
(7.23)

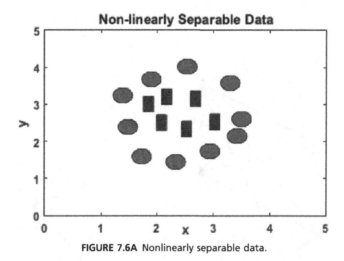

FIGURE 7.6A Nonlinearly separable data.

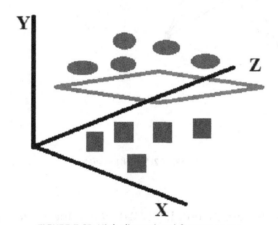

FIGURE 7.6B High-dimensional feature space.

Each (x_n, y_n) ε D influences $h(x)$ based on $||x-x_n||$.

RBF with K-centers is shown in Fig. 7.7. RBF is three layers network named as the input layer, hidden layer and output layer. First layer for input to the network. Each node in the hidden layer is a radial function; its dimensionality is the same as the input data. The output is calculated by a linear combination, i.e., the weighted sum of basis function.

For smooth interpolation in which the complexity of the data structure determines the number of basis function, some modification is required to the exact interpolation method:

1. The number of basis function K is less than a number of N data points.
2. The linear sum is included in the Bias parameters.

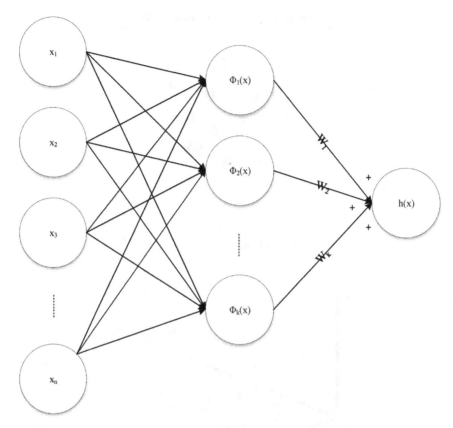

FIGURE 7.7 Rbf NN model.

RBF with K-centers is described in Eq. (7.24). N parameters $w_1, w_2 \ldots \ldots \ldots \ldots \ldots w_N$ based on N data points. Use $K << N$ centers: $\mu_1, \mu_2 \ldots \mu_K$ instead of $x_1, x_2 \ldots x_N$.

$$h(x) = \sum_{k=1}^{K} w_k \exp\left(-\frac{\|x - x_k\|^2}{\sigma_k^2} \right) \qquad (7.24)$$

RBF consists of three parameters, namely weights, centers, and standard deviation. Output layer weights (W_k) are linear parameters, which determine the height of the basis function. Centers (μ_K) of the hidden layer are nonlinear parameters. They determine the position of the basis function. The dimensions of the center are the same as the dimensions of the input layer. Centers of the hidden layer are design parameters, which defines the number of nodes in the hidden layer. Standard deviations (σ) are nonlinear parameters in hidden layers. They determine the width of the basis function.

Center of RBF: Center of RBF is determining by the clustering. Clustering is defined as the organization of the data points or objects into segment or group or cluster based on their characteristics such that there are high intraclustering similarity and high intercluster dissimilarity. There is a similarity between the objects within the cluster and dissimilarity between one cluster to another cluster.

K-means clustering: Clustering is a type of unsupervised learning. In K-means clustering, clustering is done based on attributes of data points such that data of one group is similar as possible, and data of one group differ as data of another group. The number of clusters is specified before beginning the process.

Choosing the centers: Minimize the distance between x_n and closest center μ_k. Split x_1, $x_2 \ldots x_n$ into clusters s_1, s_2, ..., s_k and assign one center to each cluster, i.e., μ_1, μ_2, ..., μ_k as shown in Eq. (7.25)

$$\text{minimize} \quad \sum_{k=1}^{K}\sum_{x_n \in s_k} \|x_n - \mu_k\|^2 \tag{7.25}$$

An iterative algorithm, Lloyd's algorithm is shown as follow:

$$\text{iteratively minimize} \quad \sum_{k=1}^{K}\sum_{x_n \in s_k} \|x_n - \mu_k\|^2 \quad w.\,r.\,t.\ \mu_k \text{ and } s_k$$

$$\mu_k \leftarrow \frac{1}{|s_k|}\sum_{x_n \in s_k} x_n$$

$$s_k = \{x_n : \|x_n - \mu_k\| \leq \text{all} \|x_n - \mu_l\|\}$$

$$\text{converge} \rightarrow \text{local minima}$$

K-means clustering algorithm

I. Get the data points.
II. Assign initial centroids.
III. Input new sample and calculate distance from each centroid. Assign the sample to the nearest cluster and update the centroid.
IV. Repeat Step 3 until all the samples have been covered.
V. Verification step involving the calculation of the distance of each sample from all the centroids and reassignment of the cluster in case of conflict.
VI. Repeat Step 5 until no samples change group.
VII. Define the threshold boundary as the mean of two neighboring centroids.

Weight of RBF NN: Weights of RBF are linear parameters. The number of cluster in RBF is much less than the number of data points; hence the number of weights are much less than data points. It means that the number unknown is much less than the number of equations. If variables are lesser than the equation, called an overdetermined system. There is no exact solution for the overdetermined system; hence the approximate solution is done by least square method or pseudoinverse method. Least squares method provides an overall solution to minimize the sum of the squares of the errors made in the result of every single equation. Input-output relation for RBF NN with K-centers is explained in Eqs. (7.26) and (7.27)

$$\sum_{k=1}^{K} w_k \exp\left(\frac{-\left\|\underline{x}_n - \underline{\mu}_k\right\|^2}{\underline{\sigma}_k^2}\right) \approx y_n \tag{7.26}$$

N equations in K $<<$ N variables.

$$\left(\begin{array}{ccc} \exp\left(\dfrac{-\left\|\underline{x}_1 - \underline{\mu}_1\right\|^2}{\sigma_1^2}\right) & \cdots & \exp\left(\dfrac{-\left\|\underline{x}_1 - \underline{\mu}_k\right\|^2}{\sigma_k^2}\right) \\ \vdots & \ddots & \vdots \\ \exp\left(\dfrac{-\left\|\underline{x}_N - \underline{\mu}_1\right\|^2}{\sigma_1^2}\right) & \cdots & \exp\left(\dfrac{-\left\|\underline{x}_N - \underline{\mu}_k\right\|^2}{\sigma_k^2}\right) \end{array} \right) \begin{bmatrix} w_1 \\ \vdots \\ w_K \end{bmatrix} = \begin{bmatrix} y_1 \\ \vdots \\ y_N \end{bmatrix} \tag{7.27}$$

If $\varphi^T \varphi$ is invertible, then the weights of RBF model is calculated as Eq. (7.28).

$$w = \left(\varphi^T \varphi\right)^{-1} \varphi^T y \tag{7.28}$$

Standard deviation (σ) of RBF NN: The spread of the Gaussian basis function is determined by gradient descent. Gradient descent is a nonlinear optimization technique. In this optimization technique, the cost function or squared error is minimizing iteratively with respect to the parameter (standard deviation). Gradient descent function updates the parameters as Eq. (7.29).

$$\sigma_{n+1} = \sigma_n - \alpha \frac{\partial (J\sigma)}{\partial \sigma} \tag{7.29}$$

7.2.6 Deep learning

Traditional machine learning models do not provide satisfactory response when data size is huge. To overcome this limitation deep learning is used. Deep learning consists of more hidden layers than feed forward neural networks. Deep learning is basically used for vast amount of data which can be structured or unstructured. The performance graph of ML and DL is shown below in Fig. 7.8.

DL is basically used to deal with real world problems. It performs feature learning means features are not needed to feed to the model manually. DL algorithm by themselves generate high order features to predict the respective object. Similar to NN, DL also consists of three main layers: Input layer, Output layer and several hidden layers. Hidden layer of DL comprises of convolution (Conv.+ ReLU) layers, pooling layers, fully connected layer and softmax/logistic layer as shown in Fig. 7.9.

Basic operations in DL involve convolution, thus it is also known as convolution neural network (CNN). In DL, input is more often in image data. Convolution layer is a technique that allows the model to extract visual features from an image in small pieces. Each neuron in convolution layer is responsible for small class of neurons in proceeding layers. The bounding box that determines the class of neurons is called a filter, also called kernels. Conceptually, a filter is moving across an image and performing a

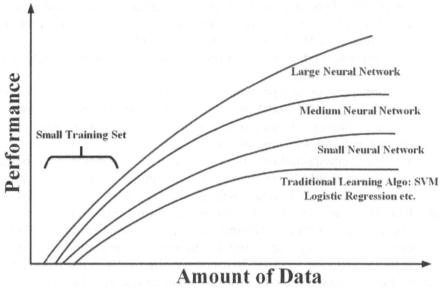

FIGURE 7.8 Performance graph of ML and DL.

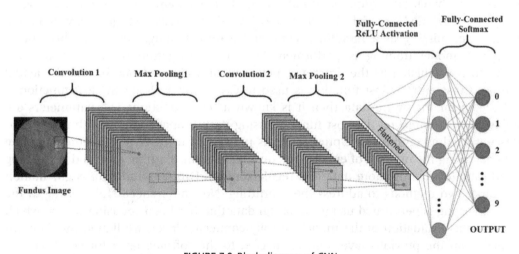

FIGURE 7.9 Block diagram of CNN.

mathematical operation (conv.) on individual reading of the image and then sends results to the corresponding neurons in the convolution layer.

In CNN, convolution is the performed by dot product of input function and kernels. After convolution, the next step of CNN is applied ReLU function as activation function which avoids the negative value. Pooling is also known as subsampling or downsampling. It reduces the volume of the image. There are three types of pooling that can be performed: Max Pooling, Min Pooling, and Average Pooling. Next layer is fully connected layer in which all the neurons of previous layer connected to all the neurons of the next layer. It is used to classify the image into different categories by training. Softmax/sigmoid layer is used after fully connected layer to classify the data into multiclass/binary class respectively.

Convolution and pooling is repeated multiple times for learning the features of input. To create deep learning model five main steps are to be followed: Gathering of data, Preprocessing, Training, Evaluation and Optimization. Selection of data is the key which entirely depends on the problem to be solve. Data can be variety of sizes. Balanced data (almost equal number of data from each class) is required to train a CNN model. Quality of data should be as good as the quantity of data. A model trained on reliable data set is more likely to deal useful predictions than a model trained on an unreliable data. There are several data sets available for CNN such as: UCI, Kaggle, Raddit, Google data set search etc. Millions of data is required to train a CNN model generally. More than 95% data are used for training. After gathering of data, preprocessing is performed. The data set is split into three subset: training, testing and validation. It cannot be split randomly. Distribution of data in Training and testing must be same. Along with parameters (weights and biases), there are many hyper-parameters such as learning rate, number of hidden layers, number of hidden nodes in hidden layers, mini batch size, number of epoch, etc. Hyper-parameters control the weights and biases of CNN model. In preprocessing, these steps are generally required: formatting, missing data handling, sampling, feature scaling, etc. Once the data is prepared, model training is performed. Forward propagation occurs and the lost function is calculated as the difference between the estimated output and the actual output. Cross-entropy lost function is used in CNN model. When the lost function is calculated for the entire data then it is known as cost function. The parameters are adjusted by minimizing the cost function using back propagation algorithm (gradient descent used as optimizer). Optimization is performed using hyper-parameter tuning (by increasing the number of epochs and by adjusting the learning rate) and addressing overfitting (by getting more data and regularization). Dropout and data augmentation are other two methods to address the overfitting. Now after successful training of the model, testing is performed using validation data that has been set aside earlier, which is known as Evaluation of the model. At fully connected layer, we flatten the data obtained from the previous layer and connected to the softmax layer for classification. Output layer occurs after softmax layer with one hot encoded.

7.3 Disease detected by machine learning techniques

7.3.1 Glaucoma and diabetic retinopathy

Glaucoma is known as optic neuropathy, which happens due to increment in intraocular pressure inside the eyeball. When intraocular pressure increases inside retina, retinal pigment epithelium cells located just beneath choroidal layer, starts damaging. It is a type of disorder, which leads to permanent loss of sight. Diabetic retinopathy (DR) is the leading cause of blindness. DR is an eye complication that develops due to long-term standing of diabetic mellitus (DM). The problem with both diseases glaucoma and DR is that they don't interfere eye sight at primary stage. They are asymptomatic in nature. They can't be recognized until reaches into advanced stage. Glaucoma and DR are detected using retinal fundus image, which consists of blood vessels, optic disc, macula, and fovea. Image is a type unstructured data. Traditional ML algorithms such as LR, DT, random forest, SVM, and RBFNN don't work on unstructured data. Hence, features form fundus image are extracted and make it into structured form before applying into traditional ML algorithm. First, preprocessing techniques are applied to improve the image quality. ROI is extracted from the image by applying segmentation techniques. Features are extracted from the ROI area. There are three types of features such as shape based features (area, perimeter, circularity, minor, and major axis), texture-based features (fine, coarse, grained, smooth, etc.), Statistical features (mean, standard deviation, entropy). These features vectors are used to train the model. RGB image directly can be feed into CNN model and CNN model extracts the features from the images.

7.3.2   Brain tumor

In India, about 5–10 cases per one lakh population was reported last year for brain tumor. Early symptoms of tumor like: headache, nausea, sleep problems, drowsiness etc. are so common, which generally mislead in detection of tumor at early stages. Using ML, tumor can be detected at early stage so that to cure in time. There are several literature available for this. Some literature have presented survey and comparison of ML techniques for brain tumor detection [34,35]. Tandel et al. have presented a comparison between ML and DL for brain tumor detection [36]. As brain disease are diagnosed by magnetic resonance images (MRI) of brains, therefore DL is the best suitable learning technique for diagnosis/prognosis of brain tumors.

7.3.3   Breast cancer

According to WHO, more than two million women were diagnosed with breast cancer among which more than six lakhs died worldwide in 2020. Severity of breast cancer increases and the infection reaches to other organs (mostly in lungs) if not diagnosed/ treated at earlier stage. The most common data set used for this is Original Wisconsin Breast Cancer data for CNN based detection. DL based breast cancer detection is done in many literature [37–39]. Apart from deep learning, some other techniques are also used for detection/prediction of breast cancer such as Naïve Bayes, Random Forest, k-nearest neighbor, etc. [40,41]. These ML techniques have reported more than 97% of accuracy for the breast cancer detection/prediction.

7.3.4   Heart disease

According to WHO, cardiac disease causes more than 17 million worldwide death each year, therefore, nowadays, it is needed to predict heart disease at earlier stage. Machine learning is the key for doing this. There are several heart disease which can be predict and detect by machine learning techniques such as: arrhythmia, artery blockage, myocardial infarction, improper functioning of valves etc. In general, neural networks, SVM and deep learning algorithms are used for classification of heart disease using various ECG database like: MITBIH, UCI, Kaggle, etc. For the feature extraction stage of heart disease detection/prediction, a number of features are available which can be extracted very easily using machine learning techniques. Some of the ECG features are: heart rate, spectral entropy, threshold cross-sample count, mean absolute value, auto regressive coefficient, etc. [18,42]. Many survey literatures are available for ML based ECG classification [43–45]. For binary classification, SVM based classifiers are used whereas for multiclass classification of ECG, neural networks are used. For ECG image data, deep learning based classifiers are used. Deep learning algorithm can also be used as 1D CNN, for classification of ECG signal data. A numerous ML technique are available for prognosis/diagnosis of heart disease.

7.3.5   Multimodal classification

Nowadays, in some disease, single parameter is not always conclusive, therefore it is required to include some other related parameters for early detection of life threatening diseases. As in the case of heart disease, it is always required to check patient's blood pressure. If ECG is normal and other parameters such as: chest X-ray, blood pressure, oxygen saturation etc. are not normal, then also patient suffers from cardiac disease. As per the need, some database are generated for this multimodal analysis for e.g., UCI database, which includes cholesterol, blood pressure, angina pain, and other 22 parameters along with arrhythmia classes. These data are the best suitable data for ML based multimodal analysis. 1-D CNN is generally used for this type of analysis [46,47].

7.4   Challenges in ML based classification for medical data

Although ML has made early detection of many life threatening diseases much easier, even then there are some challenges for using ML in medical field. Some of the challenges are as follows:

7.4.1   Data

For classification of any disease, data is the first requirement and for ML these data are used for training of ML. Therefore, the data should be reliable. For DL, huge data is required. Having medical data at this level is still challenging.

7.4.2 Selection of algorithm

Sometimes data are available, however it is difficult to decide which ML algorithm will be applicable for the precise detection or classification of disease.

7.4.3 Overfitting

If there is higher error across the testing data, then it is the case of overfitting. Overfitting can be reduced using regularization. It can also be reduced by increasing more number of data.

7.4.4 Underfitting

If training error is large, known as under fitting or high bias. In this case the algorithm has to modify. Increase the number of input variable in the algorithm. If some relevant input variables have not been considered, then relevant variable should take into account. If relevant variables have been penalize during regularization, then try to reduce the extent of regularization on those input variable or remove the regularization.

7.5 Conclusion

This chapter presented classification models of ML for medical data analysis. Mathematical modeling of various classification algorithms such as: DT, RF, LR, SVM, and NN had been discussed. The complete block diagram for CNN based classification was presented. Several life threatening disease like: cancer, cardiac disease, tumor, glaucoma, etc., which can be classified by ML/DL, were also discussed in brief. In spite of having a lot of benefits, ML based classification also faces some challenges such as overfitting, underfitting, data, etc. Machine learning has become an inevitable procedure used in every field nowadays. Use of ML in medicine and healthcare has become a boon for the patients and medical experts. In future, some more ML based techniques will be introduced to overcome the present challenges of using ML.

References

[1] World Health Organization, Global Health Estimates, Life Expectancy and Leading Causes of Death and Disability, The Global Health Observatory, 2019 [Online]. Available: https://www.who.int/data/gho/data/themes/theme-details/GHO/mortality-and-global-health-estimates%0Ahttps://www.who.int/data/gho/data/themes/mortality-and-global-health-estimates.

[2] I. Kononenko, Machine learning for medical diagnosis: history, state of the art and perspective, Artif. Intell. Med. 23 (1) (2001) 89–109.

[3] G.D. Magoulas, A. Prentza, Machine learning in medical applications, Lect. Notes Comput. Sci. 2049 (2001) 300–307. LNAI.

[4] R. Bender, U. Grouven, Ordinal logistic regression in medical research, J. R. Coll. Physicians Lond. 31 (5) (1997) 546–551.

[5] S.C. Bagley, H. White, B.A. Golomb, Logistic regression in the medical literature: standards for use and reporting, with particular attention to one medical domain, J. Clin. Epidemiol. 54 (10) (2001) 979−985.

[6] J.A. Castellanos-Garzón, E. Costa, J.L. Jaimes S., J.M. Corchado, An evolutionary framework for machine learning applied to medical data, Knowl. Base Syst. 185 (2019) 104982.

[7] E. Christodoulou, J. Ma, G.S. Collins, E.W. Steyerberg, J.Y. Verbakel, B. Van Calster, A systematic review shows no performance benefit of machine learning over logistic regression for clinical prediction models, J. Clin. Epidemiol. 110 (2019) 12−22.

[8] C. Bonte, F. Vercauteren, Privacy-preserving logistic regression training, BMC Med. Genom. 11 (Suppl. 4) (2018).

[9] S. Maji, S. Arora, Decision Tree Algorithms for Prediction of Heart Disease, vol. 40, Springer Singapore, 2019.

[10] W. Chen, S. Chen, H. Zhang, T. Wu, A hybrid prediction model for type 2 diabetes using K-means and decision tree, in: Proceedings of the IEEE International Conference on Software Engineering and Service Science (ICSESS), Vol. 2017-November, No. 61272399, 2018, pp. 386−390.

[11] M.Z. Alam, M.S. Rahman, M.S. Rahman, A Random Forest based predictor for medical data classification using feature ranking, Inform. Med. Unlocked 15 (April) (2019) 100180.

[12] S.K. Baliarsingh, W. Ding, S. Vipsita, S. Bakshi, A memetic algorithm using emperor penguin and social engineering optimization for medical data classification, Appl. Soft Comput. J. 85 (2019) 105773.

[13] A. Kalantari, A. Kamsin, S. Shamshirband, A. Gani, H. Alinejad-Rokny, A.T. Chronopoulos, Computational intelligence approaches for classification of medical data: state-of-the-art, future challenges and research directions, Neurocomputing 276 (2018) 2−22.

[14] N.P. Karlekar, N. Gomathi, S.V.M. OW-, Ontology and whale optimization-based support vector machine for privacy-preserved medical data classification in cloud, Int. J. Commun. Syst. 31 (12) (2018) 1−18.

[15] U.R. Yelipe, S. Porika, M. Golla, An efficient approach for imputation and classification of medical data values using class-based clustering of medical records, Comput. Electr. Eng. 66 (2018) 487−504.

[16] S. Das, S. Mishra, M.R. Senapati, New approaches in metaheuristic to classify medical data using artificial neural network, Arabian J. Sci. Eng. 45 (4) (2020) 2459−2471.

[17] A.N. Astafyev, S.I. Gerashchenko, M.V. Markuleva, M.S. Gerashchenko, Neural network system for medical data approximation, in: Proceedings of the IEEE Conference of Russian Young Researchers in Electrical and Electronic Engineering (EIConRus) 2020, No. 3, 2020, pp. 1483−1486.

[18] V. Patidar, R. Srivastava, B. Kumar, R.P. Tewari, N. Sahai, D. Bhatia, Arrhythmia classification based on combination of heart rate, auto regressive coefficient and spectral entropy using probabilistic neural network, in: 15th IEEE India Council International Conference (INDICON), 2018, pp. 1−4.

[19] J. Ker, L. Wang, J. Rao, T. Lim, Deep learning applications in medical image analysis, IEEE Access 6 (2017) 9375−9379.

[20] M.I. Razzak, S. Naz, A. Zaib, Deep Learning for Medical Image Processing: Overview, Challenges and the Future BT - Classification in BioApps: Automation of Decision Making, Springer, 2018, pp. 323−350.

[21] A. Tahmassebi, M.H.J. Schulte, A.H. Gandomi, A.E. Goudriaan, I. McCann, A. Meyer-Baese, Deep learning in medical imaging: FMRI big data analysis via convolutional neural networks, ACM Int. Conf. Proceeding Ser. (2018) 1−4.

[22] M. Abdar, N.Y. Yen, J.C.S. Hung, Improving the diagnosis of liver disease using multilayer perceptron neural network and boosted decision trees, J. Med. Biol. Eng. 38 (6) (2018) 953−965.

[23] K.P. Sridhar, S. Baskar, P.M. Shakeel, V.R.S. Dhulipala, Developing brain abnormality recognize system using multi-objective pattern producing neural network, J. Ambient Intell. Hum. Comput. 10 (8) (2019) 3287–3295.

[24] D. Learning, Ian Goodfellow Yoshua Bengio Aaron Courville, in: The reference book for deep learning models, 2016, pp. 1–3.

[25] M. Chstofer, Bishop, Pattern Recognition and Machine Learning, vol. 128, Springer Science Business Media, LLC, 2006 no. 9.

[26] T.M. Mitchell, Machine learning and data mining, Predict. Toxicol. 42 (11) (2005) 223–254.

[27] I. Goodfellow, Y. Bengio, A. Courville, Deep Learning, MIT Press, 2016.

[28] R.O. Duda, P.E. Hart, Pattern Classification, second ed.

[29] R. Cheruku, D.R. Edla, V. Kuppili, R. Dharavath, PSO-RBFNN: A PSO-based clustering approach for RBFNN design to classify disease data, Lect. Notes Comput. Sci. 10614 (2017) 411–419. LNCS.

[30] R. Siouda, M. Nemissi, An Optimized RBF-Neural Network for Breast Cancer Classification an Optimized RBF-Neural Network for Breast Cancer Classification, vol. 1, 2019, pp. 24–34. May.

[31] V.S. Spelmen, R. Porkodi, A review on handling imbalanced data, in: Proceedings of the International Conference on Current Trends towards Converging Technologies (ICCTCT), 2018, pp. 1–11.

[32] A.M. Karim, M.S. Güzel, M.R. Tolun, H. Kaya, F.V. Çelebi, A new framework using deep auto-encoder and energy spectral density for medical waveform data classification and processing, Biocybern. Biomed. Eng. 39 (1) (2019) 148–159.

[33] C. Jiang, Y. Li, Health big data classification using improved radial basis function neural network and nearest neighbor propagation algorithm, IEEE Access 7 (2019) 176782–176789.

[34] Z.U. Rehman, M.S. Zia, G.R. Bojja, M. Yaqub, F. Jinchao, K. Arshid, Texture based localization of a brain tumor from MR-images by using a machine learning approach, Med. Hypotheses 141 (2020) 109705, no. March.

[35] M. Aarthilakshmi, S. Meenakshi, A. Poorna Pushkala, V. Rama, N.B. Prakash, Brain tumor detection using machine learning, Int. J. Sci. Technol. Res. 9 (4) (2020) 1976–1979.

[36] G.S. Tandel, A. Balestrieri, T. Jujaray, N.N. Khanna, L. Saba, J.S. Suri, Multiclass magnetic resonance imaging brain tumor classification using artificial intelligence paradigm, Comput. Biol. Med. 122 (2020) 103804, no. March.

[37] G. Murtaza, et al., Deep learning-based breast cancer classification through medical imaging modalities: state of the art and research challenges, Artif. Intell. Rev. 53 (3) (2020) 1655–1720.

[38] T. Mahmood, J. Li, Y. Pei, F. Akhtar, A. Imran, K. Ur Rehman, A brief survey on breast cancer diagnostic with deep learning schemes using multi-image modalities, IEEE Access 8 (2020) 165779–165809.

[39] P.P. Sengar, M.J. Gaikwad, A.S. Nagdive, Comparative study of machine learning algorithms for breast cancer prediction, in: Proceedings of the Third International Conference on Smart Systems and Inventive Technology (ICSSIT). IEEE 2020, 2020, pp. 796–801.

[40] F.M. Javed Mehedi Shamrat, M.A. Raihan, A.K.M.S. Rahman, I. Mahmud, R. Akter, An analysis on breast disease prediction using machine learning approaches, Int. J. Sci. Technol. Res. 9 (2) (2020) 2450–2455.

[41] M. Amrane, S. Oukid, I. Gagaoua, T. Ensari, Breast cancer classification using machine learning, in: 2018 Electric Electronics, Computer Science, Biomedical Engineerings' Meeting (EBBT), 2018, pp. 1–4.

[42] J.P. Li, A.U. Haq, S.U. Din, J. Khan, A. Khan, A. Saboor, Heart disease identification method using machine learning classification in E-healthcare, IEEE Access 8 (Ml) (2020) 107562–107582.

[43] S. Pouriyeh, S. Vahid, G. Sannino, G. De Pietro, H. Arabnia, J. Gutierrez, A comprehensive investigation and comparison of machine learning techniques in the domain of heart disease, in: Proceedings of the IEEE Symposium on Computers and Communications (ISCC), 2017, pp. 204–207.

[44] V.V. Ramalingam, A. Dandapath, M. Karthik Raja, Heart disease prediction using machine learning techniques: a survey, Int. J. Eng. Technol. 7 (2.8 Special Issue 8) (2018) 684–687.

[45] M. Fatima, M. Pasha, Survey of machine learning algorithms for disease diagnostic, J. Intell. Learn Syst. Appl. 09 (01) (2017) 1–16.

[46] S. Münzner, P. Schmidt, A. Reiss, M. Hanselmann, R. Stiefelhagen, R. Dürichen, CNN-based sensor fusion techniques for multimodal human activity recognition, in: Proceedings of the 2017 ACM International Symposium on Wearable Computers, 2017, pp. 158–165.

[47] L. Song, et al., A deep multi-modal CNN for multi-instance multi-label image classification, IEEE Trans. Image Process. 27 (12) (2018) 6025–6038.

Regression tasks for machine learning

K.A. Venkatesh[1], K. Mohanasundaram[2], V. Pothyachi[3]

[1]SCHOOL OF MATHEMATICS AND NATURAL SCIENCES, CHANAKYA UNIVERSITY, BANGALORE, KARNATAKA, INDIA; [2]ALLIANCE SCHOOL OF BUSINESS, ALLIANCE UNIVERSITY, BANGALORE, INDIA; [3]ECONOMICS PGT, SRI S BADAL CHAND SUGAN CHORIDA VIVEKANANDA VIDAYALAYA, CHENNAI, TAMILNADU, INDIA

8.1 Introduction

To make a prediction about a real-world situation, generate sample data first, and then, by using statistical assumptions and mathematical function, a functional relationship between one or more variables is developed, which is called a statistical model. It is a collection of probability distribution on a set of all possible outcomes of an experiment. The mathematical relationship could be among the set of random variables and nonrandom variables. It is a process of applying statistical analysis and provides visualization, identifying the relationship between variables and to make predictions.

8.2 Steps in statistical modeling

1. Gather data. Sources may be from spreadsheets, databases, data lakes, or the cloud.
2. Clean the data and check for outliers and missing values using visualization.
3. Categorize as either supervised or unsupervised learning, for example, decision tree, clustering, logistic regression, classification, and regression models.

8.2.1 Regression models and classification models

- Regression model is one of predictive statistical models that analyzes the association between responses and explanatory variables. Regression models are classified as polynomial, linear, and logistic. These models are utilized in time series modeling, forecasting, and importantly in identifying the causal effect among the variables.

- Classification model is a machine learning technique, and it deploys an algorithm that analyzes the complex set of known data points. It classifies the data based on its understanding; there are a variety of such classification techniques—to name a few: nearest neighbor, random forests, decision trees, artificial neural network (ANN), and Naïve Bayes. These models are mostly used in AI [1].

8.2.2 Regression model

The regression model is used to find the cause-and-effect associations between independent and dependent variables. The most-used regression models are simple linear regression, multiple linear regression, and nonlinear regressions like logistic and polynomial. F. Galton [2] was the first to use the term regression.

8.2.3 Classification model

The classification model is an algorithm that is deployed in analyzing the complex set of known data points and appropriately classifying them. The popular models are random forest, decision trees, Naïve Bayes, and ANN.

8.3 General linear regression model

Let Y be the dependent variable (other names include response variable, outcome variable, target variable, observed variable, regressand variable, measured variable, and output variable) and X be the independent variable (other names includes regressor variable, controlled variable, manipulated variable, explanatory variable, exposure variable, and input variable). The aim of the regression model is to construct mathematical model that is represented by the dependent variable Y as a function of the independent variables X.

The mathematical and statistical models that can bring out the association between a dependent (ratio) variable and an explanatory one or more independent variables (categorical or ratio) is known as the regression model, which is

- a model that has only one explanatory variable, called a simple linear regression (SLR),
- a model that has two or more independent variables, called a multiple regression, and
- to predict the value of the response variable based on the independent variable(s) and determine how the changes in the independent variables effect the value of the dependent variable by the regression model.

The general linear regression model states that the dependent variable Y is given by

$$y = \beta_0 + \beta_1 Z_1 + \beta_2 Z_2 + \ldots + \beta_r Z_r + \varepsilon \tag{8.1}$$

[Dependent Variable/Response] = [AM (depending on the independent variables-$z_1, z_2...z_r$)] + [noise/error].

Where β_0 is the population intercept, β_1, β_2, ... β_r are population slope coefficients, and an error term is ε. From observing an event for a certain number of time periods, the response variables is captured, and these are fixed. Therefore, in such cases, the random variables are the error term and the dependent variable [3].

In general a "linear" term denotes relationship between the mean and unknown parameters $\beta_0, \beta_1, \beta_2 .... \beta_r$. The independent variables need not be present in the model in a first-order way.

Having n different observed items on the dependent variable and the corresponding items of Z_i, the entire model can be viewed as

$$Y_1 = \beta_0 + \beta_1 Z_{11} + \beta_2 Z_{12} + ... + \beta_r Z_{1r} + \varepsilon_1$$

$$Y_2 = \beta_0 + \beta_1 Z_{21} + \beta_2 Z_{22} + ... + \beta_r Z_{2r} + \varepsilon_2$$

$$Y_3 = \beta_0 + \beta_1 Z_{31} + \beta_2 Z_{32} + ... + \beta_r Z_{3r} + \varepsilon_3$$

$$Y_n = \beta_0 + \beta_1 Z_{n1} + \beta_2 Z_{n2} + ... + \beta_r Z_{nr} + \varepsilon_n \tag{8.2}$$

In model (2) the error terms must satisfy the following properties:

1. mean of the error terms is zero,
2. constant variance, and
3. if $i \neq j$ then covariance between these error terms is 0 (3).

In the matrix notation, (2) becomes

$$
\begin{bmatrix} Y1 \\ Y2 \\ Y3 \\ . \\ . \\ . \\ Yn \end{bmatrix}
=
\begin{bmatrix} 1 & z11 & z12 & . & . & . & z1r \\ 1 & z21 & z22 & . & . & . & z2r \\ . & . & . & . & . & . & . \\ . & . & . & . & . & . & . \\ . & . & . & . & . & . & . \\ 1 & zn1 & zn2 & . & . & . & znr \end{bmatrix}
\begin{bmatrix} \beta0 \\ \beta1 \\ . \\ . \\ . \\ \beta r \end{bmatrix}
+
\begin{bmatrix} \varepsilon1 \\ \varepsilon2 \\ . \\ . \\ . \\ en \end{bmatrix}
$$

or

$Y_{(nx1)} = Z_{(nx(r+1))}\beta_{(r+1x1)} + \epsilon(nxI)$

and Eq. (8.2) becomes

1. $E(\varepsilon) = 0$; and
2. $Cov(\varepsilon) = E(\varepsilon\,\varepsilon') = \sigma^2\,I$

8.4 Simple linear regression (SLR)

One has to use appropriate visualization methods such as scatterplots and statistical methods such as correlation to ascertain the linear relationships and potential strength. SLR is used to quantify the association between the explanatory variables and the response variable [4].

From a data set with "n" observations (Xi, Yi), where $i = 1,2,3 \dots$ n, a scatter plot is drawn to help understand the strength of association between the variables. To quantify the relationships, between the dependent variable and explanatory variables, SLR is used.

The functional equation is given as

$$Y_i = \beta_0 + \beta_1 X_i + \varepsilon_i$$

where

The random error (residuals) is. ε_i

The regression coefficients/parameters are β_0 and. β_1

This relation is different from the usual (mathematical one) and hence the introduction of. ε_i.

Here, the model SLR can measure the changes in the response variable value using the independent variables values X. Therefore, the prediction of dependent variable of Y_i for the known values of X_i is obtained from the SLR model $\beta_0 + \beta_1 X_i$ and the errors ε_i in the prediction. This can be stated as $E[Y_i|X_i]$, the conditional expected value [5].

8.4.1   Estimate the regression parameters

By finding the values of unknown $\beta_0 \ and \ \beta_1$ from the given set of values, one could fit the SLR. Let it be b_0 and b_1. These estimators are based on the underlying sampling distributions of the error term €, and one must understand the basic assumptions.

Underlying assumptions of linear regression model [6]

1. The error must have zero mean and the response variable must have constant variance.
2. Errors are normally distributed.
3. The errors are independent and identically distributed (i.i.d.).
4. The underlying distribution of the response variable must be normal for all values of the independent variables.

Ordinary least square (OLS) is deployed to obtain the regression coefficients. It does the expected job so that $\Sigma\left(Y_i - \widehat{Y}^2\right)$ is minimized. It provides the best linear unbiased estimator, i.e., $E\left[\beta - \widehat{\beta}\right] = 0$, where β is the population parameter and $\widehat{\beta}$ is the estimated parameter value from the sample.

To obtain the extremal value of β_0 and β_1 that will minimize the sum of squares due to error by equating to zero. On solving, we get the estimated value of β_0 as $\widehat{\beta}_0 = \widehat{Y} - \widehat{\beta}_1 X$ and the value of β_1 is given by $\beta_1 = \frac{Cov(X,Y)}{var(X)}$.

8.5 Authentication of the simple linear regression model

To authenticate the regression model and goodness-of-fit, the listed below measures are used before applying the model on the real applications:

a. the proportion of variance, coefficient of determination, $R\widehat{2}$ (squared R),
b. testing of hypothesis for regression coefficient -slope parameter β_1,
c. ANOVA (analysis of variance) for overall fit model validity,
d. error analysis to validate the underlying assumptions of the regression model, and
e. outlier analysis.

8.5.1 Squared R

Squared R is called the coefficient of determination, utilizing the available information on the independent variables to reduce the errors while predicting the dependent variable.

Based on the relationship between two types of variation, the squared R can be obtained as the variation of the dependent variable based on hat values (the obtained regression line) and the arithmetic mean.

The first variation is the sum of squares error (SSE) of the regression model.

8.5.2 Interpretation of squared R

One has to infer based on the properties of squared R. Listed below are a few properties of R:

- The value lies between 0 and 1.
- The larger values of the coefficient of determination suggests that model is a better fit; however, one must be aware of spurious regression (two sets of data without any relationship).

8.5.3 Hypothesis tests to the regression coefficients and p values

In general, one has to decide which independent variable must be in the regression model; to ascertain this fact, one has to check if the coefficient of that variable in the regression model is zero or not. In case it happens to be zero then there is no meaning in adding to the model because it will not yield any value to the dependent variable.

The corresponding null and alternative hypotheses are

$H_0 : \beta = 0$

versus

$$H_1 : \beta \neq 0.$$

The acceptance of the null hypothesis (H_0) would indicate that the independent variable does not influence the dependent variable.

To test the significance of the coefficient using t static

$$t_{n-k} = (b - \beta)/SE(b)$$

where β's estimated value and standard error are b and SE(b), respectively.

The t statistic is computed using table value based upon $n - k$ degrees of freedom, where k, n are the number of observations and parameters to be found in the model. Depending on $|t|$ value at a given level of significance, the null hypothesis is rejected or accepted. The other way to infer about the decision is based on probability (p) value; if that variable's P-value is small, then that variable must be present in the model, else that variable must be dropped from the model [6].

ANOVA table: A test for the overall fit.

Suppose the model for Y and the dependent variables are $Z_1, Z_2, ..., Z_r$, the model can be as

$$Y = \beta_0 + \beta_1 Z_1 + \beta_2 Z_2 + ... + \beta_r Z_r + \varepsilon$$

If all of the $\beta_1, \beta_2, ..., \beta_r$ are zero then the model has no explanatory power at all that is the prediction of the dependent variable Y can be always, irrespective of independent variables' values. Hence,

$H_0 : \beta_1 = \beta_2 = ... = \beta_r = 0$

versus

$$H_1 : \text{at least one of the } \beta s \text{ is not zero.}$$

One can ensure this by using the F test. This approach is popularly called the ANOVA test.

The various sources of variation can be obtained from an ANOVA table. In the regression model, we are interested to know the variation in the dependent variable.

SST is the total variation of the dependent variable, and it is given by sum of squares total. SST is equal to $\sum (Yi - \overline{Y})^2$.

The total variation has two parts:

- explained by the regression equation (SSR), and

- unexplained (sum of squared residuals); $SSE = \sum \left(Y_i - \widehat{Y_i} \right)^2$

SST − SSE = SSR.

The formal way for testing whether the unexplained variation is smaller compared to the explained variation can be achieved with F-test. Degrees of freedom (d.f.) are associated with the variation sources, and hence the variation source is associated with d.f. Mean square (MS) is defined as the ratio of either sum of squares to its degrees of freedom. The two cases of mean squares are MSR and MSE, given by

$$MSR = \frac{SSR}{r}$$

$$MSE = \frac{SSE}{(n - r - 1)}$$

Observe that the square of the standard error of estimate is nothing but MSE, that is $MSE = S_e^2$.

F-ratio is based on the mean squares and given by

$$F = \frac{MSR}{MSE} \sim F_{r,\ n-r-1} df$$

Based on the values of the F-ratio, one could state that the smaller F-ratio, the un-explained variation is large compared to the explained variation, and conclude that the regression equation has a very limited explanatory power. On the other hand, the un-explained variation is small compared to the explained variation, and can arrive that the regression model has some explanatory power.

Note: From the ANOVA table, even if one identifies that the F-ratio is extremely significant, it does not ensure the regression model is good fit in real-time problems. In such cases, one has to look into the other available measures such as R^2 and S_e.

8.5.4   Inclusion/exclusion of explanatory variable decision

This section will explain when the explanatory variables in the regression equation need to be included/excluded. Using the student's t test one has to test whether the regression coefficient of a population is zero. But it doesn't guarantee that the variable must be included or excluded just based on whether the t value is just significant or t value is insignificant. The underlying fact is that we are always trying to get the most appropriate fit possible, and according to the principle of parsimony one has to accommodate minimal number of variables.

The following is a guiding principle for adding or removing variables in a regression model:

a. Based on the t value and P-value. If the P-value of a variable at some accepted sig-nificance level, such as 0.05, this variable is a right one for exclusion.
b. If a variable's t value is more than one or smaller than one in magnitude, adjusted R^2 will increase (and S_e will decrease); when t value is smaller than one provided this variable is removed from the model. The opposite will occur when t value is larger than 1.

c. t values and P values are better ones than correlations while making decisions to add or remove variables in the model. When a high-correlated variable with the dependent variable is included along with other variables in the model it does not improve the predictability due to the presence of other variables in the model; in fact it may not be useful. This often occurs due to multicollinearity.

d. A subset of variables that are logically related in some sense, then have all variables or with no variables, is a good idea.

e. In addition, one has to use the physical or economic theory as an aid to add or remove variables.

8.6 Multiple linear regression

SLR depends on only one independent variable, but in reality, the response variable depends on more than one predicator variables. The general model of the multiple regression is given as $Y = \beta_0 + \beta_1 Z_1 + \beta_2 Z_2 + \dots + \beta_r Z_r + \varepsilon$, where X_k is the k-th predictor and β_k is the measure of the association between the respective predictor variable and the response variable. At the time of inference, β_k is the average effect on the dependent variable of a one unit increase on the predicator variable Z_k, keeping all other predicator variables fixed.

In the multiple regression analysis, one has to answer the following:

a. Is the response variable depending on at least one of the predicator variables?

b. Are all predictor variables needed to predict Y or are only a subset of predictor variables are sufficient?

c. Does the model fit the data?

d. How accurate is the predication?

As in SLR, we have to verify all $\beta_i's$ $(1 \le i \le r)$ are all zero. This can be done using hypothesis testing, as H_0 : *all regression coefficients are* 0 versus the alternate hypothesis H_1 : *at least one* β_i *is not equal to* 0.

This can be done with F-statistic.

Based on the P-value, one could say whether or not the predicator variable is meaningful in predicting the response variable. Note that there is no association between the independent variables and the response variable; the value of F-statistic would be close to unity. Dealing with F-static is somewhat tricky, that is, one has to work with number of observations and number of predictor variables. It is the right approach to use P-value and t-statistics for each predictor variable to verify the dependency of the response variable. The F-statistic method is appropriate only when the volume of predicator variables is much smaller than the volume of observations. If the number of predicted variable is too large, the model may have some false predictions; the identification of the subset of predictor variable is known as *variable selection*.

So far, the discussions fall under the assumption that all predicator variables are quantitative variables. Suppose, at least one predicator variable Z_i is qualitative (categorical/factor) with at least two levels, then by introducing dummy variable which assumes two values, that is $Z_i = 1 \; or \; 0$. Let us consider two predictor variables Z_1 and Z_2, out of these variables, Z_2 is the factor variable with two levels then the regression model is given by

$$Y = \beta_0 + \beta_1 Z_1 + \beta_2 Z_2 + \varepsilon_i = \begin{cases} \beta_0 + \beta_1 Z_1 + \beta_2 + \varepsilon_i, & Z_2 = 1 \\ \beta_0 + \beta_1 Z_1 + \varepsilon_i, & Z_2 = 0 \end{cases}$$

To address the important predictor variables, consider all possible models, that is, by trying out all possible subsets of the predictor variables set. Among the considered models, choosing the most suitable model is based on various statistics such as Mallow's C_p, Akaike information criterion (AIC), Bayesian information criterion (BIC), Akaike corrected information criterion (AIC_c), and adjusted R^2 [1].

Consider that the size of predictor variables is k and the size of observations is n.

Forward Selection: Here one has to start with null model, that is, with no predictor variables but the model has only the slope β_0. Subsequently, the linear model is augmented to the predictor variables to the model with smallest RSS, and continues this way until the prescribed stopping criterion is met.

Backward elimination: In this case, one has to start with the full model—the model with all predictor variables—and eliminate the predictor variable whose P-value is maximum. Now the (k-1) variable model is built. In this model, the predictor variable with maximum P-value is eliminated and continuing in this manner until the prescribed stopping criteria is met.

Mixed Method: As the name suggests, this model is a combination of backward elimination and forward selection. One has to start with null model and include the predictor variables one-by-one to the model based on threshold P-value. That is the predictor variable is added to the model than P-value of the predictor variable must be less the threshold P-value.

Note that the forward selection method can be applied irrespective of the number of observations and number of the predictor variables, whereas the backward elimination method is possible only when the number of observations is sufficiently larger than the number of predicator variables.

The underlying assumptions of the linear regression model are that the relation between the predictor variables and the response is linear and additive, but these assumptions need not be true in many real-time problems. Let us consider the model with three predictor variables, $Y = \beta_0 Z_1 + \beta_1 Z_1 + \beta_2 Z_2 + \beta_3 Z_3 + \varepsilon$. Here β_3 denotes the average number of units in the response variable for a unit increase in Z_3, irrespective of Z_1 & Z_2. In such situations, introduce another explanatory variable, which is the product of three variables $Z_1Z_2Z_3$ or product of two variables based on the problem. Now the model is written as $Y = \beta_0 + \beta_1Z_1 + \beta_2Z_2 + \beta_3Z_3 + \beta_4Z_2Z_3 + \varepsilon$, which is also the linear model only, that is,

$$Y = \beta_0 + \beta_1Z_1 + (\beta_2 + \beta_4Z_3)Z_2 + \beta_3Z_3 + \varepsilon$$

$$Y = \beta_0 + \beta_1Z_1 + \beta_5Z_2 + \beta_3Z_3 + \varepsilon$$

where $\beta_5 = (\beta_2 + \beta_4Z_3)$. The term that contains Z_2Z_3 is called an interaction term. Based on the need, the interaction term can be introduced. Note that whenever we use the interaction variables in the model, irrespective of P-values of the predictor variables, the variables must be included in the model.

8.7 Polynomial regression

One of the important assumptions of the linear model is the linear relationship between the independent variables and the response variable. In this case, the association between the predictor variables and the dependent variable is nonlinear, and one might consider the polynomial regression. On the identification of relationship that is not linear between predictor variable(s) and the response variable in the linear model is known as polynomial regression [1].

For a given data set, the job is to fit the data with a linear regression model, and one has to look into the following:

(a) Linearity/nonlinearity of the predictor-dependent relationship
(b) Nonconstant variance of error terms
(c) Correlation of error terms
(d) Outliers
(e) High-leverage data points
(f) Collinearity

The linear relationship can be identified as the fitted value plot. If the regression line is a straight line then it is linear. Also, one could use the residual plot to identifying the nonlinearity. In case, the relationship is nonlinear then apply transformation such as log(Z) of nth power.

In most of the cases, depending upon on the value of the response variable, the error terms' variance might vary. Using the residual plots, one could identify the outlier data points. Also, the outliers can be identified from the studentized residuals plot, and these values are obtained by dividing the each e_i by its S_e. Moreover, the same thing can be done by using leverage statistic. The leverage statistic lies between $\frac{1}{n}$ and 1. If a data point has a leverage statistic is more than $\frac{r+1}{n}$.

In case, among the predicator variables, two or more variables are highly correlated, that is linearly dependent then estimation of regression coefficient may not be possible. One might use contour plot or **variation inflation factor** (VIF) to identify collinearity. VIF for each variable is given by $VIF\left(\widehat{\beta_l}\right) = \frac{1}{1-R^2}$, where the second term in the denominator is the coefficient of determination of Z_l. If this term is approximately 1, then it indicates the presence of collinearity and hence the larger VIF.

In today's data science world, resampling techniques in regression modeling play vital roles. Mostly the data points are divided into two or three subsets, one is called training, the second is called the test data set, and the third one is the validation data set, respectively. By considering different samples from the training data set to fit the regression model with a belief that one could derive some additional information, irrespective of its expensiveness in terms of computation, the popular resampling techniques are holdout sample, three-way split, cross-validation, and bootstrapping. The holdout sample is the method where the data set is divided into two, and the training data is used to identify the model, and the rest of the test data set is used to estimate the error. Cross-validation divides the data set into multiple training and test subsets, and has different methods such as random subsampling, K-fold cross-validation, and leave-one-out cross-validation (G, D, T, & R, 2014) [1].

8.8 Implementation using R programming

The data set Carseats in MASS package in R is used.
   Code and results with discussions.
   R console.
   On installing the necessary packages (ISLR, car, and MASS).
   Type the following

➤ > View(Carseats) ## to view the data set
➤ > names(Carseats) ## to see the column names

```
➤  Loading required package: carData
➤   [1] "Sales"      "CompPrice"   "Income"    "Advertising" "Population"
➤   [6] "Price"      "ShelveLoc"   "Age"       "Education"   "Urban"
➤  [11] "US"
```

➤ > ? Carseats ## to get complete information on data set Carseats
➤ dim(Carseats) 400 rows and 11 columns

For this data set, dependent variable is Sales and consider the variables Price, Urban, and US as predictor variables.

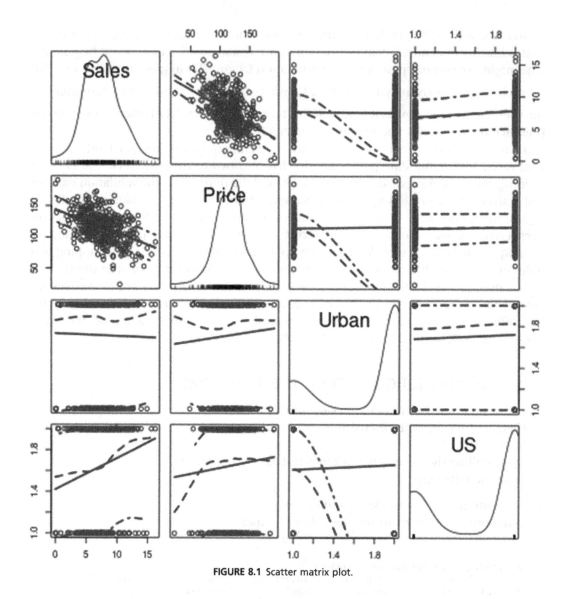

FIGURE 8.1 Scatter matrix plot.

➤ To see the relationship let us scatterplotmatrix as shown in Fig. 8.1
➤ > scatterplotMatrix(~Sales + Price + Urban + US, data=carseats)
➤ > reg.m <- lm(Sales ~ Urban + US + Price, data = Carseats)

➤ > summary(reg.m)

The output is shown as

	Min	1Q	Median	3Q	Max
➤	-6.9206	-1.6220	-0.0564	1.5786	7.0581

The coefficients are found as

	values	S-Error	t-statistic	P-value (>|t|)
➤ (β_0)	13.043469	0.651012	20.036	< 2e-16 ***
➤ β_3	-0.054459	0.005242	-10.389	< 2e-16 ***
➤ β_2	-0.021916	0.271650	-0.081	0.936
➤ β_1	1.200573	0.259042	4.635	4.86e-06 ***

➤ RSE: 2.472 on 396 DF

➤ Multiple squared R: 0.2393, Adjusted R-squared: 0.2335

➤ F-statistic: 41.52 on 3 and 396 DF, p-value: < 2.2e-16

➤ $Sales = \beta_0 + \beta_1 * Urban + \beta_2 * US + \beta_3 * Price + \varepsilon$. This is the required model. But the variables Urban and US are categorical variables and hence in the model, one has to utilize as US and Urban =1 if the store is in US and Urban, otherwise in both cases, the value is 0.

➤ Here *P*-values of all predicator variables are significant, and hence there is a relationship between predictor variables and the response variable.

➤ Suppose we consider only the Price and US as the predictor variables then the model is $Sales = \beta_0 + \beta_1 * Price + \beta_2 * US + \varepsilon$

➤ reg.model1 <- lm (Sales ~ Price + US, data = Carseats)

➤ summary(reg.model1)

➢ Residuals:

➢   Min   1Q   Median   3Q   Max

➢ -6.9269 -1.6286 -0.0574   1.5766   7.0515

➢ Coefficients:

➢   Values   S Error   t-value   P-value(>|t|)

➢ (β_0)   13.03079   0.63098   20.652   < 2e-16 ***

➢ β_1 -0.05448   0.00523   -10.416   < 2e-16 ***

➢ β_2 1.19964   0.25846   4.641   4.71e-06 ***

➢ ---

➢ Signif. codes: 0 '***' 0.001 '**' 0.01 '*' 0.05 '.' 0.1 ' ' 1

➢

➢ RSE: 2.469 on 397 DF

➢ Multiple squared R: 0.2393, Adjusted squared R: 0.24

➢ F-statistic: 62.43 on 2 and 397 DF, p-value: < 2.2e-16

➢ Now we have to identify the better one?

Looking upon the values of R-squared and RSE, both models are similar, but the latter is a little better than earlier one.

➢ Important Diagnosis [4].
  (a) residualPlots (shown in Fig. 8.2)
  (b) rsidualplots against fitted values
  (c) Marginal-Model Plots (shown in Fig. 8.3)
  (d) Added-Variable Plots (shown in Fig. 8.4)
  (e) Marginal–Conditional Plots(shown in Fig. 8.5)
  (f) QQ Plot (shown in Fig. 8.6)

For outliers, we have outlierTest() function and for leverage, we have influenceIndexPlot() function (various influence measures such as Cook's distance, DFBETA, and so on)

➢ >residualPlots(reg.model1)

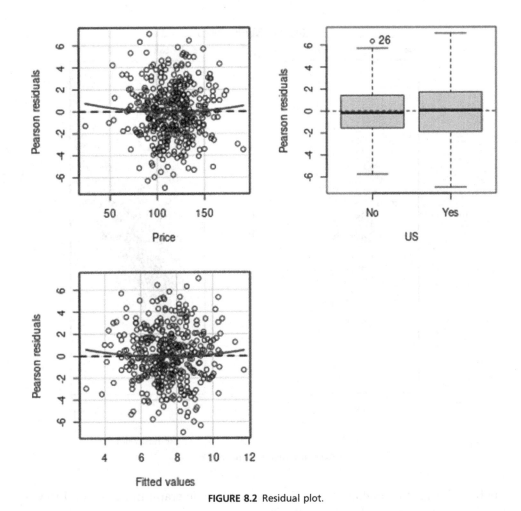

FIGURE 8.2 Residual plot.

```
➤  Loading required package: carData

➤                Test stat Pr(>|Test stat|)

➤  Price         0.6925          0.4890

➤  US
```

➤ Here the horizontal axis in the graph (first row, first column) is the predictor variable (quantitative), and vertical axis is the response variable, and the next graph is

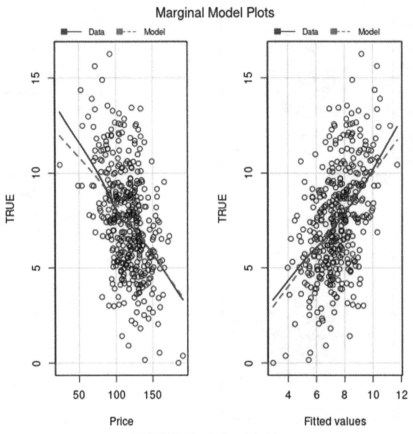

FIGURE 8.3 Marginal-model plot.

the box plot for the qualitative predictor variable. The graph in the second row is the fitted value against the residuals. The purpose of this one is that whether the linearity is met or not. Turkey's test is for nonadditivity.

➤ Tukey test   0.7087   0.4785

(c) Another version of residual plot is marginal-model plot

➤ > marginalmodelPlots(reg.model1)

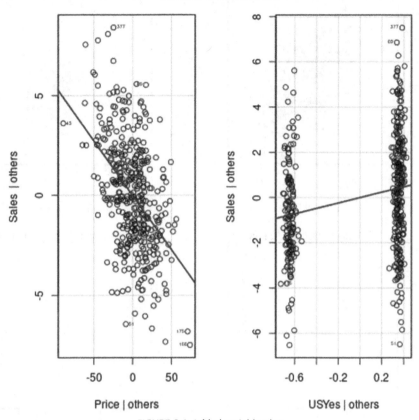

FIGURE 8.4 Added-variable plot.

In this plot the model is dashed line and thick solid line is the true one. From the graph, one could see the marginal relationship between the predictor and the response. Note that this plot is only for quantitative data.

(d) Added-Variable Plot

This graph is also, shows the partial relationship between the predictor and response (the key difference between AV plot and MM Plot is that MM Plot shows only the concerned predictor variables, not all others and AV plot is just includes all others too)

➤ >avPlots (reg.model1, id=list (n=3, cex=0.4))

(e) Marginal Conditional Plot

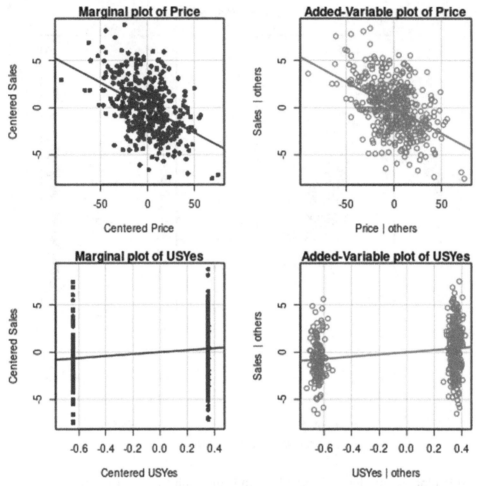

FIGURE 8.5 Marginal conditional plot.

(f) QQ plot: Quantitle-Quantile plot plotting the quantiles against studentized residuals and produces 95% confidence envelope for studentized residuals

OutlierTest: This function identifies the absolute maximum studentized residual data point and Benferroni-Corrected t-test.

InfluenceIndexPlot: Outlying points and high-leverage point has more effect on β_i. Hence, it is important to identify such data points. For the data set Carseats, the influenceindex plot is shown in Fig. 8.7

➤ >influenceIndexPlot(reg.model1, id=list=(n=3))

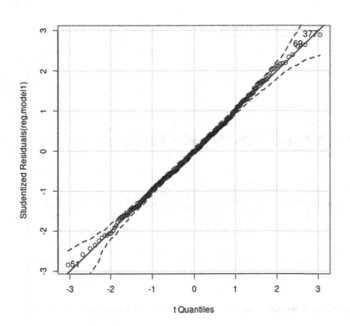

FIGURE 8.6 Qq plot.

To get all such point together, one could use the function influencePlot(reg.model1, id=list(n=3)) and to compute dfbeta's, use dfbeta(reg.model1) and plot is shown in Fig. 8.8.

```
[1] 400   11

          StudRes          Hat          CookD

26     2.5996518 0.011621599 0.026109457

43    -0.5349931 0.043337657 0.004329756

50     2.3343761 0.012552881 0.022835461

51    -2.8358431 0.004224147 0.011173381

69     2.6423636 0.005201501 0.011988429

166   -1.4214588 0.028566608 0.019755042

175   -1.2144859 0.029686718 0.015024314

368    1.7366086 0.023707048 0.024287363

377    2.8915213 0.006637175 0.018282191
```

Nonconstant Variance: The function ncvTest() is used to test the variance is constant or not.

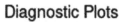

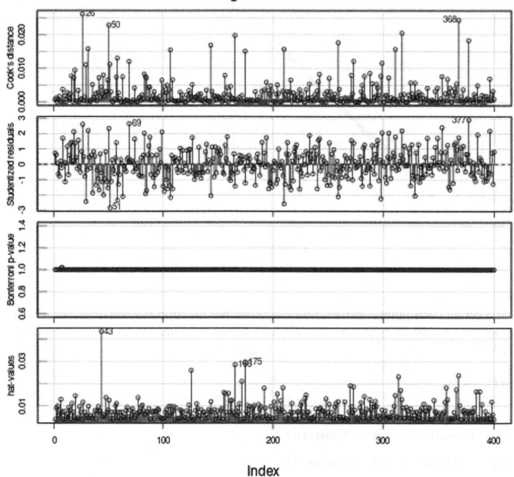

FIGURE 8.7 Influence index plot.

```
[1] 400  11

Non-constant Variance Score Test

Variance formula: ~ fitted.values

Chisquare = 1.566777, Df = 1, p = 0.21068
```

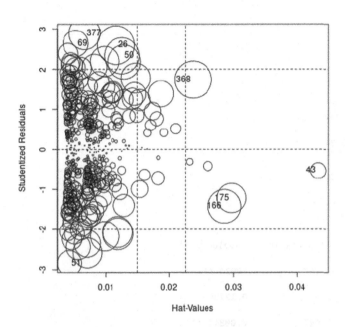

FIGURE 8.8 Influence plot.

To identify one could use the function vif(model)

Now we look into various functions in R to display the computed. $\beta'_i s, t - statistic, p - valuesandAIC, BIC$

We use auto data set in ISLR package

➤ > dim(Auto)

```
➤  [1] 392   9
```

➤ > names (Auto)

```
 [1] "mpg"          "cylinders"   "displacement" "horsepower"   "weight"

 [6] "acceleration" "year"        "origin"       "name"
```

Let us fit a linear regression where the response variable is mpg, and all other variables are predictor variables, except the variable name.

➤ >Aut<- Auto[,-9] ## remove the name column and store in Aut

>dim(Aut)
>regm<-lm(mpg~.,data=Aut) # # fitting linear model
>summary(regm) ## to get entire values

```
Call:

lm(formula = mpg ~ ., data = Aut)

Error/residual:

    Min     1Q  Median     3Q    Max

-9.5903 -2.1565 -0.1169  1.8690 13.0604




Coefficients:

              Values    S-Error   t-statstic   P-value(>|t|)

($\beta_0$)    -17.218435  4.644294  -3.707      0.00024 ***

$\beta_1$       -0.493376  0.323282  -1.526      0.12780

$\beta_2$        0.019896  0.007515   2.647      0.00844 **

$\beta_3$       -0.016951  0.013787  -1.230      0.21963

$\beta_4$       -0.006474  0.000652  -9.929     < 2e-16 ***

$\beta_5$        0.080576  0.098845   0.815      0.41548

$\beta_6$        0.750773  0.050973  14.729     < 2e-16 ***

$\beta_7$        1.426141  0.278136   5.127      4.67e-07 ***

---

Signif. codes:  0 '***' 0.001 '**' 0.01 '*' 0.05 '.' 0.1 ' ' 1

RSE: 3.328 on 384 DF

Multiple R-squared:  0.8215,    Adjusted R-squared:  0.8182

F-statistic: 252.4 on 7 and 384 DF,  p-value: < 2.2e-16
```

Instead summary function if we brief the function,
>brief(regm)

	(Intercept)	cylinders	displacement	horsepower	weight	acceleration
Estimate	-17.22	-0.493	0.01990	-0.0170	-0.006474	0.0806
Std. Error	4.64	0.323	0.00752	0.0138	0.000652	0.0988

	year	origin
Estimate	0.751	1.426
Std. Error	0.051	0.278

Residual SD = 3.33 on 384 df, R-squared = 0.821

>S(regm)

```
Call:

lm(formula = mpg ~ ., data = Aut)

Error/residual:

   Min     1Q  Median     3Q     Max

-9.5903 -2.1565 -0.1169  1.8690 13.0604




Coefficients:

               Values    S-Error    t-statstic   P-value(>|t|)

(β₀)         -17.218435  4.644294   -3.707       0.00024 ***

β₁           -0.493376  0.323282   -1.526        0.12780

β₂            0.019896  0.007515    2.647        0.00844 **

β₃           -0.016951  0.013787   -1.230        0.21963

β₄           -0.006474  0.000652   -9.929       < 2e-16 ***

β₅            0.080576  0.098845    0.815        0.41548

β₆            0.750773  0.050973   14.729       < 2e-16 ***

β₇            1.426141  0.278136    5.127        4.67e-07 ***

---

Signif. codes:  0 '***' 0.001 '**' 0.01 '*' 0.05 '.' 0.1 ' ' 1

RSE: 3.328 on 384 DF

Multiple R-squared:  0.8215,    Adjusted R-squared:  0.8182

F-statistic: 252.4 on 7 and 384 DF,  p-value: < 2.2e-16

AIC       BIC

2064.95   2100.69
```

8.9 Conclusion

This chapter introduced the concepts and various models on regression analysis and concluded with examples using the statistical software R.

References

[1] G. James, D. Witten, T. Hastie, R. Tibshirani, An Introduction to Statistical Learning, Springer Science Business Media, New York, 2014.

[2] F. Galton, Regression towards mediocrity in heredity stature, J. Anthropol. Inst. 15 (1885) 246–263.

[3] Albright & Winston, Business Analytics: Data Analysis and Decision Making, seventh ed., Cengage publications, 2019.

[4] J. Fox, Applied Regression Analysis & Generalized Linear Models, Sage Publications, 2016.

[5] R.A. Johnson, D.W. Wichern, Applied Multivariate Statistical Analysis, sixth ed., PHI, 2012.

[6] U. Dinesh Kumar, Business Analytics- the Science of Data-Driven Decision Making, Wiley India, 2017.

9

Model selection and regularization

K.A. Venkatesh[1], Dhanajay Mishra[2], T. Manimozhi[3]

[1]SCHOOL OF MATHEMATICS AND NATURAL SCIENCES, CHANAKYA UNIVERSITY,
BANGALORE, KARNATAKA, INDIA; [2]KARMAAI LIFE, BANGALORE, KARNATAKA; [3]FRANCIS
XAVIER ENGINEERING COLLEGE, TIRUNELVELI, TAMILNADU, INDIA

9.1 Introduction

Multiple regression is one of the powerful techniques in prediction. However, it has certain limitations. This chapter deals with the limitation and how to overcome such limitation. The first section of this chapter deals with the limitations of the general multiple linear regression such as the number of predictors is more than the number of observations that is too many predictors and subset selection methods. The second section is on shrinkage methods includes the Ridge and Lasso regression models, the third section is on dimension reduction, the fourth section is on dealing with higher dimensions, and the last section is with examples and coding.

First one has to understand the limitations of multiple linear regressions. The response variable in a multiple linear regression model is usually found using the ordinary least square (OLS) approach. The OLS method aims to minimize the sum of square difference between the response variable's actual and predicted value.

$$\widehat{B} = \Sigma_{i=1}^{n} \left(y_i - \widehat{y_i} \right)^2 \tag{9.1}$$

$$\widehat{B} = \Sigma_{j=1}^{n} \left(y_i - \left(B_0 + B_j X_{ji} \right) \right)^2 \tag{9.2}$$

The OLS method aims to find out the best coefficient for the data we are working with as well as it tries to minimize the error between the actual value and the predicted value of the response variable. However, there are some limitations to this action.

One of the limitations of multiple linear regression is that there are many times that some of the predictors are not important for prediction of the response variable but OLS treats all the predictors equally, which means it does not consider any independent variable more important than the other predictors. By including such types of predictors then the complexity of the model increases naturally. Just by removing such variables or by setting their coefficient to zero one could improve the performance of the model.

Statistical Modeling in Machine Learning. https://doi.org/10.1016/B978-0-323-91776-6.24001-3

Consider a simple example to understand this limitation. The problem is to find out the price of the house and there are three regressor variables: area of the house, no. of rooms in the house and the color of the house.

$$y = \text{Price of the House}$$

$$x_1 = \text{Area of the House}$$

$$x_2 = \text{Number of rooms in the House}$$

$$x_3 = \text{Color of the House}$$

Now if we use multiple linear regression model than the equation for multiple linear regression model will be

$$y = B_0 + B_1 x_{j1} + B_2 x_{j2} + B_3 x_{j3} + \varepsilon \tag{9.3}$$

Here, B_0, x_1, x_2, x_3 are the coefficients which will be decided by the OLS equation.

Just by common knowledge we can say that the color of the house does not play any important role in the price of the house but our multiple linear regression models which will not be able to predict that importance and will give the same importance to all three regressor variables present in the equation.

When the number of observations (n) are much greater than the number of predictors (p) then the OLS equation will perform well but having a small amount of bias and good accuracy result on the unseen data. In other case when the number of observations is not that much higher than the predictors $(n > p)$ in such cases the model will face the problem of over fitting, that is the model will perform poorly on the unseen data. But when the number of predictors is more than the observations $(n > p)$ in such cases the multiple linear regression models cannot perform.

Since we know the OLS vector form can be written as,

$$\widehat{B} = (X^T X)^{-1} X^T Y \tag{9.4}$$

There are more predictors in the data than the observations in such cases Ordinary Least Square method cannot perform because the equation will not be able to find the inverse of $X^T X$, as this will not result in a square matrix [1].

One could deploy something called subset selection, regularization and dimension reduction to overcome these limitations of the multiple linear regressions.

Now the overview of these methods is presented over here and will dig deep down in all these later.

9.2 Subset selection

In this approach is to find the relevant predictors which are related to the predictor in order to get the better performance of the model. There are different methods under subset selection methods.

9.2.1   Best subset selection

This is a method which is used to select the best model with the help of a subset of predictors. The OLS equation is fitted to each and every combination of p predictors. In total there will be 2^p total number of possible combinations. The algorithm for the method is as follows:

(a) First we will use a model with no predictors and will denote this by μ_0.
(b) Let x = 1,2,3, …, p predictors:
  (i) Fit all the models for the k predictors.
  (ii) Pick the best model among the models created by k predictors. We can use RSS or R^2 to predict the best model.
(c) In the end we will be selecting the best model among the best selected models for each subset of predictors using cross-validation prediction error, BIC or adjusted R^2.

   The reason why we are using adjusted R^2, BIC and cross-validation prediction error instead of R^2 and RSS is that the value of R^2 increases as the number of predictors increases but the value of RSS decreases as the number of predictors increases.

9.2.2   Stepwise selection

There are two different types of stepwise selection methods:

(a) Forward stepwise selection: In this technique we start fitting the model with the zero predictors that is we will just use intercept, then try to fit the model with the single predictor and will choose the best one amount p predictors, after that we will move to second predictor keeping the previously selected predictor and continue doing so till our performance indicator decreases.

   The method has following steps of algorithm:

  (i) Start with no predictor and name the model as μ_0.
 (ii) Let k = 0,1, …, p − 1
    (a) Consider all p-k models and with one additional predictor.
    (b) Choose the best model among these p-k models on the basis of RSS or R^2.
(iii) At the end, selecting the best model among all possible models with the help of cross-validation prediction error, BIC or adjusted R^2.

   One could clearly see that the forward selection method is having an advantage over the subset selection method since it uses a smaller number of models to provide the best fitted model and consumes less memory and time.

(b) Backwards stepwise selection: This algorithm is almost similar to the forward step-wise selection, instead that backward stepwise selection method initially uses all the predictors and removes least useful ones one-at-a-time.

The method has following steps of algorithm:

(i) Start with full model that is with all the predictors and name the model as μ_0.

(ii) Let k = p, p − 1, ..., 1
   (a) Consider all k models and for a total of k − 1 predictors.
   (b) Choose the best model among these k models on the basis of RSS or R^2.

(iii) At the end, selecting the best model out of all the models with the help of cross-validation prediction error, BIC or adjusted R^2.

9.3 Regularization

These are the frequently used techniques to overcome the problem faced due to over-fitting and correlation of the predictors. There are generally two regularization methods, namely,

(a) L1 Regularization which is also known as Lasso Regression
(b) L2 Regularization which is also known as Ridge Regression

These methods are also known as shrinkage methods and these are in detail in subsequent sections.

9.4 Shrinkage methods

There are two types of most popular shrinkage methods, namely Ridge and Lasso Regressions.

9.4.1   Ridge Regression

In multiple linear regressions sometimes, coefficients are unconstrained and can have large values resulting in a high variance. So, in order to control the variance, which is generally caused by the unconstrained coefficients, there is a need to impose a constraint on these coefficients, which is done by applying a penalty on them. One of the constraints we can apply is known as Ridge constrain, which is applied on the OLS equation.

$$(Y - XB)^T(Y - XB) + k\|B\|^2, \|B\|^2 = \sum_{j=1}^{p} B_j^2 \tag{9.5}$$

In general, the above equation is known as penalized residual sum of squares. In the following equation k is the tuning parameter whose value is always $k \geq 0$. Here the tuning parameter controls the coefficient estimates, if the value of k = 0 then Ridge regression will produce the same result as OLS, but when the $k \to \infty$ it impacts the shrinkage penalty grows and the Ridge regression coefficient estimates will approach 0. One thing to note down is that we apply the shrinkage penalty to all the coefficient parameters other than the intercept. Because PRSS is a convex function, Ridge constraint is a unique solution for B and, the solution is termed as Ridge regression estimator (RRE) because of the Ridge constraint.

To derive the RRE we will derive the predicted residual error of squares (PRESS) equation

$$(Y - XB)^T (Y - XB) + k||B||^2 \tag{9.6}$$

$$Y^T Y - 2B^T X^T Y + B^T X^T XB + xB^T B \tag{9.7}$$

Let PS(B) denotes Eq. (9.7)

So, in order to find the minimum value of B, we will differentiate Eq. (9.7) and equate it to zero.

$$\frac{\partial(PS(B))}{\partial(B)} = -2X^T Y + 2X^T XB + 2kB \tag{9.8}$$

Now equate Eq. (9.8) with respect to 0

$$-2X^T Y + 2X^T XB + 2kB = 0 \tag{9.9}$$

Solving Eq. (9.9) we will get the RRE

$$\widehat{B}(k) = (X^T X + kI_p)^{-1} X^T Y \tag{9.10}$$

Now apply equal weights to the weights of B.

9.4.1.1 Bias, mean square error, and L_2-Risk of RRE
One has to check the bias, MSE, and L_2 Risk of RRE.

First, check the bias

$$b\left(\widehat{B_n^*}(k)\right) = E\left[\widehat{B_n^*}(k)\right] - B \tag{9.11}$$

$$\beta\left(\widehat{B_v^*}(\kappa)\right) = (X + \kappa I_\pi)^{-1} XB - B \tag{9.12}$$

$$b\left(\widehat{B_n^*}(k)\right) = -k(C + kI_p)^{-1} B \tag{9.13}$$

Now look at MSE for the RRE as

$$MSE\left(\widehat{B_n^*}(k)\right) = Cov\left(\widehat{B_n^*}(k)\right) + b\left[\widehat{B_n^*}(k)\right] b\left[\widehat{B_n^*}(k)\right]^T \tag{9.14}$$

$$MSE\left(\widehat{B_n^*}(k)\right) = \sigma^2 (C + kI_p)^{-1} C(C + kI_p)^{-1} - k(C + kI_p)^{-1} BB^T (C + kI_p)^{-1} \tag{9.15}$$

Now measure the L_2 Risk of RRE as

$$R\left(\widehat{B_n^*}(k)\right) = \sigma^2 \text{tr}(C + kI_p)^{-2}C + k^2B^T(C + kI_p)^{-2}B \tag{9.16}$$

One could write as

$$\widehat{B_n^*}(k) = (X^TX + kI_p)^{-1}X^TY = W(X^TY), \quad \text{where } W = (X^TX + kI_p)^{-1} \tag{9.17}$$

and also,

$$\widehat{B_n^*}(k) = (I_p + kC^{-1})^{-1}(X^TX)^{-1}X^TY \tag{9.18}$$

$$\widehat{B_E^*}(k) = (I_p + kC^{-1})^{-1}\widehat{B_n} \tag{9.19}$$

$$\widehat{B_n^*}(k) = Z\widehat{B_n} \tag{9.20}$$

Now assume that $\lambda_i(W)$ and $\lambda_i(Z)$ be the ith eigenvalue for the W and Z

$$\lambda_i(W) = (\lambda_i + k)^{-1} \tag{9.21}$$

$$\lambda_i(Z) = \lambda_i(\lambda_i + k)^{-1} \tag{9.22}$$

where the value of $i = 1, 2, ..., p$.

Now one has to find the sum of residuals for the $\widehat{B_n^*}(k)$,

$$\text{RSS} = \left[Y - X\widehat{B_n^*}(k)\right]^T\left[Y - X\widehat{B_n^*}(k)\right] \tag{9.23}$$

$$\text{RSS} = Y^TY - 2\left[X\widehat{B_n^*}(k)\right]^TX^TY + \left[X\widehat{B_n^*}(k)\right]^TX^TX\widehat{B_n^*}(k) \tag{9.24}$$

RSS in Eqs. (9.23 and 9.24) is equal to TSS due to $\widehat{B_n^*}(k)$

Now checking the properties of L_2 with respect to Eq. (9.24),

$$R\left(\widehat{B_n^*}(k)\right) = \sigma^2 \sum_{i=1}^{p} \frac{\gamma_i}{\gamma_i + k} + k^2B^T(C + kI_p)^{-1}B \tag{9.25}$$

$$R\left(\widehat{B_n^*}(k)\right) = \text{var}(k) + b(k) \tag{9.26}$$

Fig. 9.1 explains the relationship between L_2 Risk and value of k (tuning parameter). The L_2. Risk is the combination of the variance and the bias, so here bias and variance will also have a relationship with the value of k.

Since the variance is monotonic decreasing function of k and bias is a monotonically increasing function of k. Find the derivative of variance and bias for $k = 0$ as

$$\lim_{k \to 0} \frac{d \text{ var}(k)}{d(k)} = -2\sigma^2 \sum_{j=1}^{p} \frac{1}{\gamma_j} \tag{9.27}$$

$$\lim_{k \to 0} \frac{d \text{ bias}(k)}{d(k)} = 0 \tag{9.28}$$

Since variance is giving negative derivative when $k = 0$ for an orthogonal X^TX but with ill-conditioned X^TX variance will tend to go ∞.

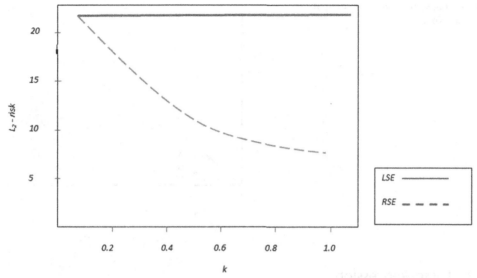

FIGURE 9.1 Relation between L_2 Risk and Tuning Parameter k.

While bias is flat and 0 at k = 0. So, from these it is very much clear that if the value of k > 0 then there will definitely be some increase in the bias but at the same time there will be a huge drop in variance [1].

9.4.1.2 Graphical representation of RRE

Fig. 9.2 shows the geometrical representation of the Ridge regression.

Here assumed that there are just two predictors and have coefficients as to B_1 and B_2.

$$B^{\hat{}} = \text{Least Square Solution}$$

The minimum value of sum of squares will be achieved at $\widehat{B}$. The small ellipse is the locus of points in the B_1. The circle which is located at the center of the origin is tangent to the smaller ellipse at $\widehat{B}^*$ and is known as the Ridge estimate. Since we know that $\widehat{B}^*$ is the shortest vector which will provide the minimum value of the sum of square anywhere on a small ellipse.

There are some general properties of the Ridge regression which we will see here:

1. The sum of squares of residuals is minimized by the $\widehat{B}^*$ which is located at the origin and B^* is the radius. The sum of the square of residual is an increasing function of k
2. The length $\widehat{B}^*$ is the decreasing function of k and is also the radius of the circle. With the increasing value of k, the length of the radius will decrease.
3. The angle γ between Ridge solution $\widehat{B}^*$ and the gradient vector g is a decreasing function of k. The value of γ steadily decreases with an increase in the value of k.

FIGURE 9.2 Geometrical representation of
Ridge Regression.

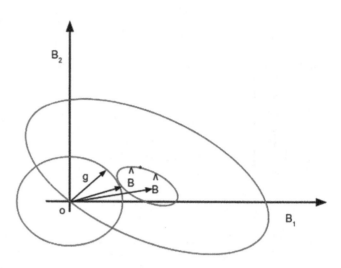

9.4.2   Lasso Regression

Another shrinkage method we can often use to overcome the problems related to the OLS is Lasso regression. Lasso regression does perform almost similar to the Ridge regression but not exactly similar. Ridge regression uses penalty to shrink all the co-efficients toward zero but that does not make any of the coefficients exactly zero, while penalty in the Lasso regression makes some of the variables exactly zero which will help in eliminating the coefficients which finally helps in removing some of the variables. Lasso regression not only shrinks the coefficients toward zero but it helps in variable selection too. The term sparse is used for a model which is having a very few nonzero coefficients. Therefore, the key property of the L1-constraint is that it has the ability to yield sparse solutions.

The Lasso coefficients minimize

$$\sum_{i,j=1}^{n} \left(y_i - (B_0 + B_jX_{ji})\right)^2 + k||B_j|| \tag{9.29}$$

The equation is known as the Lasso equation, which can also be written in another form Lagrangian form

$$\text{minimize}\left\{\frac{1}{2N}||Y - XB||_2^2 + X||B||_1\right\} \tag{9.30}$$

where $k \geq 0$.

Typically, before performing the Lasso regression we standardize all the predictors so that they are cantered and have unit variance. If we do not do so then Lasso will depend on the units present in the predictors and will not give a proper outcome. But if the predictors are in the same unit then there will be no need to standardize the predictors.

Geometry of Lasso Regression: The main reason for setting some of the coefficients to zero and help in eliminating them is the geometry of the Lasso regression.

Here in Fig. 9.3, the assumption is that there are two predictors used and the criterion. $\sum_{i,j=1}^{n} (y_i - B_j X_{ji})^2$ equals the quadratic function

$$\left(B - \widehat{B_0}\right)^T X^T X \left(B - \widehat{B_0}\right) \tag{9.31}$$

The residual sum of squares has elliptical contours having center at the full least square estimates. The constraints for the Lasso regression are the diamond $|B_1| + |B_2| \leq t$. The Lasso solution is the first place where contours touch the diamond and occasionally this process will occur at the corners of the diamond, resulting in a zero coefficient.

Now a big question arises how to estimate the value of k (tuning parameter). For this there are two methods:

1. Cross-validation: Suppose

$$Y = n(X) + \varepsilon \tag{9.32}$$

where $E(\varepsilon) = 0$ and $var(\varepsilon) = \sigma^2$.

The mean-squared error of estimates $\widehat{n}(X)$ can be defined as

$$MSE = E\{(\widehat{n}(X) - \widehat{n}(X)\}^2 \tag{9.33}$$

The prediction error for $\widehat{n}(X)$ is -

$$PE = \{Y - \widehat{n}(X)\}^2 = ME + \sigma^2 \tag{9.34}$$

FIGURE 9.3 Regression with two predictors.

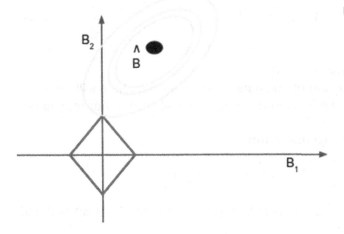

In order to estimate this best value for k, one has to create training and test data points from the given data set by splitting them up at random, and estimating performance on the test data. Generally, the data set is divided into a number of groups $g > 1$ where g belongs to number of groups, more specifically to divide the data set into 5 to 10 groups. Now the PE is measured for all the values of k in each data set and tested for the respective test data set. The value of k yielding the lowest PE is selected.

2. The second method for estimating the value of k is Stein's unbiased estimate of risk. By assuming z is a multivariate normal random vector having mean as μ and variance as I. Let $\hat{\mu}$ be the estimator for μ and $\hat{\mu} = z + g(z)$.

Then,

$$E_\mu ||\hat{\mu} - \mu|| = p + E_\mu \left(||g(z)|| + 2 \sum_{i=1}^{p} \frac{dg_i}{dz_i} \right) \tag{9.35}$$

By using this result and apply this to the Lasso estimator equation.

From the above equation one could derive the formula for approximately unbiased estimate of risk as:

$$R\left\{ \hat{B}(Y) \right\} \approx \hat{T}^2 \left\{ p - 2\#\left(j; \left| \hat{B}_j^0 / \hat{T} \right| < \gamma \right) + \sum_{j=1}^{p} \max\left(\hat{B}_j^0 / \hat{T} |, \gamma^2 \right) \right\} \tag{9.36}$$

where $\hat{T} = \sigma/\sqrt{N}$ and $\hat{B}_j(\gamma) = \text{sign}\left(\widehat{B_j^0} \right) \left(|\widehat{B_j^0}/\hat{T}| - \gamma \right)^+$.

Now an estimate of γ can be obtained using the minimizer of $R\left\{ \hat{B}(Y) \right\}$

$$\hat{\gamma} = \text{argarg min}\left[R\left\{ \hat{B}(Y) \right\} \right] \tag{9.37}$$

From this estimate of Lasso parameter can be obtained as

$$\hat{k} = \Sigma\left(|\widehat{B_j^0}| - \gamma \right) \tag{9.38}$$

9.4.2.1   Computation of the Lasso solution

A Lasso equation is a convex program of quadratic type problems. There will be many sophisticated quadratic problem methods to solve the Lasso algorithm but we will see coordinate descent procedure in detail.

For our simplicity we will use Lagrangian form:

$$\text{minimize}\left\{ \frac{1}{2N} \sum_{i=1}^{N} \left(y_i - \sum_{j=1}^{p} B_j x_{ij} \right)^2 + k||B_j|| \right\} \tag{9.39}$$

Here the assumption is that x_{ij} and y_i are standardized and whose mean is 0 and variance is 1.

The two cases one with single predictors and other with multiple predictors are considered here:

1. Single predictors: Here assume that the data set just one predictor i.e., p = 9. The equation will become

$$\text{minimize}\left\{\frac{1}{2N}\sum_{i=1}^{N}(y_i - Bx_i)^2 + k||B||\right\} \qquad (9.40)$$

According to the standard approach, find the first order derivative of Eq. (9.40) and set it equal to 0 in order to minimize the solution. Since, the absolute value of the function |B| does not have any derivative at B = 0. So, we will proceed by using the equation and get

$$\hat{B} = \frac{1}{N}<x,y> -k \ \text{if} \ \frac{1}{N}<x,y>> k$$

$$\hat{B} = 0 \ \text{if} \ \frac{1}{N}<x,y> = k \qquad (9.41)$$

$$\hat{B} = \frac{1}{N}<x,y> +k \ \text{if} \ \frac{1}{N}<x,y> < k \qquad (9.42)$$

Eqs. (9.41)−(9.42) can be written as:

$$\hat{B} = S_k\left(\frac{1}{N}<x,y>\right) \qquad (9.43)$$

where $S_k(x) = \text{sign}(x)(|x| - k)_+$ which is nothing but the soft-thresholding operator. Soft-thresholding operator translates its argument x toward zero by the amount k, and sets it to zero if $|x| \le k$.

2. Multiple Predictors: Using the univariate problem one has to solve the Lasso problem with multiple predictors using a simple coordinate wise scheme. For multivariate predictors p > 1 we repeatedly cycle through the predictors in some fixed order (say j = 1, 2,..., p), where at the jth step, update the coefficient B_j by minimizing the objective function while keeping all other coefficients at their current value. The Lasso equation can be written in terms of objective function as

$$\frac{1}{2N}\sum_{i=1}^{N}\left(y_i - \sum_{z\neq j}x_{iz}B_z - x_{ij}B_j\right)^2 + k\sum_{z\neq j}|B_z| + k|B_j| \qquad (9.44)$$

here the solution for every B_j can be described in terms of partial residuals.

$r_j^i = (y_i - \Sigma_{z\neq j}x_{iz}B_z)$. Now observe that jth coefficient is updated in terms of partial residuals as:

$$\hat{B}_j = S_k\left(\frac{1}{N}<x_j, r^j>\right) \qquad (9.45)$$

So, the update of the thresholding will be given by

$$\widehat{B}_j = S_k\left(\widehat{B}_j + \frac{1}{N} < x_j, r^j >\right) \tag{9.46}$$

The overall algorithm functions by updating the soft-threshold continuously in a cyclical way which will keep on updating the coordinates of $\widehat{B}_j$ along the way [2,3].

9.5 Dimensional reduction

In this section, deals with the one of the most important algorithms used in dimensional reduction, Principal Component Analysis (PCA). When a data set that has a high number of predictors then the first instance is to reduce the number of variables without losing much information about the data. The main idea behind the PCA is to get rid of high dimensionality from the data set, by removing the variables which are highly correlated among them and retaining the number of variables that will give important information about the data set. The following is attained by getting another set of variables, principal component(s); these principal components will not be correlated among them and will cover most of the information about the data set.

Definition and Derivation: Suppose a vector X with n random variables and interested in the variance of n variables and correlation among the variables. Unless the value of n is very small, looking at the variance of each variable and looking at the correlation among each variable is not possible as it might create some confusion. Assume that there are 50 variables then looking at the variation of 50 variables is not so easy and looking at correlation 1225 [$^1/_2$ (50(50−1))] is not a good idea. So, an alternative approach to this can be using a few variables from the 50 variables. But the main concern before picking a few variables is that without losing any important information from those variables and another concern is how many variables are to be chosen. For such concerns, PCA comes in handy and plays an important role. It is possible, to obtain a two-dimensional representation of the data which can capture most of the variations, in such cases plotting the observations in low-dimensional space is possible. PCA provides a tool to find a low-dimensional representation of a data set that contains as much as possible of the variation. The broad idea is that each of the n variables is living in p-dimensional space, but not all of them are equally important. So, PCA helps us in getting small number of dimensions which are most important to explain the most of the variation present.

The step in the PCA includes having a linear function $a_1^T x$ of the elements of x which is having a maximum variance, where a_1^T is a vector of constants:

$$a_1^T x = a_{11}x_1 + a_{12}x_2 + \dots + a_{1p}x_p \tag{9.47}$$

$$a_1^T x = \sum_{j=1}^{p} a_{1j} x_j \tag{9.48}$$

Similarly, one has to look at another PC, uncorrelated with $a_1^T x$ and similarly for every k PC, having maximum variance being correlated to $a_1^T x, a_2^T x, a_3^T x, a_{k-1}^T x$. We can find up to p PCs but in general, most of the variation in the data is covered by m variables, where $m \ll p$.

To derive the form of Principal Components, the first PC $a_1^T x$ the vector of a_1 maximizing $\text{var}(a_1^T x) = a_1^T \Sigma a_1$. To achieve the maximum value, one has to impose a normalization constraint.

To maximize $a_1^T \Sigma a_1$ and subject to $a_1^T a_1 = 1$, the standard approach is to use Lagrange multipliers.

Maximize,

$$a_1^T \Sigma a_1 - \lambda(a_1^T a_1 - 1), \tag{9.49}$$

λ = Lagrange Multiplier. In order to get the maximum out of this, differentiate Eq. (9.49) wrt a_1 and equate to 0.

$\Sigma - \lambda a_1$, now equating this to 0

$$\Sigma a_1 - \lambda a_1 = 0$$

Or

$$(\Sigma a_1 - \lambda a_1) a_1 = 0 \tag{9.50}$$

I_p = Identity matrix of size p*p. Here λ is the eigenvalue of a_1 is the eigenvector. Now one has to decide which of the p eigenvectors will give us $a_1^T x$ with the maximum value of variance, in order to obtain, maximize

$$a_1^T \Sigma a_1 = a_1^T \lambda a_1 = \lambda a_1^T a_1 = \lambda \tag{9.51}$$

So one has to provide the maximum value to the λ in order to get the p where one could obtain $a_1^T x$ with maximum variance.

Thus, a_1 is the eigenvector having the largest eigenvalues of $\text{var}(a_1^T x) = a_1^T \sum a_1 = \lambda_1$, which is the largest eigenvalue.

Now look at the second PC $a_2^T x$, which maximizes $a_2^T \sum a_2$ which is uncorrelated to the first PC $a_1^T x$ or one could say that both are having correlation as 0. But

$$\text{cov}[a_1^T x, a_2^T x] = a_1^T \Sigma a_2 = a_2^T \Sigma a_1 = a_2^T \lambda_1 a_1^T = \lambda_1 a_2^T a_1 = \lambda_1 a_1^T a_2 = 0$$

Hence,

$$a_1^T \sum a_2 = 0$$

$$a_2^T \sum a_1 = 0$$

$$\lambda_1 a_2^T a_1 = 0$$

$$\lambda_1 a_1^T a_2 = 0$$

One could use any of the above to specify that the correlation between $a_1^T x, a_2^T x$ is 0. Arbitrarily choose any of the above equations but one point to remember here is that already the system undergone to the normalization constraints, the quantity to be maximized is -

$$a_2^T \Sigma a_2 - \lambda(a_2^T a_2 - 1) - B a_2^T a_1 \tag{9.52}$$

Here λ and B both are Lagrange multipliers. In order to find the maximum value, one has to differentiate with respect to a_2 the above equation and equate 0.

$$\sum a_2 - \lambda a_2 - B a_1$$

Now equate the resulting equation to 0.

$\sum a_2 - \lambda a_2 - B a_1 = 0$, multiplying by a_1^T to both side of the equation

$$a_1^T \sum a_2 - a_1^T \lambda a_2 - B a_1^T a_1 = 0 \tag{9.53}$$

From past equations, it is evident that the first two terms are 0 and $a_1^T a_1 = 1$.

Therefore, $\sum a_2 - \lambda a_2 = 0$, so λ is again an eigenvalue of a_2 is the corresponding eigenvector.

Again.

$\lambda = a_2^T \sum a_2$ So the value of λ needs to be as large as possible. Here one has to make an assumption that the system does not have repeated eigenvalues.

Similarly, it can be shown for the third, fourth, pth PCs, the vector of coefficients a_3, a_4, a_p are the eigenvectors, and $\lambda_3, \lambda_4, \lambda_p$ are the eigenvalues for the $\sum \lambda_3, \lambda_4, \lambda_p$ will be the third, fourth, and pth largest value of eigenvalue respectively.

9.5.1   How many principal components?

This is one of the biggest challenges faced during the use of PCA. How many PCs should be considered in order to get the best result? A few ideas about some of the methods one can use in order to choose what should be the number of principal components (PCs).

1. Cumulative Percentage of Total Variation: This is the frequently used informal method to check the appropriate number of PCs. In this method, select the appropriate (cumulative) percentage of variance selected PCs should contribute like 85% or 95% say v^*. So whenever the variation of first m PCs exceeds v^* to obtain the number of the required PCs.
2. Size of variances of PCs: The rule in the current section is used only when using a correlation matrix. The idea behind the size of variances of PCs is that if all elements of x are independent, then the number of PCs will be the same as the, and all have unit variances in the case of a correlation matrix. Thus, any PC having variance <1 contains less information than that of the original variables and so that variable is not important so here retain only those PCs whose variance >1

3. The Scree Graph: The scree graph method provides a plot of eigenvalue and the number of the variables which helps in deciding at which eigenvalue the lines joining the plotted points are 'steep' to the left side of a number of variables, and 'not steep' to the right. This value of a number of variables defines an 'elbow' in the scree graph, is then taken to be the number of PCs that will be having the maximum information of the data provided [3].

9.6 Implementation of Ridge and Lasso Regression

Here we will take a data set and compare the result of Multiple Linear Regression, Ridge Regression, and Lasso regression. The data set is about predicting the price of the car, having 205 observations and 26 variables where 1 is the dependent variable and 25 are independent variables. The data type of 16 variables is integers or float types while the rest 10 are categorical variables.

Before creating the model of multiple linear, Ridge, and Lasso regression one has to scale down the data (Fig. 9.4).

For all the categorical variables, on-hot encoding method is utilized. The data is split as: 80:20, 80% will be included for the training and 20% will be for testing the data. Apply all three types of models and compare all of them on the basis of R^2, root mean square error (Fig. 9.5).

Code for Multiple Linear Regressions for the considered data is.

In the first line shows the linear regression model on the training data set. In the second line shows predicted output for the test data set. In the last line, shows the formula to compute of RMSE and R^2, the corresponding output is shown as (Fig. 9.6):

```
    wheelbase   carlength    carwidth   carheight   curbweight  enginesize  boreratio      stroke
1  -1.6866429  -0.4254799  -0.8427194  -2.0154834  -0.01453071  0.07426712  0.5178038  -1.8348856
2  -1.6866429  -0.4254799  -0.8427194  -2.0154834  -0.01453071  0.07426712  0.5178038  -1.8348856
3  -0.7068655  -0.2309477  -0.1901008  -0.5422002   0.51362457  0.60257108 -2.3990076   0.6842711
   compressionratio  horsepower     peakrpm    citympg  highwaympg  fueltype_gas  aspiration_turbo
1        -0.2876448   0.1740567  -0.2623181  -0.6449741  -0.5447253     0.327995        -0.4681493
2        -0.2876448   0.1740567  -0.2623181  -0.6449741  -0.5447253     0.327995        -0.4681493
3        -0.2876448   1.2614484  -0.2623181  -0.9506844  -0.6899381     0.327995        -0.4681493
   doornumber_two  carbody_hardtop  carbody_hatchback  carbody_sedan  carbody_wagon  drivewheel_rwd
1        1.127628        -0.201025         -0.7183239     -0.9361825     -0.3717679        1.299649
2        1.127628        -0.201025         -0.7183239     -0.9361825     -0.3717679        1.299649
3        1.127628        -0.201025          1.3853389     -0.9361825     -0.3717679        1.299649
   enginelocation_rear  enginetype_dohc  enginetype_l  enginetype_ohc  enginetype_ohcf  enginetype_ohcv
1           -0.1215691      -0.06984303    -0.2487426       -1.607428       -0.2802896       -0.2595728
2           -0.1215691      -0.06984303    -0.2487426       -1.607428       -0.2802896       -0.2595728
3           -0.1215691      -0.06984303    -0.2487426       -1.607428       -0.2802896        3.8336909
   enginetype_rotor  cylindernumber_five  cylindernumber_four  cylindernumber_six  cylindernumber_three
1        -0.1407246           -0.2375383           0.5365603          -0.363249           -0.06984303
2        -0.1407246           -0.2375383           0.5365603          -0.363249           -0.06984303
3        -0.1407246           -0.2375383          -1.8546322           2.739503           -0.06984303
   cylindernumber_twelve  cylindernumber_two  price
1            -0.06984303          -0.1407246  13495
2            -0.06984303          -0.1407246  16500
3            -0.06984303          -0.1407246  16500
> |
```

FIGURE 9.4 Scaled data.

FIGURE 9.5 Code line for Multiple Linear Regression.

```
linear_model = lm(price ~ ., data = train_mydata)

predict_linear = predict(linear_model, newdata =mydata_xtest)

root_mean_error = mean((abs(y_test-predict_linear)))^0.5
SSE <- sum((predict_linear - y_test)^2)
SST <- sum((y_test - mean(y_test))^2)

R_square <- 1 - SSE / SST
```

```
> print(root_mean_error)
[1] 39.96243
> print(R_square)
[1] 0.9231386
>
```

FIGURE 9.6 Value of root_mean_error and R_square.

So, the value of R^2 for the multiple linear regression model is 0.92 and RMSE is 39.96.

The code used for Ridge regression is to model to predict the outcome for the test data. By using the function cv.glmnet, that is to do cross-validation and glmnet is used for the ride, Lasso regression, where parameter alpha will define Ridge or Lasso, by using alpha = 0 will call Ridge regression while the value alpha = 1 will call Lasso regression. To extract the values of R^2 and RMSE, the code is given as (Fig. 9.7):

As it is clearly visible that there has been a small increase in the value of R^2 and there has been a decrease in the value of RMSE, hence, there is definitely an improvement in the performance (Figs. 9.8 and 9.9).

The code for Lasso Regression is:

Now extracting the value of R^2 and RMSE, using the code as (Fig. 9.10):

The value of RMSE is minimum and R^2 is maximum in the case of Lasso Regression in comparison to all three models (Fig. 9.11).

```
ridge <- cv.glmnet(x = x_train,y =y_train, alpha = 0,type.measure = "mse"
                   ,family ="gaussian", lambda = seq(0.01,100,by=0.01))


predicted_ridge <- predict(ridge, s=ridge$lambda.1se, newx = x_test)

root_mean_error = mean((abs(y_test-predicted_ridge)))^0.5
SSE <- sum((predicted_ridge - y_test)^2)
SST <- sum((y_test - mean(y_test))^2)

R_square <- 1 - SSE / SST
```

FIGURE 9.7 Code line for Ridge Regression.

```
> R_square
[1] 0.9282358
> root_mean_error
[1] 38.81294
>
```

FIGURE 9.8 Result of R_square and root_mean_error.

```
lasso <- cv.glmnet(x = x_train,y =y_train, alpha = 1,type.measure = "mse"
                ,family ="gaussian", lambda = seq(0.0001, 10, 0.01))


predicted_lasso <- predict(lasso, s=lasso$lambda.1se, newx = x_test)


root_mean_error = mean((abs(y_test-predicted_lasso)))^0.5

SSE <- sum((predicted_lasso - y_test)^2)
SST <- sum((y_test - mean(y_test))^2)

R_square <- 1 - SSE / SST
```

FIGURE 9.9 Code line for Lasso Regression.

```
> R_square
[1] 0.9293858
> root_mean_error
[1] 38.3338
```

FIGURE 9.10 Value of R-square and root_mean_square.

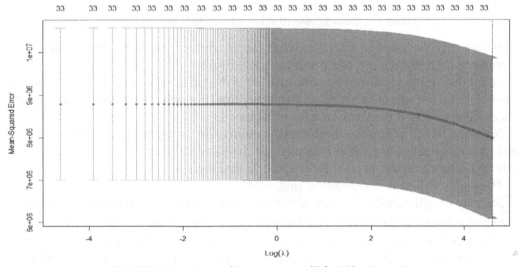

FIGURE 9.11 Mean-Squared Error versus Log(λ) for Ridge Regression.

Now compare the value of lambda and mean-square error for Ridge and Lasso Regression both.

As the value of Log(λ) increases the value of Mean-Squared Error decreases. The value at the top 33 is showing the number of the variables at each value Log(λ). Similarly, one could identify for Lasso Regression as shown in Fig. 9.12.

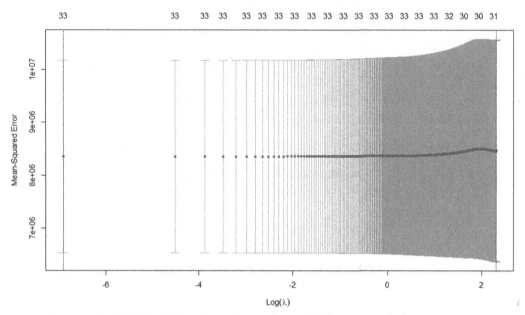

FIGURE 9.12 Mean-Squared Error versus Log(λ) for Lasso Regression.

Notice that the value of Log(λ) reaches the maximum the value of variables starts decreasing for a while but it again increases afterward.

So for the overall scenario, the performance of the Lasso regression is best among the three of them. But this does not mean that every time one will obtain best result from the Lasso regression as it depends on a lot of factors.

Now one has to create the three models namely linear regression, Ridge regression, and Lasso regression in Python language for the same data set, under the assumption that no preprocessing is needed for the considered data set.

The code for Linear Regression is (Fig. 9.13):

After training the model one has to test the model on the test data set and predict the values of output and then study the performance of model, using RMSE and R^2.

The code and output for extracting the values RMSE and R^2 on test data (Fig. 9.14).

As it is clearly visible that RMSE on a linear model is 73 and R2 is 0.43. Now build the Ridge regression and study at the performance on Ridge Regression. The Ridge Regression gets the best value for penalty by using cross-validation. The corresponding code is shown as (Fig. 9.15):

FIGURE 9.13 Linear Regression code line in Python.

```
#Creatng object of Linear Regression
ln_reg = LinearRegression()

#Fitting the model on training dataset
ln_reg.fit(X_train,Y_train)
```

```
LinearRegression()
```

```
#Predicting the values on test data
pred = ln_reg.predict(X_test)
print("The value of Root Mean Square Error for Linear Regression is {}.".format(mean_absolute_error(Y_test,pred)**0.5))
print("The value of R square for Linear Regression is {}.".format(r2_score(Y_test,pred)))
```

```
The value of Root Mean Square Error for Linear Regression is 73.1910839860037.
The value of R square for Linear Regression is 0.43323434093820756.
```

FIGURE 9.14 Value of root_mean_square_error and R_square.

```
#Applying Ridge Regression
ridge= Ridge()

#Setting the values for parameter
parameters = {'alpha':np.arange(0.001,10,0.01)}

#Using GridSearchCV
ridge_regressor = GridSearchCV(ridge,parameters,scoring='neg_mean_squared_error', cv =5)

#Fitting the model on training set
ridge_regressor.fit(X_train,Y_train)
```

FIGURE 9.15 Ridge Regression code file in python with grid search.

The code for test the output on test data and check the evaluation parameters RMSE and R^2 is as (Fig. 9.16):

The value of RMSE reduced to 49.95 and R^2 increased to 0.868. It is evident that Ridge Regression has performed well in comparison to Linear Regression.

The code for Lasso Regression shown as (Fig. 9.17):

The code for test the output on test data and check the evaluation parameters RMSE and R^2 is as (Fig. 9.18):

The value of RMSE is 55.37 and R^2 is 0.811, Lasso regression has performed better than Linear regression but has not performed better than Ridge regression.

```
#Predicting Values for test dataset
pred_ridge = ridge_regressor.predict(X_test)
print("The value of Root Mean Square Error for Ridge Regression is {}.".format(mean_absolute_error(Y_test,pred_ridge)**0.5))
print("The value of R square for Ridge Regression is {}.".format(r2_score(Y_test,pred_ridge)))
```

```
The value of Root Mean Square Error for Ridge Regression is 49.95621424551953.
The value of R square for Ridge Regression is 0.8680828610051492.
```

FIGURE 9.16 Value of root_mean_square_error and R_square.

```
#Lasso Regression
lasso- Lasso()

#Setting the value of Lambda
parameters - {'alpha':np.arange(0.001,10,0.01)}

#Using GridSearchCV
lasso_regressor = GridSearchCV(lasso,parameters,scoring='neg_mean_squared_error', cv =5)

#Fitting model on training set
lasso_regressor.fit(X_train,Y_train)
```

FIGURE 9.17 Code line for Lasso regression in python with grid search.

```
#Predicting values on test data
pred_lasso = lasso_regressor.predict(X_test)
print("The value of Root Mean Square Error for Lasso Regression is {}.".format(mean_absolute_error(Y_test,pred_lasso)**0.5))
print("The value of R square for Lasso Regression is {}.".format(r2_score(Y_test,pred_lasso)))
```

```
The value of Root Mean Square Error is 55.37135502396852.
The value of R square is 0.811985299857219.
```

FIGURE 9.18 Value of root_mean_square_error and R_square.

9.7 Conclusion

This chapter introduces the importance of regularization and the two widely used shrinkage methods and ends with the implementation of Lasso and Ridge regression using Python.

References

[1] A.K. Md, E. Saleh, B.M. Mohammad Arashi, Golam Kibria Theory of Ridge Regression Estimation with Applications, Wiley, 2019.

[2] D.W. Marquardt, R.D. Snee, Ridge Regression in practice, Am. Statistician 29 (No. 1) (Feb., 1975) 3–20.

[3] G. James, D. Witten, T. Hastie, R. Tibshirani, An Introduction to Statistical Learning, Springer Science Business Media, New York, 2014.

10

Data clustering using unsupervised machine learning

Bhanu Chander[1], Kumaravelan Gopalakrishnan[2]

[1]DEPARTMENT OF COMPUTER SCIENCE AND ENGINEERING, INDIAN INSTITUTE OF INFORMATION TECHNOLOGY, KOTTAYAM, KERALA, INDIA; [2]DEPARTMENT OF COMPUTER SCIENCE AND ENGINEERING, PONDICHERRY UNIVERSITY, PONDICHERRY, INDIA

10.1 Introduction

The recent advances in technologies like mobile, web, social, cloud, artificial intelligence (AI) makes an easy way to design and develop intelligent mechanical systems, distributed network procedures with big-data management. This computer-generated creation has created a massive amount of data which accelerating the adoption of AI-based techniques. Here the automation of data gathering and recording implied an overflow of information about numerous dissimilar types of systems [1–3]. To overcome these data-related issues, researchers developed various approaches for organizing and data modeling. Most of these approaches are motivated to form their application domains, and the definitions of these approaches are closely related to AI and ML-based techniques. In general, ML techniques design to imitate and adapt human behavior in each iteration, then learn and use learned experience for upcoming sessions.

Clustering is one of the emerging approaches in data gathering and data management. Grouping of data objects plays an essential role in various domains. For example, students who are studying engineering have some indications and groups according to their academic branch. Similarly, the students who have not contain similar subjects will not place in that group. The students grouped as a branch will train accordingly, while at the same time, students who do not belong to the branch will train differently. So, a professor can easily teach the branch students based on the appropriate syllabus. It can easily say from the above example that it is easy and immediately places that object into the appropriate group with the same label when find any labeled data object. If the label of objects provides in advance, it is easy to make a group. However, it is not an easy task in real-time applications to know the labeling information in advance.

In such cases, it is better to group objects based on some similarities. These labeled and unlabeled data objects play a crucial role in the analysis of data for numerous problems. In general, the label of objects specified in advance is termed supervised classification, while labels not attached with objects in advance are called unsupervised classification. The primary objective of analyzing the study of classification problems is to design or develop an algorithmic technique that can easily forecast an unknown object's class, which is not labeled. Here the algorithmic techniques are called a classifier. Instances or patterns commonly denote the objects in the classification procedure, and a pattern contains several features or attributes [2–4]. The fact referees the precision of classification as to how many testing patterns it has classified acceptably. Some pioneers' supervised classification procedures can originate in neural networks, support vector machine (SVM), fuzzy sets, regression, particle swarm optimization (PSO), Bayesian classifiers, rough sets, decision tree, etc.

Here, the unsupervised classification denotes clustering, which deals with crucial data classes without knowing the class labels. The goal of data clustering is to find a group of objects that are similar to other objects. Such models are closely related to defining a bunch of simplified data properties that deliver in-built descriptions of a data set's appropriate characteristics. In recent days, most researchers have shown interest in unsupervised-based clustering models for data analysis since getting labeled data from real-world applications is a challenging task. So, the clustering approaches are generally more demanding than supervised techniques; however, they produce more perceptions about composite data [5]. Since clustering procedures that contain numerous parameters function in high-dimensional spaces, procedures have to cope with noisy, imperfect, and tested data; their performance can vary substantially for different applications and categories of data.

The data mining experts proposed numerous clustering approaches based on data types and applications, and it is still challenging to choose an appropriate clustering approach for a given data set. Most of them focused on artificial and customarily distributed data containing different classes, features, various objects for separation, and mixed construction of the complicated groups. The main reason for using artificial data is that it can be easy to change the properties mentioned earlier in the data set. These features allow the clustering procedures to be complete and strictly evaluated in an enormous number of conditions and contribute to quantifying the performance's sensitivity concerning minor variations in the data.

10.2 Techniques in unsupervised learning

ML consists of three learning schemes: supervised, unsupervised, and reinforcement learning. The training samples are provided with correct labels in supervised learning, which means the ground truth is available. In unsupervised learning, no prior knowledge of the data is available, and they only need collected data to train a model. Reinforcement learning is practical when an agent takes movements in an environment and learns from the opposite acting policy [3–5]. Here, both supervised and unsupervised models are employed when the system's output does not influence the inputs. Deep learning (DL) is considered a subpart of ML, widely applied in various research domains; however, most of them use supervised learning. However, the training model requires many labeled samples to attain good generalization based on the numerous amounts of trained data. Labeling on the samples is a very time-consuming and resource-constrained process; moreover, labeling can be done manually. Hence, there is a need for methods that can quickly train with unlabeled data samples with less time. For this, need to apply unsupervised learning, then find the fine-tuning model with the help of labeled and supervised learning. As mentioned above, the upcoming DL models are expected to be obsessed by the progression of sophisticated and detailed unsupervised learning approaches. Supervised learning approaches have a well-defined objective, which means the output topographies are previously known, just like the class labels in sorting. In unsupervised, the intention is to realize unknown constructions, comparisons, and alliances in the data. Clustering is when the objective is to generate groups of data samples (clusters) based on some kind of likeness measure among the data samples [5,6]. The variance among classification and clustering is that clustering is carried out in an unsupervised style, so no class labels are provided. Occasionally, even the number of clusters is also not identified a priori.

In pattern recognition issues, training data consisting of a set of input data instances uses with no information on target data standards. As mentioned above, the main objective of unsupervised learning issues is to notice groups of similar examples within the data, known as clustering. Otherwise, recognize the data distribution, which is known as density estimation. K-means clustering, KNN (k-nearest neighbors), hierarchal clustering, neural networks, principle component analysis (PCA), independent component analysis (ICA), a priori algorithm, and singular value decomposition are coming under the unsupervised learning techniques. A broad range of discussions and proofs are already available on the aforementioned techniques so here not elaborating much about them. However, various issues, reasons for employing unsupervised techniques in various domains along with the applications are discussed.

10.2.1   Issues with unsupervised learning

1. Based on available data instances and computation perspective, unsupervised learning is more challenging than supervised learning tasks.
2. There is no proper justification for knowing that the obtained results are correct and meaningful since we are working on unlabeled or no answerable data labels.
3. On what basis the data expert has to look at the results.
4. Mathematical function to define the objective of clustering approach.
5. It is not possible to observe precise information for data sorting, and we cannot get exact information concerning data sorting, in addition, the output data used in unsupervised learning is labeled and not acknowledged.
6. In some cases, simulation accuracy will be less since the input data is not recognized and not labeled by expert people in advance, like supervised learning. It means that the designed mechanism needs to do this itself.
7. The supernatural classes do not constantly link to informative classes.
8. The operator desires to spend time for data understanding then label the modules which follow that sorting.
9. Computational complications due to a huge size of training data.

10. Longer training epochs.
11. Higher risk of imprecise consequences.
12. Needs Human involvement to authorize output variables.
13. Nonexistence of transparency into the origin on which data clustered.

10.2.2 Why is unsupervised learning needed despite these issues?

1. In general, explaining massive data sets is very expensive, so we can explain few examples automatically by explaining few labels.
2. In some cases, bot possible to recognize how many/what classes the data is separated.
3. With the help of clustering, model might improve some vision into the data structure before designing a classifier.
4. Unsupervised learning approaches nearly found all types of unknown and useful patterns in the given data.
5. Most of the unsupervised models assist in finding features that can be valuable for classification.
6. Unsupervised models are mostly applied for real-time applications; hence every input data is scrutinized and labeled in the existence of designed learners and data experts.
7. It is easier to get unlabeled information from a computer than labeled data, which wants manual interference.
8. Clustering inevitably splits the data set into groups based on their comparisons.
9. Anomaly discovery can notice infrequent, rare data points in a given training data set, which is suitable for fraudulent verdict dealings.
10. Latent variable representations are extensively applied for data preprocessing, which reduces the number of structures in a data set or decomposes the data set into various mechanisms.

10.2.3 Applications of unsupervised learning

ML-based approaches have developed a communal way to progress a product user practice then test quality statement outlines. Unsupervised learning (USL) delivers an investigative path to understand data, consenting industries to categorize patterns in vast sizes of data rapidly. Some real-world appliances of unsupervised learning mentioned below [1,3–8]:

1. News Sectors: Google News makes use of USL to classify numerous articles on a similar story from several online news channels. For instance, news of terrorists related to ISI.
2. Computer vision: The highest use of visual perception tasks like object recognition and face recognition.

3. Medical imaging: In the field of USL it offers important topographies to medicinal imagery devices, like image classification, exposure, and separation; these are primarily applied in radioscopy and pathology to analyze affected patients rapidly and precisely.
4. Anomaly detection: USL-based clustering and classification models can examine over large quantities of data then notice uncharacteristic, different data points inside a data set. These irregularities can increase responsiveness to security breaches, machine faults, decision-making, and human errors.
5. Customer personas: Defining client personas makes it calmer to understand mutual behaviors and clients' purchasing traditions. USL consents businesses to shape healthier buyer identity profiles, allowing administrations to bring into line their product more suitably.
6. Recommendation Machines: With previous purchase behavior data, USL can benefit learn data trends to progress more dynamic cross-selling tactics.

10.3 Unsupervised clustering

The main objective of unsupervised clustering is to abstract or expose the in-depth hidden patterns from the given input data features along with their structures and hierarchies from the given data samples. Data exposer or abstraction approaches make an easy way to understand the given data and fundamental procedures to use from the above sections clustering models designed on similarity or distances measurements. Similarity measurement is the most popular clustering method because it produces the district topographies as clusters. Feature extraction is another method that effectively finds the principal guidelines of the feature space [4–8]. Unsupervised-based deep clustering models make a stronger assumption on the nature of data than natural labeling with automatic generation of multiple labels. Unsupervised deep clustering models try to reconstruct the input of the network on its output. Here the reconstruction error is expected to extract a diplomatic representation of the given input data from which it can reconstruct with minimum possible error [9–11].

In some cases, two neural networks can parallelly employ: generator—it creates the data almost similar to the input data of the network, and a discriminator—it discriminates the generated samples from the real-ones. Here, both generator and discriminator run parallel, where the generator is trained to supply data points that can fool the discriminator; on the other hand, the discriminator is trained in a unique way to differentiate between synthetic and natural data since it knows which samples come for the generator. Clustering models find the hidden shapes from unlabeled input data in clusters, and the resultant controlled data is called data-concept. Clustering is broadly applied in various domains such as networking, anomaly detection, classification issues, cloud computing, and wireless sensor networks.

10.3.1   Hierarchical clustering

Hierarchical clustering designed in an iterative distributed manner with both top-down or bottom-up approaches. There is no need for prior distribution on several clusters, so it makes faster clustering computations. It creates cluster structures in the shape of the dendrogram. Here, it leaves presumed as samples and the root of the structure is considered the cluster containing all the samples. Hierarchical clusters developed in two schemes: agglomerative (leaves-to-root or bottom-to-top) and divisive clustering (root-to-leave or top-to-down). The agglomerative schemes form clusters opening with a solitary object, then mix these microgroups into larger clusters until the complete substances lie in a solitary cluster, or until more favorable circumstances are fulfilled [8–12]. The divisive scheme breaks the cluster of every object into small clusters until each object formulae a cluster its own or until it pleases optimistic assumption.

The hierarchical clustering approaches are additionally categorized into single-linkage, complete-linkage, average linkage based on similarity measures or linkages. In single-linkage clustering, the link among any two clusters is made by a single element pair the closeness to each other. Here the distance among two clusters is calculated by the adjacent distance from any member of single cluster to any member of added clusters. It also describes the comparison of two clusters with the help of minimum pairwise variations among the elements of the two clusters. In complete-linkage clustering approaches, the distance among two clusters is determined by longest from any member of 1 cluster to any member of the added cluster or else similarity of two clusters as the maximum of the pairwise dissimilarities among the elements of the two clusters [8–13]. Average linkage clustering is also a kind of minimum discrepancy scheme; the distance among two clusters is defined by the average distance from one cluster member to any other cluster member. Lastly, the clusters can also be given centroids computed based on the given samples fit into the clusters similar to the clusters in the k-means algorithm. The centroid clustering approach describes cluster similarity with the variation measure among the centroids of the clusters.

Balanced Iterative Reducing and Clustering Using Hierarchies (BIRCH): BIRCH is the combination of the number of data objects in a cluster (n), the sum of the linear attributes and sum of squares of the characteristic, and the place where these all tuples are stored is acknowledged as cluster features (CF). The main enthusiasms of BIRCH lie in two characteristics, the capability to contract with massive data sets then heftiness to anomalous values.

Clustering Using Representatives (CURE): It deals with high volume databases; it takes clusters of various forms and is robust against anomaly behavior data. Its performance is decent with 2-D data sets; moreover, CURE has better cluster quality than BIRCH.

ROCK: ROCK models do not consider any kind of distance function and mostly employed on categorical data sets, which is the same as the agglomerative cluster models. It based on the sum of associations among two records; links capture the sum of additional records that match each other.

CHAMELEON: In CHAMELEON based models, the clusters will amalgamate only if the interconnectivity then intimacy among two clusters are high comparative to the interior connectivity of the clusters then intimacy of items inside the clusters calculated.

10.3.2 Partitional clustering

Partitional clustering is a particular class of clustering procedures; it decomposes data into a set of separate clusters [13–15]. For n observations, the clustering procedure divides data into k < n clusters. Partitional clustering approaches divided into k-means clustering and mixture models.

K-Means Clustering: K-means clustering broadly applied for classification issues, where it considers the statical vector as an input to presume the labeling representations. It takes n number of observations and creates k number of clusters where each observation is close to the adjacent cluster. Cluster means to consider as a measurement to determine the cluster mean value. In the k-means process, the given data set is classified over a user-defined (*k*) number of clusters. The key objective is to express *k* centroids intended for every cluster.

$$K - \text{means} = \sum_{j=1}^{k} \sum_{i=1}^{n} \left\| x_i^{(j)} - c_j \right\|^2$$

From the equation $\left\| x_i^{(j)} - c_j \right\|^2$ is the possible distance among a data point x_i and the cluster center c_j. The distance measurement among the data points and cluster centroids is estimated then that data point is the closest centroid. Moreover, every data point can only relate to one cluster. Every iteration in k-means clustering contains two phases: In phase one, whole training data related to one of the clusters, coming to second phase location of a cluster centroids will be updated. These two phases will run till the satisfying condition is encountered. The update of the centroids of clusters is carried out after each element of the training set has been associated with one of the clusters. The new cluster centroids can be calculated similarly to a center of mass for entire samples related to that specified cluster. Mostly Euclidean and Mahalanobis distance measurements are used in the clustering process.

The k-means cluster performance is greatly affected based on the appropriate collection of initial parameters, such as the number of clusters or the initial position of centroids. However, k-means clustering is computationally expensive since it needs to calculate the distance of every data sample and every centroid. In addition, the basic structure of the algorithm will change if new data is added.

Fuzzy c-means (FCM): In FCM, there is a chance that one point could fit to dual or more clusters dissimilar to k-means, where just a single clustering is allowed to each point. The procedure of fuzzy c-means bases on the minimization of the resulting objective function.

$$J_m = \sum_{i=1}^{N} \sum_{j=1}^{c} \mu_{ij}^m \left\| x_i - c_j \right\|^2$$

Here N is the quantity of samples, C is the quantity of clusters, x_i is the ith sample, where $i \in \{1, 2 \ldots N\}$, c_j is the jth cluster centroid, where $j \in \{1, 2, \ldots, C\}$, μ_{ij} is the grade of attachment of xi in cluster j, $\|.\|$ is any measurement distance, and m is a coefficient to control fuzziness $1 \le m \le \infty$.

Mixture Models: Mixture models make use of statistical implications and assumptions, world efficiently for univariate and multivariate data sets. Both models have dissimilar parameters from the respective data observations, leading to a clustering set of observations. Mixture models are highly utilizing for pattern recognition, medical image processing, big-data analysis, and computer vision. Mixture-based clustering models effectively handle the weaknesses of heuristic and agglomerative clustering models. These models are considered the most efficient sensor node classification methods in a large-scale network.

10.3.3   Latent variable models for clustering

Latent variable models are entirely statistical approaches that permit relatively composite distributions in controllable joint distributions over a prolonged variable space [12–16]. For instance, fundamental variables of a procedure in higher dimensional space consider as latent variable models. Latent variables primarily used for dimensionality reduction, factor analysis, and data analysis.

Mixture Distribution: Mixture distribution is one of the significant latent variable models used for guesstimating the underlying density function. Notably, it delivers an overall outline for density estimation with the simpler parametric distributions. The expectation-maximization (EM) procedure is one of the mixture distributions models employed to estimate through maximization of the log-likelihood.

Factor Analysis: Factor analysis (FA) is a density guesstimate model utilized for combined operations like filtering and dimensionality reduction. Compared to other existing dimensionality reduction models, factor analysis uses fixed variance Gaussian noise models in conventional settings.

Blind Signal Separation: It identifies and separates autonomous source signs from diverse input indications with very slight info around the mixing procedure. PCA and ICA are the two extensively used blind signal separation models in all forms of multi-dimensional data dispensation.

10.3.4   Dimensionality reduction

As the result of significant growth in various computer technologies, the collected data has many dimensions, and these dimensions further contain countless potential interrelated dimensions. However, researchers found that the inherent dimensionality of the real-world data features is less than the total amount of dimensions. Hence, there is a need for a model that can discover the basic design of the causal data by abstracting inherent dimensions. At the same time, it just verified that the absolute values are not changed, which is also called the curse of dimensionality. Dimensionality reduction is an unsupervised task because it can create a new cumulated feature for all the available features [10–16]. The research community designed numerous approaches to the reduction of data dimensions some of them are as follows: sparse representation—which signifies the data vectors in linear mixtures of minor base vectors, independent representation—makes an effect of disconnecting the source of causal the data distribution so the illustrations will remain statistically independent. In addition, numerous nonlinear dimensionality reduction approaches developed and focused on avoiding repression of linear data dimensionality.

ISOMAP: A nonlinear dimensionality reduction method makes the fundamentals for low dimensional geometric information for a given data set. It primarily uses the geometric distances along with Dijkstra's algorithm to estimate low-dimensional shortest path representation.

Generative Topographic Model (GTM): GTM expresses the nonlinear latent variable plotting from small to high-dimensional spaces. GTM expresses as the reference vectors in data-space, and GTM enhances the log-likelihood, then the resultant probability describes the density in data-space.

Principal Curves: It is a data summarizing method where nonparametric curves permit over the mid of a multidimensional data set that summarizes the data sets. Here the smooth arcs reduce the average distance among two data-points in the existence of Gaussian noise.

Nonlinear Multidimensional Scaling (NMDS): It works as FA, where a multivariate standard dispersal is preassumed, then the correlation among two dissimilar objects shows as a correlation matrix. It mainly applied to three dimensioned data sets for better visualization and abstract the multidimensional data.

10.3.5   The search-based clustering approaches

Search-based approaches primarily employed to get the optimal rate of the distance or the objective function. These kinds of approaches categorize into stochastic then deterministic search practices. Here, stochastic search practices develop a rough optimal explanation. In addition, greatest part of the stochastic practices are evolutionary-based tactics. The remained search practices considered as deterministic search practices produce a best result through performing comprehensive records, and deterministic tactics are characteristically greedy descent methods. The stochastic search practices will be sequential or parallel like simulated annealing (SA), whereas evolutionary tactics are fundamentally parallel. SA methods design to evade or improve from solutions that resemble the objective functions' local targets, which train to accept some likelihood as a newfangled solution for the subsequent iteration of lower quality as restrained by the standard function.

10.3.6   Bayesian clustering

Bayesian clustering is designed based on probabilistic strategies, where the subsequent dispersal of the given data is learned based on the preceding likelihood distribution. It has two versions: parametric and nonparametric. The main dissimilarity among these versions is the dimensionality of parameter space. For instance, if the model has fixed sizes in the parameter space, then the applied version is acknowledged as Bayesian parametric or Bayesian nonparametric.

10.3.7   Spectral clustering

Spectral clustering designed by Donath for the purpose of graph partitioning through studying graphs with systematic approaches of linear algebra. It is an emergent practice based on graph clustering, which contains cluster points with eigenvectors resultant from the given data. Here, the training data represent in a comparison graph, an undirected graph with the training samples as the vertex. Then, the edges connect with a weight of the similarity between two vertices' where they attach. However, the triumph of spectral clustering is that it does not make strong assumptions on clusters. Spectral clusters proficiently implemented with many data sets; if a similarity graph is selected, then try to resolve linear problem, there are no problems on local minima of getting stuck in local minima or resuming the process numerous times with dissimilar commencements. Spectral clustering cannot assist as a black-box assumption which inevitably notices the suitable clusters in prearranged data set. However, it can consider as an authoritative tool that can produce decent results if used sensibly.

10.4 Taxonomy of neural network-based deep clustering

Deep learning clustering uses deep neural networks to learn the feature illustrations suitable for clustering tasks [17–22]. In deep clustering, the optimization function or the loss function fabricate with net loss and clustering loss.

$$\text{Deep cluster loss function (L)} = \lambda L_{\text{Network-loss}} + (1 - \lambda) L_{\text{Cluster-loss}}$$

Here $\lambda \in [0, 1]$ is a hyper-parameter employed to balance the network-loss and cluster-loss. In addition, network loss uses to learn possible structures then evade minor solutions, and the cluster-loss improves the possible points to form groups.

10.4.1   Autoencoder (AE) based deep clustering

In general, autoencoding (AE) is one of the neural networks which aims to minimize the reconstruction loss and is mainly used to build unsupervised data representations [17–19]. AE consists of two functions: encoder: the function $f_\varnothing(x)$ which plots the given input data to a latent representation (h); decoder: prepares a reconstruction (r) which is $g_\theta(h)$. Here, the reconstructed representation (r) needs to be an identical to input x. When the distance among the two variables is mean-square-error, then the optimization objective in mathematical formation will be written as follows:

$$L_{RE} = \|x_i - g_\theta(f_\varnothing(x_i))\|^2$$

some deep cluster-based representative methods are mentioned below:

Deep Clustering Network (DCN): DCN stands as a popular approach, combining AE with a k-means algorithm. In the initial stage model, pre-train with AE then works to optimize both reconstruction and clustering losses. Because of computation performances, DCN has lower complexity with minimal recourses used. *Deep Continuous Clustering (DCC):* DCC designed based on AE-deep clustering to solve the intrinsic limitation of traditional clustering approaches like center-based, hierarchical and diverged based clustering. Deep-clustering. Optimizations of mentioned models contain discrete reconfigurations of the objective, which needs constant updates of clustering parameters. DCC formation contains strong constant objective and no preceding information of cluster number. *Deep Embedding Network (DEN):* It is employed to abstract the practical features for clustering, where Deep-AE is used to learn appropriate features from raw-sensed data. Then build the local strength of original data. Moreover, it integrates the group's sparsity constraints to diagonalize the responsiveness of representations. *Deep Multimanifold Clustering (DMC):* DMC fabricated on the working style of the multimanifold framework. It improves a joint loss function: locality preserving—it makes the learned representations meaningful with AE reconstruction-loss and cluster-oriented objectives. Proximity representation of every cluster centroid to make the representation cluster-friendly and discriminative.

10.4.2 CDNN based deep clustering

Convolutional deep neural network (CDNN) based uses the clustering loss value to train the networks like CNN, DBN, and FCN [18–21]. In CDNN-based procedures, if the data is mapped into tight clusters, it suffers from the risk of gaining dull feature space, resultant in a minor clustering loss.

Deep Nonparametric clustering (DNC) improves unsupervised feature learning with DBN for accurate cluster scrutiny. Then runs nonparametric models to make the appropriate number of clusters and labels all the existing training data. At last, it considers the skilled progression to enhance the parameters from the DBN top-layer. Discriminative boosted clustering (DBC): Its pretrains the AE, then fine-tunes the network with the help of cluster assignment loss. DBC based approaches are mainly applied for image analysis. *Deep Adaptive Image Clustering (DAC):* DAC design from the basic assumptions from the relationships among the pairwise images with binary values. Its optimization objective is binary pairwise classification. DAC is a single-stage convolutional net to cluster the binary value images. It adopts an efficient adaptive learning process. In every iteration, a pairwise image with the appropriate similarities is select on a secure network, and then the framed net is proficient by the chosen labeled examples. Lastly, images are clustered based on the results of the largest response of label topographies. *Clustering Convolutional Neural Network (CCNN):* CCNN is a well-organized then constant deep clustering model employed on high-volume image data sets. CCNN randomly chooses the k samples and prepares a primary pretrained model to be abstract the high-level features and respective cluster centroids. Every mini-batch k-means is accomplishing to update developments of examples and cluster centroids, whereas stochastic gradient descent is utilized to keep informed the designed CCNN model's parameters. Here the mini-batch k-means tremendously diminishes the computation, and memory expenses, includes an original iterative centroid updating process that evades drift error made by the feature inconsistency among two consecutive reiterations. Only the top k-samples with the least distance to their equivalent centroids are selected from each iteration to update the network parameters and improve updates' consistency. It is only the deep cluster approach that can effectively deal with millions of images.

10.4.3   Variational AE-based deep clustering

From the above discussion, both AE and CDNN grounded deep clustering approaches have produced inspiring results compared to traditional clustering approaches. In addition, designed approaches explicitly designed for clustering and there failed to expose the real underlaying construction of data, preventing them from being stretched to additional tasks like generating or predicting missing values [21,22]. Since we are working on large data sets, the expectations designed based on independent assumptions reduce the dimensionality, but there is no theoretical evidence that they learn feasible representations. With the invention of variational autoencoder (VAE), we can extract the hefted and motivate many variants. VAE measure as a reproductive variant of AE, and it imposes the latent code of AE to follow a well-defined distribution. It uses reparameterization-based neural networks to fit the conditional posterior to augment the variational inference objective via stochastic gradient and standard backpropagation. At last, the most significant variance among standard AE and VAE is that VAE imposes a probabilistic prior distribution over the hidden representation.

Variational Deep Embedding (VaDE): A VaDE tuned to best use the probability of the specified samples. The VaDE framed as follows:

$$VaDE = E[logp(x|z)] - D_{KL}(q(z, \ c|x) \, ||p \, (z, c))$$

From the equation, the first term indicates the reconstruction-loss and the second term indicates Kullback-Leibler (KL) deviation from the mixture-of-Gaussians (MoG) preceding p(z, c) to the variational subsequent value q(z, c|x), which contemplate as the clustering-loss Lc.

10.4.4   GAN-based deep clustering

Generative adversarial network (GAN) is a famous deep generative prototypical that effectively makes adversarial alterations among pairs of neural networks. GAN generally attempts to plot a sample z from a previous distribution p(z) to the data-space. However, the discriminatory net attempts to calculate the likelihood where input is an actual example from the data dispersal rather than an example produced through the generative system.

Deep Adversarial Clustering (DAC): DAC is a specially designed adversarial AE-based deep clustering model. VaAE inspires DAC; here in DAC, the adversarial AE practices an adversarial training method to contest the gathered subsequent latent representation with the former distribution. Importantly the objective function of DAC contained with general AE reconstruction objective function, adversarial objective, and Gaussian mixture probability. *Categorial Generative Adversarial Network (CatGAN):* CatGAN is a generalized GAN framework. It consists of unsupervised discriminative, which categorizes the data instances into a priori selected categories.

10.5 Cluster evolution criteria

10.5.1 Similarity measurements

For many real-time applications, formations of clustering are a necessary procedure. The evolution of clustering tests, accuracy, and validity of the clusters is also an important task. It should be noted that clusters should be verified whether the formed clusters with a particular technique show supreme comparison between the objects in the similar clusters then minimum likeness between those in added clusters. A measurement of data instances of a cluster is vital to building unsupervised learning procedures [14–23]. If the distances among the two objects are closer to each other, it is assumed as the basis of likeness. Researchers proposed numerous evaluation criteria, some of them mentioned below.

10.5.1.1 L_p and L_1 based distance measurements
Minkowski distances: This is also acknowledged as Lp norm; it is formalized as shown below.

$$D_{Mink}(x, \ y) = \sqrt[n]{\sum_{i=1}^{n} |x_i - y_i^p|}$$

Here p is the positive Euclidean distance value, and if we change the values of p, we get different Lp norm variants.

Manhattan (MD): It is also recognized as L1 norm and rectilinear distance, designed based on the sum of the absolute variances among the differing standards in vectors.

$$MD(x, \ y) = \sum_{i=1}^{n} |x_i - y_i|$$

Chebyshev (CD): It is also called maximum value distance and chessboard distance. It gives the best results when the two objects are definite as dissimilar if they have changed in one dimension. It is a measurement system well-defined on a vector space where the distance among a pair of vectors is extreme of their alteration with any direct dimensions.

$$CD(x, \ y) = \max |x_i - y_i|$$

Euclidean (ED): ED is one of the popular distance measurements and is also acknowledged as the L2 norm. It is a rearrangement to the Pythagorean statement, and then the distance signifies the origin of the sum-of-the-square of alterations among the conflicting standards in vectors.

$$ED(x, y) = \sqrt{\sum_{i=1}^{n} |x_i - y_i|^2}$$

Sorensen distance (SD): SD measurement mainly used to get or express relationships among two data instances; it has another name acknowledged as Bray–Curtis. SD is the adapted version to the Manhattan metric, where the summated variances among the attribute's values of the vectors x, y are identical by their summed attributes values. SD measures values between 0 and 1.

$$SD(x, y) = \frac{\sum_{i=1}^{n} x_i - y_i}{\sum_{i=1}^{n} x_i + y_i}$$

Lorentzian distance (LD): LD kind distance measurements are characterized by the standard log of the complete change among two vectors. LD is delicate to minor deviations because the log scale enlarges the minor range then compresses the high range.

$$LD(x, y) = \sum_{i=1}^{n} \ln(1 + x_i - y_i|)$$

From the LD equation, ln is the natural algorithm depend on the application, and one is to make sure the non-negativity property then avoids a log of zero from the equation.

Mean Character Distance (MCD): MCD famous as Average Manhattan and the Gower distance.

$$MCD(x, y) = \frac{\sum_{i=1}^{n} x_i - y_i}{n}$$

Non-Intersection Distance (NID): NID is the counterpart to the joint comparison and attains by deducting the intersection comparison from 1.

$$NID(x, y) = \frac{1}{2} \sum_{i=1}^{n} x_i - y_i$$

Canberra distance (CanD): CanD is a weighted distance variation of MD. The complete variance between the attribute standards of the vectors x and y is separated through the sum of the complete attribute standards before summating. CanD is very sensitive for the small changes and employed only for the positive values. Thus, CanD characteristics turn out to be more ostensible in high-dimensional space, mainly with a cumulative amount of variables.

$$CanD(x, y) = \sum_{i=1}^{n} \frac{|x_i - y_i|}{|x_i| + |y_i|}$$

10.5.2 Internal quality criteria

Designing clusters is a vital procedure; however, testing both the legitimacy and precision of the clusters shaped by any technique is also necessary. The designed method must test since the clusters formed through a particular approach show minimum similarity between the objects in the similar cluster then maximum comparison between the objects in other clusters. In general, internal quality criteria [24–26] measures intercluster separability, compactness, and intracluster consistency. Some of them are mention below:

Jaccard distance (JacD): JacD measures the differential among sample sets, and it is obtained by subtracting the Jaccard coefficient from:

$$\text{JacD}(x,\ y) = \frac{\sum_{i=1}^{n}(x_i - y_i)^2}{\sum_{i=1}^{n}x^2 \sum_{i=1}^{n}y_i^2 \sum_{i=1}^{n}x_i y_i}$$

Hellinger distance (HeD): HeD is a metric introduced in 1909 by Hellinger to measure the similarity among two probability distributions.

$$\text{HeD}(x,\ y) = \sqrt{2\sum_{i=1}^{n}\left(\sqrt{x_i} - \sqrt{y_i}\right)^2}$$

10.5.2.1 Sum of squared error (SSE)

$$\text{SSE} = \sum_{K=1}^{K}\sum_{x_i \in C_k}^{x} x_i - \mu_K^2$$

Here C_k is the set of examples in cluster K; μ_K is mean of cluster K.

Chord distance (ChoD): ChoD is an extension to the Euclidean distance and is mostly employed for clustering continuous data. It states the length of the chord linking two standardized points inside a hypersphere of radius 1.

$$\text{ChoD}(x,\ y) = \sqrt{2 - 2\frac{\sum_{i=1}^{n}x_i y_i}{\sum_{i=1}^{n}x^2 \sum_{i=1}^{n}y^2}}$$

Squared chord distance (SCD): In SCD, the summation-of-squares of square-root differences at every point be taken along both vectors, which increase the differences for more different feature.

$$\text{SCD}(x,\ y) = \sum_{i=1}^{n}\left(\sqrt{x_i} - \sqrt{y_i}\right)^2$$

10.5.2.2 Scatter criteria

$$\text{Defined for the Kth cluster}: \ S_k = \sum_{x \in C_k}(x - \mu_K)(x - \mu_K)^T$$

10.5.2.3 Condorcet's criterion

Applied for ranking issues and defined as follows:

$$\sum_{x_j \in C, \, x_k, x_{je} C} S(x_j, x_k) + \sum_{x_j \in C, \, x_k, x_{je} C} d(x_j, x_k)$$

From above $S(x_j, x_k)$ and $d(x_j, x_k)$ demonstrate the likeness then distance of the vectors x_j and x_k.

C-criterion: which is an extended version of Condorcet's criterion and well-defined as:

$$\sum_{x_j \in C, \, x_k, x_{je} C} (S(x_j, x_k) - \gamma) + \sum_{x_j \in C, \, x_k, x_{je} C} (\gamma - S(x_j, x_k))$$

Here γ is the threshold value.

Edge Cut Metrics: An edge cut minimization problem was valuable in solving clustering problems. The clustering superiority is measured as the residual edge weights percentage to the overall precut edge weights. Finding the optimal rate for various mathematical models becomes simple with edge cut minimization when no limits on the magnitude of the clusters.

10.5.3 External quality standards

External quality standards used to verify and contest the structure of the cluster to a prearranged sorting data instance [24–29].

Mutual information-based: For n data instances-based cluster using $C = (C_1, C_1 \ldots \ldots C_g)$ then stating to the target feature z whose field is dom (z) = $(C_1, C_1 \ldots \ldots C_k)$ then the formal description as follows:

$$C = \frac{2}{n} \sum_{l=1}^{g} \sum_{h=1}^{k} n_{l,h} \, log_{g,k} \, (n_{l,h} \cdot n \, / \, n_{.,l} \cdot n_{l,.})$$

From the above equation $n_{l,h}$ defines the amount of data examples that are in a cluster C_l then C_h.

Rand index is an easy norm used to calculate how similar the clusters are to a benchmark data classification.

$$Rand \, Index = TP + TN/TP + FP + FN + TN$$

From the equation: TP defines the quantity of true-positives, FP defines the quantity of false-positives, TN indicates the true-negatives, and FN indicates false-negatives. When the two instances perfectly agree, then it shows 1 or else 0.

F-Measure: While computing the rand index, both the FP and FN may similarly weigh because of an undesirable feature of estimated clustering appliances. The F-measure addresses this anxiety then use for the steadiness of false negatives by weighting the recall parameter.

$$F\text{-Measure} = \frac{(\eta^2 + 1) \cdot P \cdot R}{\eta^2 \cdot P + R}$$

Here P is the precision-rate, and R recall-rate. Recall has no influence on weighting recall parameter $\eta = 0$ when growing η assigns an increasing quantity of weight to recall in the last F-measure.

Fowlkes-Mallows index: The Fowlkes-Mallows index governs the likeness among the clusters gained afterward the clustering procedure. The higher-value indicates more likeness among the clusters.

$$FM = \sqrt{\frac{TP}{TP + FP} \cdot \frac{TP}{TP + FN}}$$

10.5.4 Clustering loss

In cluster learning, cluster-friendliness of the learned representations are vital functions which recognized as clustering-loss [19–29].

k-Means loss: K-Means loss is broadly applied in many clusters research operations. It ensures that the new available representation is closely related to k-means friendly function where every data point is closely circulating to the cluster centers. Neural network-based k-means loss can be express as:

$$L = \sum_{i=1}^{N} \sum_{k=1}^{K} y_{ik}\, x_i - C_k^2$$

Here, x_i is an embedded data-point, C_k is a cluster center and y_{ik} is a Boolean variable for conveying x_i with C_k. Minimalizing this loss concerning the network parameters guarantees that the distance among every data point and its allocated cluster center is small. Hence, employing k-means would result in better clustering quality.

Balanced assignments loss: These kinds of lose functions are employed along with other losses to make balanced cluster assignments, mainly applied in k-means, sparsity losses etc. Formalized balanced assignments loss can write as follows:

$$L_{bas} = KL(G\|U)$$

Here, bias indicates the balanced assignment loss G is a probability distribution of assigning a point to every available cluster, U indicates the uniform distribution.

Group sparsity loss: the fundamentals of group sparsity loss collected from spectral clustering where block diagonal comparison atmosphere is deflated for feature representation learning. For instance, a data point x_i Moreover, the loss function is represented as:

$$L_{gsl} = \sum_{i=1}^{n} \sum_{g=1}^{G} \lambda_g\, x_i$$

$$\lambda_g = \lambda \sqrt{n_g}$$

Here n_g is the group size, and λ indicates the constant value.

Cluster classification loss: These losses found while updating the clusters with fake data classes for classification loss and model evolution, which inspire expressive feature extraction in every available network layer.

Agglomerative clustering loss: As mentioned in the above sections, agglomerative clustering combines pair of clusters with maximum attraction in every step till it reaches a positive stopping criterion. In neural networks, the agglomerative clustering loss is calculated as follows: first selects the finest affinity, next trains to further optimize, and then updates network switches to optimize the affinity of the newly merged cluster pairs. Likewise, changing the network layer by layer and optimizing the network constraints with this loss function would make a clustering space more appropriate for agglomerative-based clustering.

10.5.5   Nonclustering loss

The nonclustering loss is an independent measurement of the clustering algorithm, and it usually enforces a desired constraint on the learned model. Some of the nonclustering models mention below [19,26–29]:

Self-Augmentation loss: It forces both representations of the original sample and their expansions.

$$L = -1/N \sum_n s(f(x), (T(x)))$$

In equation: x is the standard or original sample, $f(x)$ represents the value created by the model, s is the similarity measure, and T indicates the argumentation function.

No nonclustering loss: No additional nonclustering functions were used, and the clustering loss only controls the system model. For the most clustering losses, the absenteeism of a nonclustering loss can have a hazard of worse representations/consequences, or tentatively even in trouble clusters formation.

Autoencoder (AE) reconstruction loss: As discussed in the above sections, AE works with two phases: encoder and decoder. In AE-clustering, once the training is done, the decoder part is not needed further, then encoder $=$ maps input to the latent-space. In AE, reconstruction-loss assumed as a distance measure d_{AE} $(x_i, f(x_i)))$, here, x_i is input, $f(x_i)$ indicates the corresponding reconstruction loss. The mean-squared-error of two variables can formalize as:

$$L = d_{AE}\left(x_i, f((x_i)) = \sum_i ||x_i - f(x_i)||^2\right)$$

This loss function assurances that the learned representation conserves essential information from the initial one, which is why reconstruction is possible.

10.6   Applications of clustering

1. **Image Segmentation**: Hierarchical and k-means clustering approaches are highly apply in image segmentation. Segmentation of MRI imageries is a critical job in numerous medicinal appliances like surgical preparation and irregularity

recognition. In general, MRI image segmentation aims to divide an input image into important structural parts according to particular image properties. In some cases, it can formulate as a clustering problem. A set of feature vectors gained over transformation image dimensions then pixel locations clustered into some structures.

2. **Bioinformatics**: In recent times, genome sequencing and DNA projects have improved with gene expression data clustering. The applications of bioinformatics work based on comparable genes or proteins that share alike patterns or prime sequenced structures.

3. **Object Recognition**: clustering in object recognition groups the view of data objects where every object to be recognized will be characterized in a library of various images of the object.

4. **Character Recognition**: Clustering in character recognition detects the lexemes in given hand-written text for writer independent hand-writing recognition. The system accomplishments mostly depend on the user's acceptance rate, but requires a large amount of training data.

5. **Information retrieval** (IR): It is related to the automatic storage and recovery of essential documents. Many educational institutes employed the IR schemes in their library systems to provide easy access to books, articles, and journal-related documents.

6. **Data mining and big data**: It is extracting knowledge from massive databases. Data mining techniques are beneficial to relational, spatial, large, shapeless, and transactional databases. Extensive data mining is considered one of the emergent topics. Since the size of the data sets which are beyond the model capacity, The volume of data that is beyond.

7. There are numerous data mining approaches available, like basketball association coaches noticing movements, style, strengths, and outlines play of specific players and classifying kids' patterns in the foster care system. Clustering will assist in aggregating similar data instances gathered from unformatted examples. Hadoop is one of the big data processing tools.

8. **Spatial Data Analysis Clustering**: Dealing with large spatial data sets like medical equipment, image data exploration, satellite images, and geographical information systems. Spatial data clustering is helpful to exciting abstract features, then identify the shapes, which exist in real-world data sets. The clustering procedure helps to recognize spatial data by analyzing data process inevitably.

9. **Business:** As the world changes toward digitalization, clustering is pretty exciting in various business fields. Clustering helps market researchers and business experts to analyze and predict customer choices and requirements. Clustering is helpful for market segmentation, product position, and new product design.

10. There are numerous appliances on clustering in computer systems are already implemented and currently in use for various services. Intrusion exposure and internet-based traffic classification are the essential operations with clustering approaches. Genetic and BIRCH algorithms provide efficient abstracted data instances, giving high accuracy and fast classification compared to remained cluster models.

11. **Data Reduction**: In this digital world, every second of new data is added into our systems; due to the usage of various social and internet-related sites, a considerable amount of data is heaped every second. So, data reduction or compression is one of the essential tasks for handling extensive data; moreover, its processing turns out to be very problematic. Here clustering approaches will assist in compressing the data information by grouping essential features into clusters. After forming clusters, we can select the info or set-of-data that is suitable for our task.

10.7 Feature selection with ML for clustering

With recent developments in computer networks, Internet of Things, mobile communications, and other modern technologies, the world continuously generates enormous volumes of raw data at first-rate speed, like image, text, audio, and video. Most of these data gained from social media with the growth of the cloud computing. These informatic data often have the features of high scopes, which postures a high rated challenge for data examination then decision-making. In real-time applications, high-dimensional data scrutiny is a challenge for many researchers in ML, DL, as well as data mining domains. Feature selection (FS) is a practical approach that produces easy ways to resolve high-dimensional data by eliminating inappropriate, extraneous, and redundant data; this will reduce computation and communication time, then progress learning precision to simplify improved understanding for the learning model [30]. FS is the best theoretical and practical process for gaining a subset from novel features based on positive FS principles, which chooses functional features from the training data. FS models built based on statistical, manifold, rough-sets, and information theory. FS applies in various research domains like text-mining, fault diagnosis, image recognition, intrusion detection, natural language processing, image retrieval, biometric data analysis, etc. In particular, feature collection and abstraction are the two approaches to reduce the data dimensionality; however, feature abstraction desires to transmute the unique data into features with dense pattern recognition abilities.

FS techniques based on the given trained data such as labeled, partially labeled, and unlabeled are alienated into supervised, unsupervised, and semisupervised approaches [30–32]. From the literature work (1) based on the learning relationships, FS approaches can be alienated into a filter, embedded, and wrapper models; (b) based on evolution criteria, FS classified into information measure, Euclidian distance, correlation, dependence, and reliability; (c) based on search approaches, FS classified into hybrid models, backward, forward and random models; (d) based on output type, feature rank, and subset selection models. Here we discuss the unsupervised feature selection models that are unsupervised filter and wrapper, FS models.

10.7.1 Unsupervised filter model

These models choose the features affording to the nature of given data structures. Researchers give equal importance to clustering and learning models for FS, thus decreases the proposed model complexity. Unsupervised feature selection approaches use the statistical background theorems for the training and evolution since they are highly adaptable and appropriate for high-dimensional data sets.

Authors of [31] projected an unsupervised FS-based competitive learning system to categorize the samples then determine the quantity of clusters, after that divide the input features into several feature subsets. In addition, authors also make threshold values to fix the candidate features, at last, the correlation differentiation among candidate and selected feature designed. In Ref. [32] presented an unsupervised FS approach to measuring the likeness among features using maximum information density index value. From Ref. [33] designed an unsupervised FS; for each feature, authors calculated the maximal information coefficient to hypothesis an attribute distance matrix, then applied k-Mode for automatic cluster recognition. In Ref. [34] designed a novel model based on data inside a class that should be similar, Laplacian score employed for evolution [35]. authors employed features using a GA and Sammon's stress function, thereby conserving the topology structure of the standard data in the condensed feature space.

10.7.2   Unsupervised wrapper model

Unsupervised wrapper models mostly applied clustering algorithms to fine-tune the validity of FS. Here, the feature subset with the finest clustering performance will study as the last ideal feature subset. Meanwhile, every feature subset must evaluate by the cluster model, which makes the model with high complexity in space and runtime. In some cases, Unsupervised wrapper models could divide into local and global models depends on the FS applied on multiple clusters or single clusters.

Authors in Ref. [36] employed sequentially forward and backward searches to discover best feature subsets. The authors used optimal feature subset for model clustering evolution. In Ref. [37] designed a novel conceptual information hierarchical cluster model where unsupervised FS uses to search optimal feature subset, and the search continues until the novel features are selected [38]. designed sample-based clustering with PSO on genetic factor data. The authors use the k-means algorithm for subset evolution. In Ref. [39] proposed an object evolution role to choose the best feature subset, the Bayesian model employed to find the optimal number of clusters [40]. developed an advanced k-means clustering model for FS. Feature weighting is employed to guide on selection of essential features.

10.7.3   Challenges

FS consider as an optimization solution that could do through an exhaustive search. Hence, researchers broadly apply the heuristic search-based models along with polynomial time complexity.

1. **Extreme data for ML:** In general, extreme data consists of large-scale groups with high-dimensional, biased labels and various data mining fields. A high number of researchers just focused on imbalanced classification problems. There is a need for a novel FS for high-dimensional data sets and works differently on different data sets. At the same time, FS models focused on the precision with imbalanced data

sets, along with scalability and stability. The FS with better scalability gives the best results with large data sets. As well as FS will not show more changes while adding and reducing data features is called stability. FS with promising stability evaluation approaches need, improved feature subset comparison measure then inconsiderateness to the size variance among feature subsets. In Ref. [41] authors examined various techniques on imbalanced data sets and designed SMOTE, ADASYN, and random forest with Near miss techniques for accurate formation of data sets. Authors compared the proposed model with logistic, regression and decision trees, and the results shows that the proposed model ADASYN technique has performed good prediction with appropriate balance.

2. **Ensemble feature selection:** Ensemble FS collects base classifiers talented by diverse feature gained by FS. It can enhance the FS stability on different data sets; random forest and random subspace are the best examples. Generalized additive-based models consider as the alternative for ensemble FS approaches when working with micro and gene data sets. Authors of [42] presented a novel collective model base on gene selection via information scheme [43], employed empirical breadth-first-search to discover more ideal gene subsets trained with SVM. Ensemble FS intentions to lessen the impact of high dimensions data sets on learning procedures whereas collecting various base classifiers that suit classification difficulties. But, the performance of collective approaches is unbalanced since the feature subsets are separated randomly.

3. **Dynamic feature selection:** In traditional FS models, data instances are static, which does not change the entire process. However, the data instances change over time in recent real-time applications, so the obtained FS changes constantly. FS with dynamic or continuous data features acknowledged as online or dynamic FS. Dynamic FS works effectively with unknown features and achieves redundancy analysis with feature-to-feature correlation. Appropriate stability, a real-time processing scheme, and first-rate algorithm design are some tips to improve the dynamic FS process.

4. **Feature selection and deep learning**: Usually, DL associates with low-level features, then construct the high-level feature after that learns distributed illustration of data. DL as the subpart of ML, with numerous hidden layers broadly employed in pattern recognition, video game intelligence, image processing, speech recognition and information understanding. The inappropriate data features might consume a great-part of properties in the model training process. A model training, the weight of features will be zero, it expresses FS as a feature reconstruction error, and the reconstruction error uses to select the appropriate input feature. FS-based DL approaches need to be at advanced levels like removing unrelated and noisy features from the novel features, with only valuable features to train the DL model.

10.8 Classification in ML: challenges and research issues

Several tools are available to describe the learning classifier system. Some of them are mentioned below.

Precision and Recall: Most of the classification and clustering models evaluated with various metrics, precision, and recall are important. Precision measures the amount of appropriately classified NFRs concerning the number of NFRs regained. The formula of precision defined as follows precision (p) = true-positive/(true-positive + false-positive). Recall defined as the percentage of NFRs those are properly classified and demonstrated as: Recall (R) = true-positives/(true-positives + false-negatives).

Both precision and recall are employed together, and there is differentiation among them. Precision makes sure that the retrieved information is genuinely applicable, whereas recall focuses on repossessing all associated requirements. However, there is still no idea which measurement value is essential for estimating classification results. Some researchers focused on attaining high recall since eliminating inappropriate features is more accessible than finding every missing data. Some other researchers argued that precision measurement was essential than recall for automatic classification. In some cases, researchers achieved high recall and precision values by retrieving all relevant features accurately.

F-Score: F-score is the harmonic mean of precision and recalls measurement values, defined as F-score. It widely employed for information retrieval tasks, combined with both precision and recall values. in some cases, F-score is utilized as an indicator for performance evolution and comparison and utilized as the measure when we don't know the evolution measure to apply or apply like in ROC.

Confusion Matrix: The confusion matrix is one of the famous evolution metrics for classification as well as clustering approaches. Usually, it shows the classification results, which show the percentage of accurate classifications and inaccurate classifications. The confusion matrix produces valuable results for further improvement of the designed model.

Accuracy: Accuracy or precision is well-defined as the measurement for the number of accurate classifications divided by the absolute quantity of classifications. Formalized as accuracy = (true-positive + true-negative)/(true-positive + true-negative + false-positives + false-negative). It has two variants: transudative—estimates the accuracy of predicting unlabeled data instances in the training period, whereas inductive accuracy predicts the unseen test data.

The AUC Curve: AUC is the pictorial representation of classifiers performance, where the X-axis shows false-positive-rat and the Y-axis shows true-positive-rate. It shows the accuracy of the performance concerning new instances; at the same time, AUC suffers from less bias associated with other trials like F-Score.

10.9 Key findings and open challenges

1. ML-based approaches perform very well when they have preprocessed data and achieve nearly 75% detection and classification accuracy.
2. From the literature work, unsupervised algorithms obtained better results with real-time applications than supervised algorithms.

3. Most of the ML models show better results when they have high-priority features rather than phrases. This mechanism is very much helpful for classification and identification.
4. In some special cases, unsupervised models also need preprocessed data sets, like identifying appropriate features for ML.
5. In addition, model designers do not explain model working procedures in most cases since they do not express how the model work. However, those metrics are not very well explained, like why they utilize for a particular evolution approach.
6. Researchers proposed various evolution metrics based on theory and practice, but those metrics are not explained clearly, like why they are used for certain evolution methods.
7. Nonexistence of training data sets is one of the key challenges handled by the present researchers. The research community prearranged some data sets, but they are in some specific format and not used in every problem.

10.10 Conclusion

Clustering high-dimensional data is a compound task which involves the optimal solution among many dissimilar approaches. The objects previously labeled are positioned in supervised classified clusters while those not labeled are positioned in unsupervised classified clusters. This chapter discusses taxonomy of clustering along with the usage of unsupervised clustering approaches and DL based unsupervised clustering models. The taxonomy clearly demonstrates the numerous approaches with their features, advantages and disadvantages. Various cluster distance measurements, internal and external evolution measurements, clustering and nonclustering losses are clearly discussed. In addition, chapter deliver some unsupervised feature selection directions utilized for cluster formation. The discussed information of this chapter can help as a valued reference for researchers who are attentive in unsupervised clustering.

References

[1] P. Berkhin, A survey of clustering data mining techniques, in: Grouping Multidimensional Data, Springer, Berlin, Germany, 2006, pp. 25–71.

[2] B. Chander, Deep learning network: deep neural networks, in: Neural Networks for Natural Language Processing, IGI Global, 2020, pp. 1–30.

[3] B. Chander, Feature selection techniques in high dimensional data with machine learning and deep learning, in: Handbook of Research on Automated Feature Engineering and Advanced Applications in Data Science, IGI Global, 2021, pp. 17–37.

[4] B. Chander, Clustering and Bayesian networks, in: Handbook of Research on Big Data Clustering and Machine Learning, IGI Global, 2020, pp. 50–73.

[5] Q. Wang, C. Wang, Z.Y. Feng, J.F. Ye, Review of K-means clustering algorithm, Electron. Des. Eng. 20 (7) (2012) 21–24.

[6] M. Goyal, S. Aggarwal, A review on K-mode clustering algorithm, Int. J. Adv. Res. Comput. Sci. 8 (7) (2017).

[7] H. Wang, D.-Y. Yeung, Towards Bayesian Deep Learning: A Survey, 2016 [Online]. Available, https://arxiv.org/abs/1604.01662.

[8] C. Li, F. Kulwa, J. Zhang, Z. Li, H. Xu, X. Zhao, A review of clustering methods in microorganism image analysis, in: Information Technology in Biomedicine, 2021, pp. 13–25.

[9] V. Mehta, S. Bawa, J. Singh, Analytical review of clustering techniques and proximity measures, Artif. Intell. Rev. 53 (8) (2020) 5995–6023.

[10] S. Singh, S. Srivastava, Review of clustering techniques in control system: review of clustering techniques in control system, Procedia Comput. Sci. 173 (2020) 272–280.

[11] N.M. Mahfuz, M. Yusoff, Z. Ahmad, Review of single clustering methods, IAES Int. J. Artif. Intell. 8 (3) (2019) 221.

[12] I. Bonet, A. Escobar, A. Mesa-Múnera, J.F. Alzate, Clustering of metagenomic data by combining different distance functions, Acta Polytech. Hung. 14 (3) (2017).

[13] U. Von Luxburg, A tutorial on spectral clustering, Stat. Comput. 17 (4) (2007) 395–416.

[14] B. Yang, X. Fu, N.D. Sidiropoulos, M. Hong, Towards k-means-friendly spaces: simultaneous deep learning and clustering, International Conference on Machine Learning. PMLR 6 (4) (2017) 3861–3870.

[15] K. Qader, M. Adda, M. Al-Kasassbeh, Comparative analysis of clustering techniques in network traffic faults classification, Int. J. Innov. Res. Comput. Commun. Eng. 5 (4) (2017) 6551–6563.

[16] A.I. Károly, R. Fullér, P. Galambos, Unsupervised clustering for deep learning: a tutorial survey, Acta Polytech. Hung. 15 (8) (2018) 29–53.

[17] E. Min, X. Guo, Q. Liu, G. Zhang, J. Cui, J. Long, A survey of clustering with deep learning: from the perspective of network architecture, IEEE Access 6 (2018) 39501–39514.

[18] J. Schmidhuber, Deep learning in neural networks: an overview, Neural Network. 61 (2015) 85–117.

[19] E. Aljalbout, V. Golkov, Y. Siddiqui, D. Cremers, Clustering with deep learning: Taxonomy and new methods, arXiv preprint arXiv:1801.07648 (2018) [Online]. Available: https://arxiv.org/abs/1801.07648.

[20] C.-C. Hsu, C.-W. Lin, CNN-based joint clustering and representation learning with feature drift compensation for large-scale image data, IEEE Trans. Multimed. 20 (2) (2018) 421–429.

[21] A. Makhzani, J. Shlens, N. Jaitly, I. Goodfellow, B. Frey, Adversarial autoencoders, arXiv preprint arXiv:1511.05644 (2015). Available: https://arxiv.org/abs/1511.05644.

[22] J. Guérin, O. Gibaru, S. Thiery, E. Nyiri, CNN Features Are Also Great at Unsupervised Classification, 2017. Available: https://arxiv.org/abs/1707.01700.

[23] V.S. Prasatha, H.A.A. Alfeilate, A.B. Hassanate, O. Lasassmehe, A.S. Tarawnehf, M.B. Alhasanatg, H.S.E. Salmane, Effects of Distance Measure Choice on Knn Classifier Performance-A Review, 2017 arXiv preprint arXiv:1708.04321.

[24] E. Aljalbout, V. Golkov, Y. Siddiqui, M. Strobel, D. Cremers, Clustering with Deep Learning: Taxonomy and New Methods, 2018 arXiv preprint arXiv:1801.07648.

[25] D. Chen, J. Lv, Z. Yi, Unsupervised multi-manifold clustering by learning deep representation, in: Workshops at the AAAI Conference on Artificial Intelligence, 2017.

[26] J. Xie, R. Girshick, A. Farhadi, Unsupervised deep embedding for clustering analysis, in: International Conference on Machine Learning (ICML), 2016, pp. 478–487.

[27] M.Z. Rodriguez, C.H. Comin, D. Casanova, O.M. Bruno, D.R. Amancio, L.D.F. Costa, F.A. Rodrigues, Clustering algorithms: a comparative approach, PLoS One 14 (1) (2019) e0210236.

[28] M. Usama, J. Qadir, A. Raza, H. Arif, K.L.A. Yau, Y. Elkhatib, A. Al-Fuqaha, Unsupervised machine learning for networking: techniques, applications and research challenges, IEEE Access 7 (2019) 65579–65615.

[29] A. Saxena, M. Prasad, A. Gupta, N. Bharill, O.P. Patel, A. Tiwari, C.T. Lin, A review of clustering techniques and developments, Neurocomputing 267 (2017) 664–681.

[30] J. Cai, J. Luo, S. Wang, S. Yang, Feature selection in machine learning: a new perspective, Neurocomputing 300 (2018) 70–79.

[31] N. Vandenbroucke, L. Macaire, J.G. Postaire, Unsupervised color texture feature extraction and selection for soccer image segmentation, in: Proceedings of International Conference on Image Processing, 2000, pp. 800–803.

[32] P. Mitra, C. Murthy, S.K. Pal, Unsupervised feature selection using feature similarity, IEEE Trans. Pattern Anal. Mach. Intell 24 (2002) 301–312.

[33] P.Y. Zhou, K.C. Chan, An unsupervised attribute clustering algorithm for un- supervised feature selection, in: Proceedings of IEEE International Conference on Data Science and Advanced Analytics (DSAA), 2015, 2015, pp. 1–7.

[34] X. He, D. Cai, P. Niyogi, Laplacian score for feature selection, in: Proceedings of Advances in Neural Information Processing Systems, 2015, pp. 507–514.

[35] A. Saxena, N.R. Pal, M. Vora, Evolutionary methods for unsupervised feature selection using Sammon's stress function, Fuzzy Inf. Eng. (2015) 229–247.

[36] M. Devaney, A. Ram, Efficient feature selection in conceptual clustering, in: Proceedings of International Conference on Machine Learning (ICML), 1997, pp. 92–97.

[37] J. Gennari, Concept formation and attention, in: Proceedings of the Thirteenth Annual Conference of the Cognitive Science Society, 1991, pp. 724–728.

[38] P. Deepthi, S.M. Thampi, Unsupervised gene selection using particle swarm optimization and k-means, in: Proceedings of the Second ACM IKDD Conference on Data Sciences, 2015, pp. 134–135.

[39] S. Vaithyanathan, B. Dom, Model selection in unsupervised learning with applications to document clustering, in: Proceedings of International Conference on Machine Learning, 1999, pp. 433–443.

[40] J.Z. Huang, J. Xu, M. Ng, Y. Ye, Weighting method for feature selection in k-means, in: Computational Methods of Feature Selection, 2008, pp. 193–209. CRC Press.

[41] T. Goswami, U.B. Roy, Classification accuracy comparison for imbalanced datasets with its balanced counterparts obtained by different sampling techniques, in: ICCCE 2020, Springer, Singapore, 2021, pp. 45–54.

[42] H. Liu, L. Liu, H. Zhang, Ensemble gene selection for cancer classification, Pattern Recognit. 43 (2010) 2763–2772.

[43] S.L. Wang, X.L. Li, J. Fang, Finding minimum gene subsets with heuristic breadth-first search algorithm for robust tumor classification, BMC Bioinf. 13 (2012) 178.

11

Emotion-based classification through fuzzy entropy-enhanced FCM clustering

Barbara Cardone[1], Ferdinando Di Martino[1,2], Sabrina Senatore[3]

[1]UNIVERSITÀ DEGLI STUDI DI NAPOLI FEDERICO II — DIPARTIMENTO DI ARCHITETTURA, NAPOLI, ITALY; [2]CENTRO INTERDIPARTIMENTALE DI RICERCA "ALBERTO CALZA BINI", UNIVERSITÀ DEGLI STUDI DI NAPOLI FEDERICO II, NAPOLI, ITALY; [3]UNIVERSITÀ DEGLI STUDI DI SALERNO, DIPARTIMENTO DI INGEGNERIA DELL'INFORMAZIONE ED ELETTRICA E MATEMATICA APPLICATA, SALERNO, ITALY

11.1 Introduction

The development of digital platforms for microblogging and the spread of product review websites, e-commerce, and social media feeds a continuous stream of thoughts and opinions. Statical data analysis models confirm well-established evidence, that is that the social data stream is a real new oil, whose potential could have an important effect on the global market. Capturing the essence of "the chattering" on the web is crucial in many decision-making processes: companies monitor the public behavior of users toward their items, analyzing the human reaction toward ongoing marketing strategies; the influence of users on other users in expressing an opinion about products/events. For example, the current global pandemic COVID-19 is characterized by an incremental presence of opinionated data, often polarized by misinformation published in the usual web channels, that may induce changes in vaccination intent. A recent study [1] investigated the impact of (mis)information on population considering sociodemographic factors (i.e., age, gender, education, employment type, race, and political affiliation), daily time spent on social media, and sources of information on COVID-19 pandemic.

Social events and political activities can influence collective opinion and sentiments; at the same time, real-time events, such as accidents, earthquakes, or other natural crises, crime are often kept posted by users that broadcast them on social web platforms, by continuous monitoring of the situation's evolution.

The uncontrolled Internet-generated content in several digital formats, coupled with the technological enhancements in the communication infrastructure, device miniaturization, and the development of networking platforms represent the face of the same coin: an uncontrollable growth of the data volume and availability.

Many studies in the literature face the "opinionated" text mining for grabbing relevant information from these data sources by analyzing the feeling and judgments about targets of interest in the digital content. Analyzing user-generated content to understand user preferences and moods is crucial in fields like market analysis, business, political consensus study, and even nowadays, with the recent pandemic, vaccination campaigns.

Sentiment Analysis (SA) has become an umbrella term concerning the investigation of feelings, moods, and emotions conveyed in the text. In a sense, SA refers to the sentiment polarity that varies between positive, neutral, and negative, drawn from the natural language of the written text; sometimes, it focuses specifically on capturing emotions, e.g., joy, hate, rage, and sadness expressed in the text. The SA can become challenging when it aims at discriminating objective (neutral) from subjective (opinionated) texts, especially when the polarity from text is sough toward a specific target, such as an item, a topic, a service, an event [2]. Moreover, the automatic identification of a wide range of emotional nuances in the natural language text is definitive in social, economic, and marketing strategies. Understanding, for example, the feeling of frustration of users dissatisfied with their purchase could allow the selling company to react proactively, initiating new marketing strategies targeted at relieving possible deriving losses.

The proposed work introduces an emotion-based classification method from a social media data stream, such as Twitter, to detect the main emotions triggered by a social trend. The case study is analyzed by exploiting an extension of the fuzzy c-means (FCM) clustering algorithm, that exploits the fuzzy entropy measure to assess the data distribution and overcome the FCM's sensitivity to the random cluster initialization. The remainder of the paper is organized as follows. Section 11.2 introduces a literature overview about sentiment analysis, emotion extraction approaches, in the Twitter's domain. Section 11.3 provides an overview on the main emotion-based models from the literature. The theoretical background on FCM as well as the fuzzy entropy measure and the proposed entropy based FCM model are presented in Section 11.4, while Section 11.5 describes the approach for emotion detection from an Italian reference Twitter data set. Finally, Section 11.6 compares the results achieved by the entropy-based FCM with the classical FCM algorithm, showing that the proposed method converges faster and provides promising classification performance, evaluated by the common metrics such as accuracy, precision, and F1-score. Section 11.7 outlines the concluding remarks.

11.2 Related work

The advent of social computing paved the way for the analysis of large amounts of data collected as a massive data stream from social networks. With the rise of microblogging websites, such as Twitter, people have been able to share their opinions and feelings about various focal events and pressing issues on such social networks. The continuous flow of user-generated content creates unprecedented opportunities to gain insights into the major trends of the global community, especially when aimed at capturing the opinions and the emotions of the web users.

The need to detect emotions and sentiments conveyed through Twitter messages has led to several research works. Understanding human psychology and mutual influence in the user behaviors are crucial to drive marketing strategies [3,4]: companies study the user attitudes toward their items, services, based on comments left on the online social network; social and political events influence and are influenced by collective opinion, as well as social issues and disasters can have a strong impact on the collective thinking. One of the main issues is the difficulty of processing the vast volumes of unstructured textual content to extract relevant information. Sentiment Analysis tasks are often closely affected by the vagueness of natural language, that can make the sentiment polarity and emotion identification very complex especially when some issues of polysemy, ambiguity, sarcasm, or irony appear in the text processing. Moreover, in short messages such as tweets, idiomatic expressions and sarcasms are common habits of people's cultures, and slang expressions are common habits of people [5]. Sentiment Analysis was studied at different levels of granularity: sentiment polarity was detected at the level of the whole document, a specific sentence, and a single aspect (i.e., the entity described in the text). A text analysis applied at the sentence level allows to detect more targeted sentiments from documents [6], especially if they refer to a given target, i.e., entity, person, item, event [7], providing higher accuracy than the overall document sentiment [8].

The complexity of accurately extracting feelings and emotions from the text to capture the meaning of the actual sentence promotes hybrid approaches [9,10] that often represent promising solutions in terms of performance. Several SA approaches exploit annotated corpora and define disambiguation rules [4]. Twitter SA studies include both supervised methods [11–14] and lexicon-based [15–17] methods. Supervised methods are mainly classification approaches that adopt Natural Language Processing tasks to clean the text from noisy words. Lexicon-based methods exploit lexicon and thesaurus with sentiment orientation to determine the general sentiment trend of a given text, such as SentiWordNet [18]. Distinguishing the sentiment polarity may not provide sufficient information about the actual feeling expressed in the text, especially when this study is employed in activities that requires accurate evaluation such as in market and social trend analysis.

Understanding emotion and opinion involves the deep integration of expertise with knowledge acquired by domains such as psychology, linguistics, cognitive science, sociology, and ethics [19]. Affective computing represents an emergent research area bringing computer science with psychology and cognitive science-oriented fields, with the goal of making machines capable of capturing, inferring, and interpreting human emotions [20]. Several approaches in the literature have investigated this domain, for example, a system for automatic emotions extraction from news headlines is presented in Ref. [21] obtained by constructing a large, annotated data set with emotions. Rule-based approaches, probabilistic and machine learning methods have been intensely used for emotion extraction [22–25], especially deep learning methods employed for topic detection and sentiment classification [26]. Word embedding models like Glove [27], Word2Vec [28], Fast-Text [29], and BERT [30] have been deployed along with lexicons to increase the sentiment and emotion classification performance. Text classification applications with CNNs have also attracted increasing attention from research communities; in Ref. [31] a deep convolutional character-level embedding (Conv-char-Emb) neural network model is proposed for SA on unstructured data; in Ref. [32] on the other hand, a CNN model with multiple convolution kernels accomplishes the emotion classification model in microblogs, addressing the typical issues of short messages such as slangs, abbreviation, and especially, the difficulty of extracting accurate features from short texts.

Hybrid solutions in the literature combine machine learning and linguistic models [22,33] for social stream classification. The main drawback of these systems is the high computational demand, which can greatly affect the algorithm performance in terms of memory usage and execution time.

The approaches described are attractive solutions for processing big data, which means that a huge amount of data must be available to apply ML-based techniques. However, in some specific application domains, the data sets are small, and sometimes, it is necessary to enrich them with additional data that is not easy to find, resulting in expensive or time-consuming activities. Unsupervised or semisupervised learning approaches represent viable alternatives to process data especially when no annotation is available, a typical situation in real-world scenarios. Indeed, one of the main issues is the availability of large unlabeled data, which requires automatic methods for labeling, based, for example, on predictive approaches to assign labels to unlabeled data using predictive approaches [34] or training multiple supervised models with the unlabeled data, previously augmented with predictions from earlier iterations of the unsupervised models. Our approach assumes significance in this context: it is unsupervised, suitable for working on data sets from real word that are generally unlabeled.

11.3 Emotion-based models

In recent decades, emotion sensing is becoming increasingly crucial in the study and understanding of human social behavior. Sentiments and emotions deeply influence the human experience, they are often described by a straight dichotomic classification: positive versus negative, pleasant versus unpleasant, activated versus deactivated, etc. [35,36]. Ranging from Darwin's concept of emotion, described as expressive and as physiological states for living and learning to Ekman's universality of facial expressions of emotion, the impact of emotions on human behavior is the focus of many studies. Noteworthy was Picard's contribution in affective computing (1997), stating that machines could be truly intelligent and interact naturally with humans, if computers were able to identify, interpret, and express emotions. Picard showed that the Turing test could solve the dilemma of whether or not a machine could think, and the hidden Markov model could show transitions from one emotional state to another, given some observations over time. Marvin Minsky's conception of emotions provides an alternative interpretation, arguing that the emotional state results from turning on some resources (that make up the human mind) and turning off others. Minsky argued that emotions are just different "ways of thinking" to deal with issues from the world to fuel human intelligence.

11.3.1   Ekman's emotions

Although the ability to identify the basic emotions through facial expression seems to be universal among humans [37], Paul Ekman's research meets the Darwin's evolutionary theory about the role of emotions in the evolution to facilitate rapid communication of danger or safety [38]. Basic emotions influence and are influenced by the primary tasks of life; they are crucial for living and learning; for instance, feelings like fear and rage, can lead to survival in danger by influencing living beings to flee, run for their lives or fight to defend themselves.

Ekman recognized six universal emotions: sadness, fear, disgust, anger, surprise, and happiness; they come from the same cross-cultural facial expressions that describe basic emotions and depend neither on age nor race. According to Ref. [39], these emotions are described by considering unrelated dimensions: pleasant-unpleasant, tension-relaxation, and excitation-calm although some studies have shown that the last two dimensions overlap.

Ekman considered only the pleasant-pleasant dimension and the active-passive scale to adequately address the distinction between emotions. In addition, he characterized new, more complex or compound emotions by combining elementary emotions together. Many later models build on Ekman's model of emotion, by incorporating multiple nuances into the emotions obtained as a combination of the primary ones. One of the best known is Plutchik's wheel of emotions.

11.3.2   The Plutchik's main emotions

Plutchik's multidimensional model is derived from the psychoevolutionary theory of emotions, extending Ekman's primary emotions with as many secondary emotions. Emotions are part of an evolutionary process involving events, perceptions, moods, and actions aimed at preserving behavioral homeostasis. According to Plutchik's idea, humans earn life experience by harnessing primary emotions, paired as distinct opposite emotions. Like Ekman, Plutchik argued that these primary human emotions are purely primitive and culturally independent; moreover, their evolution allows species to survive. He identified eight basic, primary emotions, placed in a circular model: trust, joy, surprise, fear, anticipation, sadness, disgust, anger, and asserted that there is an opposite to each primary emotion. Plutchik's wheel of emotions is shown in Fig. 11.1: opposite emotions are placed one in front of the another, while the adjacent emotions, when they merge, can generate new and more complex emotions. Each primary emotion is, in turn, divided into subgroups corresponding to the secondary and tertiary emotions in the wheel representation, as they take different intensities and polarities [35]. The intensity of an emotion becomes greater as approaching the center of the wheel, while it decreases

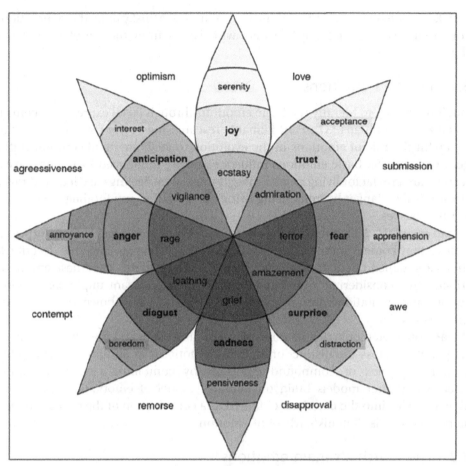

FIGURE 11.1 Primary and secondary emotions in the Plutchik's wheel-shaped representation. *Free license: https://commons.wikimedia.org/wiki/File:Plutchik-wheel.svg.*

as moving away from it. The fusion of primary emotions can produce further emotions, which are located in the outer circle of the wheel. An example of blended emotions can be remorse, resulting from the combination of sadness and disgust, or love from the fusion of joy and trust. The theory behind Plutchik's model concerns the different origins associated with the primary and secondary emotions. While primary emotions control the human instincts, secondary or mixed emotions are from a social reaction to the stimuli of mental states or triggered by cognitive actions. Plutchik's wheel of emotions summarizes, in a simple way, the wide range of human emotions, emphasizing that they represent the result of "mixtures" and various events.

11.3.3   The Hourglass of emotion

A reinterpretation of Plutchik's wheel of emotions is the Hourglass of Emotions [40], an affective-based model composed of four affective dimensions. Each dimension consists of sentic levels, i.e., the activation levels of emotional states of the mind, whose intensity is strictly related to the activated level. The model reflects Marvin Minsky's conception of emotions, which describes the emotions as the result of turning on/off some specific resources within our mind. For instance, the state of anger turns on certain resources that allow humans to respond quickly, while dismissing other resources that usually require more prudent behavior.

The model covers an affective classification splitting the emotions into independent but concomitant affective dimensions and, through the combination of different levels of activation, define the resulting, mixed emotional states of our mind. This model considers four affective dimensions: Pleasantness, Attention, Sensitivity, and Attitude (Fig. 11.2). Each dimension includes distinct sentic levels, which measure the strength of the perceived emotion (expressed in the range [−3, 3]). The levels are labeled with 24 basic emotions (six for each of the affective dimensions). The sentic level describes an emotional state which can be more or less strong, depending on the value of the position occupied in the corresponding dimension. In Fig. 11.2, for example, the Pleasantness dimension encloses two opposite emotions, joy and sadness, located respectively, at sentic levels 2 and −2.

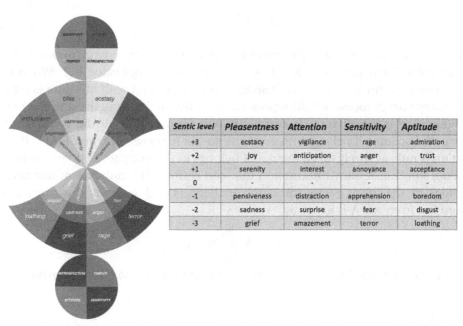

Sentic level	Pleasentness	Attention	Sensitivity	Aptitude
+3	ecstacy	vigilance	rage	admiration
+2	joy	anticipation	anger	trust
+1	serenity	interest	annoyance	acceptance
0	-	-	-	-
-1	pensiveness	distraction	apprehension	boredom
-2	sadness	surprise	fear	disgust
-3	grief	amazement	terror	loathing

FIGURE 11.2 The hourglass of emotions [40]. *A part of this figure is from https://commons.wikimedia.org/wiki/File: Hourglass-of-emotions.jpg. This file is licensed under the Creative Commons Attribution-Share Alike 4.0 International license.*

The transition between different emotional states can vary according to the intensity value of an affective dimension that ranges from zero (emotional emptiness) to unit value (heightened emotionality). The mapping of this emotional space yields the typical hourglass shape of this model.

11.4 Theoretical background

11.4.1 Fuzzy entropy and fuzziness

Let X be a universe of discourse and $F(X) = \{A: X \to [0, 1]\}$ be a collection of all fuzzy sets defined on X. Then, let x be an element of X and $A(x)$ be the membership degree of x to the fuzzy set $A \in F(X)$. A continuous function e: $[0,1] \to [0, 1]$ is called *fuzzy energy function* [41,42] and it is characterized as follows.

- $e(0) = 0$
- $e(1) = 1$
- e is monotonically increasing.

Let us notice that the higher the value $e(x)$, the greater the certainty that object x belongs to the fuzzy set A.

Let X be a discrete set with cardinality m, $X = \{x_1, ..., x_m\}$, the energy measure E of the fuzzy set A is given by Refs. [41,42]:

$$E(A) = K \sum_{i=1}^{m} e(A(x_i)) \tag{11.1}$$

where K is a positive constant. The minimal value of the fuzzy energy measure is 0 and the maximal value is Km. With $K = 1/m$, the maximal energy value is 1. When $E(A)$ is equal to one all elements $x_1, ..., x_m$ belong to the fuzzy set A with absolute certainty.

The fuzzy energy measure $E(A)$ represents the information contained in the fuzzy set A. If $F(A) = 1$ this information is complete as it implies that all the elements of the universe of discourse X belong with certainty to the fuzzy set A. When $F(A) = 0$ this information is null, as surely no element of X belongs to the fuzzy set A, therefore the fuzzy set A does not transfer any information that can connote the universe of discourse X.

Besides to the fuzzy energy measure, in Refs. [41,42] is defined a fuzzy entropy measure the fuzzy set A is given by:

$$H(A) = K' \sum_{i=1}^{m} h(A(x_i)) \tag{11.2}$$

where the continuous function h: $[0,1] \to [0,1]$ describes the *fuzzy entropy function* where:

- $h(1) = 0$
- $h(u) = h(1 - u)$ for each $u \in [0,1]$
- h is monotonically increasing in $[0, 1/2)$

- h is monotonically decreasing in $[1/2, 1]$

The positive constant K' is set to the value $1/m$ to guarantee the normalization of the fuzzy entropy index in the interval $[0,1]$.

The fuzzy entropy (11.2) measures the fuzziness of the fuzzy set A. The minimal value 0 of H(A) is given when A is a crisp set (the membership degree of each element of X to A is 0 or 1); the maximum value, i.e., 1 is given with $u = 1/2$.

In [42], De Luca and Termini present the fuzzy entropy function, defined as follows.

$$h(u) = \begin{cases} 0 & \text{if } u = 0 \\ -u\ln(u) - (1-u)\ln(1-u) & \text{if } 0 < u < 1 \\ 0 & \text{if } u = 1 \end{cases} \tag{11.3}$$

The fuzzy entropy function assumes the maximum value, i.e., equal to 1, with $u = 1/2$; and it is known as Shannon's function.

11.4.2   FCM and weighted FCM algorithms

The FCM algorithm is one of the most known partitive clustering algorithms [43,44]. It is widely employed in several domains, such as information retrieval, image segmentation, medical imaging, etc. In literature, partitive clustering algorithms such as K-means, FCM are used as a baseline and compared for data classification [45–47]. Empirical evidence shows the overwhelming of FCM compared to K-Means, in terms of performance results (i.e., cluster accuracy), even though it needs more computation time than K-Means clustering.

The FCM algorithm is well suited for multidimensional data analysis: in fact, one of its strengths is its computational complexity, which is linear with respect to the size of the input data. Nevertheless, FCM is sensitive to the presence of noise and outliers and to random initialization. A further disadvantage is that the FCM algorithm needs to know a priori the number of clusters, so sometimes it is necessary a preprocessing stage to assess some validity indices in order to find the best number of clusters. In the remaining of the section, some theoretical background about FCM and its weighted extension are given.

Let $\mathbf{P} = \{\mathbf{p}_1, ..., \mathbf{p}_N\} \subset R^n$ be composed of N data points; the jth element of $\mathbf{P}$ is given by the vector $\mathbf{p}_j = (p_{j1}, ..., p_{jn})^{-1}$ in the n-dimensional space R^n of the features.

The FCM algorithm [43,44] detects as points in the n-dimensional space R^n, the centers of C clusters, where C is the prefixed number of clusters. FCM uses the Euclidean metrics to compute the distances.

Let $\mathbf{V} = \{\mathbf{v}_1, ..., \mathbf{v}_C\} \subset R^n$ be the set composed of the centers of the C clusters. Let $\mathbf{U}$ be the *partition matrix*, a $C \times N$ matrix whose elements u_{ij} are the membership degree of the jth data point $\mathbf{p}_j$ to the ith cluster whose center is $\mathbf{v}_i$.

FCM aims at finding the cluster centers and the partition matrix, by minimizing the following objective function:

$$J(\mathbf{U}, \mathbf{V}) = \sum_{i=1}^{C} \sum_{j=1}^{N} u_{ij}^m d_{ij}^2 = \sum_{i=1}^{C} \sum_{j=1}^{N} u_{ij}^m \left\| \mathbf{p}_j - \mathbf{v}_i \right\|^2 \tag{11.4}$$

where $d_{ij} = \|p_j - v_i\|^2$ is the Euclidean distance between the center v_i of the ith cluster and the jth data point p_j and the fuzzifier parameter $m \in [1,+\infty)$ is a constant which impacts on the FCM clustering performance: the higher the value of m, the higher the fuzziness level of the clusters. Generally, the value of m is set to 2. The following equations describe the imposed constraints.

$$\sum_{i=1}^{C} u_{ij} = 1 \quad \forall \ j \in \{1, \dots, N\} \tag{11.5}$$

$$0 < \sum_{j=1}^{N} u_{ij} < N \quad \forall \ i \in \{1, \dots, C\} \tag{11.6}$$

Eq. (11.5) ensures that the membership degrees sum of a data point to all the clusters is equal to 1, Eq. (11.6) instead, guarantees that the membership degree of at least one data point to a cluster is greater than zero.

The solutions for **U** and **V** are obtained applying the Lagrange multipliers, described as follows.

$$v_i = \frac{\sum_{j=1}^{N} u_{ij}^m p_j}{\sum_{j=1}^{N} u_{ij}^m} \quad i \in \{1, \dots, C\} \tag{11.7}$$

and

$$u_{ij} = \frac{1}{\sum_{k=1}^{C} \left(\frac{d_{ij}}{d_{kj}}\right)^{\frac{2}{m-1}}} \quad i \in \{1, \dots, C\}, \ j \in \{1, \dots, N\} \tag{11.8}$$

FCM initially assigns the membership degrees u_{ij} randomly; during each iteration the cluster centers are calculated by Eq. (11.7), then the membership degree components are calculated by Eq. (11.8). The partition matrix obtained at the tth iteration is compared with the one calculated in the previous iteration, to check whether to perform a new iteration or terminate the process. The termination criterion to check if the process can be stopped is:

$$\left|\mathbf{U}^{(t)} - \mathbf{U}^{(t-1)}\right| < \varepsilon \quad i = 1, \dots, C; \ j = 1, \dots, N \tag{11.9}$$

where $\varepsilon > 0$ is a parameter assigned a priori to stop the iteration process and

$$\left|\mathbf{U}^{(t)} - \mathbf{U}^{(t-1)}\right| = \max_{\substack{i=1,\dots,C \\ j=1,\dots,N}} \left\{\left|u_{ij}^{(t)} - u_{ij}^{(t-1)}\right|\right\} \quad i = 1, \dots, C; \ j = 1, \dots, N \tag{11.10}$$

A pseudocode of the FCM algorithm is shown in Algorithm 11.1.

Algorithm 11.1. Pseudocode of FCM algorithm

Set m, ε, C
Initialize randomly the partition matrix components u_{ij} i = 1,...,C j = 1,...,N
Repeat
  Calculate v_i i = 1,..., C by (11.7)
  Calculate u_{ij} i = 1,..., C j = 1,..., N by (11.8)
Until $\left|U^{(t)} - U^{(t-1)}\right| > \varepsilon$

An extension of the FCM algorithm is the weighted FCM algorithm (for short, wFCM): compared to FCM, whose data point contributes equally to find the final optimal clustering, it assigns a weight to each data point, that can affect differently, in discovering the optimal solution.

The wFCM algorithm minimizes the following objective function:

$$J_w(U,V) = \sum_{i=1}^{C} \sum_{j=1}^{N} w\left(\mathbf{p}_j\right) u_{ij}^m d_{ij}^2 = \sum_{i=1}^{C} \sum_{j=1}^{N} w\left(\mathbf{p}_j\right) u_{ij}^m \left\| \mathbf{p}_j - \mathbf{v}_i \right\|^2 \qquad (11.11)$$

Applying Lagrange multipliers, the cluster centers is expressed as follows:

$$\mathbf{v}_i = \frac{\sum_{j=1}^{N} w\left(\mathbf{p}_j\right) u_{ij}^m \mathbf{p}_j}{\sum_{j=1}^{N} w_j u_{ij}^m} \quad i \in \{1, \dots, C\} \qquad (11.12)$$

where the weight $w(\mathbf{p}_j)$ describes the degree of influence of the jth object to find the cluster centers; the value $w(\mathbf{p}_j)$ assigned to the jth data point $\mathbf{p}_j$ is calculated in any iteration of the algorithm, as a function of $\mathbf{x}_j$.

A pseudocode of the wFCM algorithm is shown in Algorithm 11.2.

Algorithm 11.2. Pseudocode of wFCM algorithm

> *Set* m, ε, C
> *Initialize* randomly the partition matrix components u_{ij} i = 1,...,C j = 1,...,N
> **Repeat**
> | *Calculate* $w(p_j)$ j = 1,...,N
> | *Calculate* v_i i = 1,...,C by (11.12)
> | *Calculate* u_{ij} i = 1,...,C j = 1,...,N by using (11.8)
> **Until** $\left| U^{(t)} - U^{(t-1)} \right| > \varepsilon$

As introduced above, one of the main drawbacks of the FCM algorithm is the random initialization of the cluster centers that can yield to being trapped in local minima or to determining global minima after many iterations, at the expense of the running time.

11.4.3 Entropy-based FCM algorithm

In Ref. [48], the fuzzy entropy function (11.3) allows setting the data point weights in the wFCM algorithm. The fuzzy entropy function calculated on $u_{i,j}$, the membership degree of the jth data point to the ith cluster is given by:

$$h\left(u_{ij}\right) = \begin{cases} 0 & \text{if } u_{ij} = 0 \\ -u_{ij}\ln\left(u_{ij}\right) - \left(1 - u_{ij}\right)\ln\left(1 - u_{ij}\right) & \text{if } 0 < u_{ij} < 1 \\ 0 & \text{if } u_{ij} = 1 \end{cases} \qquad (11.13)$$

The weight of the jth data point is calculated by applying the formula:

$$w\left(\mathbf{p}_j\right) = 1 - \frac{1}{C}\sum_{i=1}^{C}h\left(u_{ij}\right) \qquad (11.14)$$

where the second term is the average value calculated on the membership degree of the jth data point to each cluster. Due to the constraint of Eq. (11.6), the weight $w(\mathbf{p}_j)$ is a value within the range [0, 1]. The maximum value $w(\mathbf{p}_j) = 1$ is reached if the data point belongs with a membership degree of one to a single cluster and 0 to all others. The minimum value $w(\mathbf{p}_j) = 0$ is reached when the data point belongs to any cluster with the same membership degree $1/C$.

In Ref. [48] is defined the mean fuzziness of the clusters $\overline{H}$ given by:

$$\overline{H} = \frac{1}{C}\sum_{i=1}^{C}\left(\frac{1}{N}\sum_{j=1}^{N}h\left(u_{ij}\right)\right) \qquad (11.15)$$

where the term in parenthesis represents the fuzziness of the ith fuzzy cluster, defined following formula (11.2). $\overline{H}$ ranges in $[0, h_{max}]$; it is equal to 0 when the fuzzy clusters become crisp sets, and $h_{max} = h(1/C) \leq 1$ when all the objects assume a membership degree to each fuzzy cluster equal to $1/C$.

In Ref. [6] a novel weighted FCM algorithm, called Entropy weighted FCM (for short, EwFCM) fixes the problem of the random initialization of the cluster centers.

In EwFCM, a preprocessing step is performed to identify the optimal cluster centers. Initially, the centers of the C clusters are randomly assigned; at each iteration, the average fuzziness value of the clusters is calculated (Eq. 11.15). If the difference between the mean fuzziness value calculated in the current iteration and the one calculated in the previous iteration is below a fixed threshold η, the preprocessing phase stops and then the FCM algorithm is started to detect the final clusters. The preprocessing phase also ends when a prefixed maximum number of iterations i_{max} is achieved. The pseudocode of EwFCM is shown in Algorithm 11.3.

Algorithm 11.3. Pseudocode of EwFCM algorithm

Set m, ε, C
Initialize randomly the partition matrix components u_{ij} $i = 1,...,C$ $j = 1,...,N$
$n_{iter} \leftarrow 1$ // iteration number set initially to 1
Repeat   //preprocessing phase
  Calculate $w(p_j)$ $j = 1,...,N$ by (11.14)
  Calculate v_i $i = 1,...,C$ by (11.12)
  Calculate u_{ij} $i = 1,...,C$ $j = 1,...,N$ by (11.8)
  Calculate $\check{H}^{(t)}$ by (11.15)
  $n_{iter} \leftarrow n_{iter} + 1$
Until $\left|\check{H}^{(t)} - \check{H}^{(t-1)}\right| > \eta$ OR $n_{iter} = i_{max}$

Calculate u_{ij} $i = 1,...,C$ $j = 1,...,N$ by (11.8)   // used the optimal cluster centers detected
Repeat   //execute FCM setting the cluster centers calculated in the preprocessing phase
  Calculate v_i $i = 1,...,C$ by (11.7)
  Calculate u_{ij} $i = 1,...,C$ $j = 1,...,N$ by (11.8)
Until $\left|U^{(t)} - U^{(t-1)}\right| > \varepsilon$

In a nutshell, the EwFCM algorithm requires two iterative processes (compared to FCM): a preprocessing phase targeted at calculating the optimal centers of the initial clusters, and then the typical iterations of FCM, exploiting the computed setting parameters (cluster center values) determined in the preprocessing phase. In addition, the EwFCM algorithm allows improving the FCM performance. In fact, in Ref. [48] the authors showed, from comparative analysis, that EwFCM, besides improving the classification performance of FCM, also reduces its running time.

11.5 Logical design model

The data flow diagram describing the proposed emotion classification method of social stream messages is sketched in Fig. 11.3. The orange-colored rectangles are the functional components, while the parallelogram shape describes input/output of functional components.

The corpus of documents was built starting from the microblogging stream messages, specifically, Twitter's messages was collected and processed.

An initial parsing task filters out unnecessary data from the textual content: text from tweets is usually processed to remove noise words (stop words, slangs, abbreviations, etc.). Due to the short message nature of tweets, they are aggregated by word hashtag to form an individual document describing a social trend, according to the selected hashtag. A tweet containing multiple hashtags is replicated in every document containing one of those hashtags.

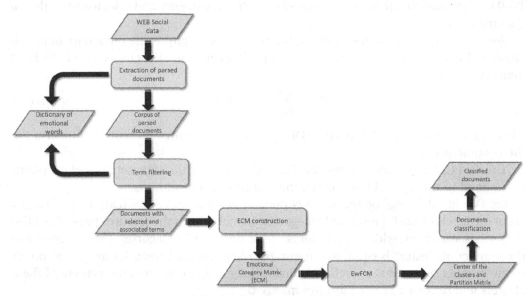

FIGURE 11.3 The dataflow diagram of the design model.

Once the documents are built, the text parsing is accomplished: each term in the documents is reduced to its stemmed form (for instance, the term happy and the derivative ones such as happiness, happily, are reduced to a root word happi).

At that point, stemmed terms are grouped according to the emotions that thy express or are associated with. To this purpose, a *Dictionary of Emotional Words* was built ad hoc: precisely, this dictionary is composed of all the words that describes emotions and are grouped according to their affinity with some selected basic emotions, called *emotional categories*. These categories represent an emotion classification as described by the emotion-based models introduced in Section 1.3. The choice of emotion model is at the operator's discretion; for example, if Plutchik's emotion classification is used to build the dictionary, each emotion label (primary and secondary, as shown in Fig. 11.1) provides the root term of an emotional category, that can be enriched by synonyms or related terms with similar meaning. Just to give an example, the stemmed term *happi* from the text parsing is associated with root term identified by the emotional category labeled *Joy* from the Plutchik's emotion model.

The *Term Filtering* component accomplishes such task: it associates each stemmed term from text with the emotional category which the term belongs. The occurrences of the individual terms associated with an emotional category of the dictionary, contribute to build the Emotional Category matrix (shortly, ECM), by providing the entry values of the matrix cells. This matrix will be input into the *EwFCM* component that accomplishes the homonym algorithm, described in Section 1.4.

The *ECM Construction* component generates an ECM matrix, whose cell values based on the known Term Frequency, Inverse Document Frequency index (for short TF-IDF). Precisely, each element ECM_{ij} of the matrix ECM is obtained by summing the values of TF-IDF calculated for all terms associated with the jth category and selected from the ith document.

Formally, if d_i is the ith document in the corpus and Dict(j) is the dictionary of the jth emotional category composed of a set of related terms, the element ECM_{ij} of the ECM matrix is given by

$$ECM_{ij} = \sum_{t \in \text{Dict}(j)} TF - IDF(t, d_i) \tag{11.16}$$

where t is a term in Dict(j) and $TF - IDF(t, d_i)$ is the relevance measure of the term t in the document d_i.

The EwFCM component uses the ECM matrix as a training set. The data points correspond to the rows of this matrix, the features to the columns.

Let $D = \{d_1, d_2, ...d_N\}$ be the set of N documents, and let $CAT = \{cat_1, cat_2, ...cat_C\}$ a set of the labels of the C emotional categories, that also form the feature space. The ECM matrix has size $N \times C$ with N documents and C emotional categories. Let us notice that the number of clusters is equal to the number of emotional categories and corresponds to the number of features as well. The EwFCM algorithm discovers the centers of the C clusters and returns a $C \times N$ partition matrix **U**.

The *Document Classification* component achieves a mapping between the clusters obtained from the EwFCM and the emotional categories. It associates each cluster with the emotional category corresponding to the label of feature with the highest value in the cluster center coordinates. In other words, each document is allocated to the emotional category corresponding to the cluster where the document belongs to with the highest membership degree.

11.6 Experimental results

A series of tests were carried out; the goal of these tests is to show the effectiveness of the EwFCM, our entropy-enhanced version of FCM, comparing the performances of EwFCM and FCM. The known classification metrics, namely, accuracy, precision, recall, and F1-score were used to evaluate the performance of the proposed method. These measures are based on the evaluation of the following parameters:

TP (True Positive) is the number of data correctly assigned to the class.
TN (True Negative) is the number of objects correctly not assigned to the class.
FP (False Positive) is the number of objects wrongly assigned to the class.
FN (False Negative) is the number of objects wrongly not assigned to the class.
The four classification measures are given as follows.

$$\text{accuracy} = \frac{\text{TP} + \text{TN}}{\text{TP} + \text{TN} + \text{FP} + \text{FN}} \times 100 \tag{11.17}$$

$$\text{precision} = \frac{\text{TP}}{\text{TP} + \text{FP}} \times 100 \tag{11.18}$$

$$\text{recall} = \frac{\text{TP}}{\text{TP} + \text{FN}} \times 100 \tag{11.19}$$

$$\text{F1 score} = \frac{2 \times \text{recall} \times \text{precision}}{\text{recall} + \text{precision}} \times 100 \tag{11.20}$$

Performance was also evaluated with respect to the number of iterations required for algorithm convergence and algorithm execution time.

Two experiments were conducted. The first one was performed on the well-known classification data set *Wine*, from the UCI Machine Learning repository.[1] The second data set was composed of a collection of tweets arranged in trend-based documents (as described in Section 1.5), classified according to prevailing emotions.

The values assigned in the configuration setup are as follows:

- fuzzifier parameter m = 2
- stop iteration error ε = 0.2
- maximum number of iterations = 50.

[1]https://archive.ics.uci.edu/ml/datasets.php.

In Table 11.1, the classification performance obtained on the Wine data set is reported. The data set contains information on 178 types of wines from three different crops, classified based on 13 numerical features that characterize their chemical composition. Each data point is labeled with the class corresponding to its crop.

The results after executing the two algorithms shows that EwFCM algorithm always overwhelms the traditional FCM algorithm, showing how the entropy measure strongly influences on the algorithm, speeding up its convergence, as evidenced in Table 11.2 where the iteration number of EwFCM is lower than FCM, and consequently the running time.

The results obtained for other data sets from the UCI Machine Learning classification confirm that EwFCM has better classification performances than FCM; furthermore, EwFCM achieves convergence faster than FCM.

The other experiment is performed on a Twitter data set presented in Ref. [49]. The data set consists of more than 1400 documents, which were built by processing a stream of 450 1000 tweets, collected during the time period May 2018–July 2018, in the cities of Washington, New York and London.

As stated in Ref. [49], each document is composed of about 350–400 tweets, including the same hashtag that represents the document trend. Initially, the feature set was composed of 107 words, that composed the emotional categories describing *pleasant feelings*; then they were reduced to eight features, by considering only the emotional categories that, in this experiment, are labeled as *Open, Happy, Alive, Good, Love, Interest, Positive* and *Strong*.

Table 11.1   Data set wine – FCM versus EwFCM – classification performance.

Measure	Algorithm	Class 1	Class 2	Class 3	Mean
Accuracy	FCM	91.33%	87.79%	89.35%	89.49%
	EwFCM	97.75%	96.63%	97.75%	97.38%
Precision	FCM	83.33%	87.88%	82.61%	84.61%
	EwFCM	94.92%	98.48%	93.88%	95.76%
Recall	FCM	93.22%	81.69%	79.17%	84.69%
	EwFCM	100.00%	92.86%	95.83%	96.23%
F_1 score	FCM	88.00%	84.67%	80.85%	84.65%
	EwFCM	97.39%	95.59%	94.85%	95.94%

Table 11.2   Data set wine - FCM versus EwFCM: iterations and running time.

	Iterations	Running time (s)
FCM	12	0.19
EwFCM	10	0.14

The *Dictionary of Emotional Words* was built by considering these height emotional categories as root terms and populated by words encountered in the text with similar emotional meaning.

Table 11.3 shows the classification measures obtained executing FCM and comparing the classification results with the ones in the data set. The accuracy, precision, recall and F1-score are evaluated for each emotional category, as reported in the table. Then, for each measure, the average value (*Mean value*) calculated on all the emotional categories is given in the last column.

Similar evaluation has been reported in Table 11.4 for the EwFCM algorithm. It turns out that the performance results of EwFCM are better than FCM: for all emotional categories, the measures values are higher than those assessed on FCM, confirming than the entropy measure allows us to control the random initialization of the cluster centers and improve the clustering results.

Analyzing the average values, let us notice that the differences between the value of each performance measure obtained by running EwFCM and the correspondent one obtained by running FCM is equal to about 8%, for all the performance metrics. Finally, a comparison between FCM and EwFCM, in terms of number of iterations and running time is shown in Table 11.5.

These results show that the EwFCM-based classification of documents into emotional categories overwhelms the FCM performance, yielding better classification accuracy, precision and running time.

Table 11.3 Twitter data set - FCM performance.

Measure	Emotion category								Mean value
	Open	Happy	Alive	Good	Love	Interested	Positive	Strong	
Accuracy	50.86%	50.79%	50.79%	49.53%	49.14%	47.09%	47.30%	47.47%	49.12%
Precision	41.50%	42.16%	42.12%	40.32%	40.65%	38.04%	38.30%	38.17%	40.16%
Recall	43.84%	41.67%	39.53%	40.47%	43.84%	38.34%	40.47%	36.67%	40.60%
F1-score	42.64%	41.91%	40.79%	40.40%	42.19%	38.19%	39.36%	37.40%	40.36%

Table 11.4 EwFCM results on the data set built on Twitter.

Measure	Emotion category								Mean value
	Open	Happy	Alive	Good	Love	Interested	Positive	Strong	
Accuracy	61.83%	59.06%	59.06%	60.27%	59.81%	54.76%	56.65%	55.20%	58.33%
Precision	50.78%	49.42%	49.59%	50.34%	48.55%	45.51%	47.33%	45.57%	48.39%
Recall	51.98%	49.02%	49.17%	48.41%	51.53%	46.27%	48.60%	44.85%	48.73%
F1-score	51.37%	49.22%	49.38%	49.36%	50.00%	45.89%	47.96%	45.21%	48.55%

Table 11.5   Twitter data set — FCM versus EwFCM: iterations and running time.

	Iterations	Running time (s)
FCM	19	146.13
EwFCM	14	124.77

11.7 Conclusion

This paper proposes an emotion classification of social stream messages to detect the predominant emotions expressed by web users from a social trend. Although several approaches for emotion classification exist in the literature, the entropy-weighed FCM highlights that the fuzzy entropy measure allows for accelerated convergence when compared to the classical FCM algorithm, since, by exploiting the data distribution in the n-dimensional space, it overcomes the sensitivity of the FCM to random cluster initialization.

The comparative analysis performed on both the Wine data set of the UCI Machine Learning repository and our Twitter data set shows that the classification, performed using an entropy-based FCM algorithm, overwhelms the performance of the classical FCM, and highlights that the proposed method converges faster (i.e., lower iteration number), with better classification performance, as evidenced by the metrics such as accuracy, precision, and F1-score.

References

[1] S. Loomba, A. de Figueiredo, S.J. Piatek, et al., Measuring the impact of COVID-19 vaccine misinformation on vaccination intent in the UK and USA, Nat. Human Behav. 5 (2021) 337–348.

[2] E. D'Andrea, P. Ducange, A. Bechini, A. Renda, F. Marcelloni, Monitoring the public opinion about the vaccination topic from tweets analysis, Expert Syst. Appl. 116 (2019) 209–226, https://doi.org/10.1016/j.eswa.2018.09.009. ISSN 0957-4174.

[3] J. Islam, R.E. Mercer, L. Xiao, Multi-channel convolutional neural network for twitter emotion and sentiment recognition, in: Conference of the North American Chapter of the Association for Computational Linguistics: Human Language Technologies, Volume 1, Association for Computational Linguistics, 2019, pp. 1355–1365, https://doi.org/10.18653/v1/N19-1137.

[4] L. Yue, W. Chen, X. Li, W. Zuo, M. Yin, A Survey of Sentiment Analysis in Social Media, Knowl Inf Syst, 2018, https://doi.org/10.1007/s10115-018-1236-4.

[5] V. Gupta, S. Gurpreet, S. Lehal, A survey of text mining techniques and applications, J. Emerg. Technol. Web Intell. 1 (1) (2009) 60–76.

[6] B. Pang, L. Lee, S. Vaithyanathan, Thumbs up? Sentiment classification using machine learning techniques, in: Proceedings of the Conference on Empirical Methods in Natural Language Processing, Philadelphia, Pennsylvania, USA, 2002, pp. 79–86.

[7] S.M. Mohammad, P. Sobhani, S. Kiritchenko, Stance and sentiment in tweets, ACM Trans. Internet Technol. 17 (3) (2017).

[8] H. Kanayama, T. Nasukawa, Textual demand analysis: detection of users' wants and needs from opinions, in: 22nd International Conference on Computational Linguistics − Volume 1 (COLING '08), Association for Computational Linguistics, USA, 2008, pp. 409−416.

[9] D. Wang, S. Zhu, T. Li, Sumview: a web-based engine for summarizing product reviews and customer opinions, Expert Syst. Appl. 40 (1) (January 2013) 27−33.

[10] X. Fu, W. Liu, Y. Xu, L. Cui, Combine HowNet lexicon to train phrase recursive autoencoder for sentence-level sentiment analysis, Neurocomputing 241 (2017) 18−27.

[11] S. Kiritchenko, X. Zhu, S.M. Mohammad, Sentiment analysis of short informal texts, J. Artif. Intell. Res. 50 (March 2014) 723−762.

[12] N.F.F. da Silva, E.R. Hruschka, E.R. Hruschka, Tweet sentiment analysis with classifier ensembles, Decis. Support Syst. 66 (Oct. 2014) 170−179.

[13] Z. Jianqiang, C. Xueliang, Combining semantic and prior polarity for boosting twitter sentiment analysis, in: Proc. IEEE Int. Conf. Smart City/SocialCom/SustainCom (SmartCity), December 2015, pp. 832−837.

[14] Z. Jianqiang, Combing semantic and prior polarity features for boosting twitter sentiment analysis using ensemble learning, in: Proc. IEEE Int. Conf. Data Sci. Cyberspace (DSC), June 2016, pp. 709−714.

[15] M. Thelwall, K. Buckley, G. Paltoglou, Sentiment strength detection for the social Web, J. Am. Soc. Inf. Sci. Technol. 63 (1) (2012) 163−173.

[16] G. Paltoglou, M. Thelwall, Twitter, MySpace, Digg: unsupervised sentiment analysis in social media, ACM Trans. Intell. Syst. Technol. 3 (4) (2012) 1−19.

[17] A. Montejo-Rááez, E. Martínez-Cámara, M.T. Martíín-Valdivia, L.A. Ureña-López, A knowledge-based approach for polarity classification in Twitter, J. Assoc. Inf. Sci. Technol. 65 (2) (2014) 414−425.

[18] S. Baccianella, A. Esuli, F. Sebastiani, Sentiwordnet 3.0: an enhanced lexical resource for sentiment analysis and opinion mining, in: Proc. 17th Conf. Int. Lang. Resource Eval, 2010, pp. 2200−2204.

[19] J.D. Westaby, D.L. Pfaff, N. Redding, Psychology and social networks: a dynamic network theory perspective, Am. Psychol. 69 (3) (2014) 269−284. https://doi.org/10.1037/a0036106.

[20] S. Poria, E. Cambria, R. Bajpai, A. Hussain, A review of affective computing: from unimodal analysis to multimodal fusion, Inf. Fusion 37 (2017) 98−125. https://doi.org/10.1016/j.inffus.2017.02.003. ISSN 1566-2535.

[21] C. Strapparava, R. Mihalcea, Learning to identify emotions in text, in: Proceedings of the 2008 ACM Symposium on Applied Computing, SAC 08, ACM, New York, NY, USA, 2008, pp. 1556−1560. http://doi.acm.org/10.1145/1363686.1364052.

[22] R. Bandana, Sentiment analysis of movie reviews using heterogeneous features, in: 2nd International Conference on Electronics, Materials Engineering & Nano-Technology (IEMENTech), Kolkata, 2018, pp. 1−4. https://doi.org/10.1109/IEMENTECH.2018.8465346.

[23] L. Barbosa, J. Feng, Robust sentiment detection on Twitter from biased and noisy data, in: Proceedings of the 23rd International Conference on Computational Linguistics, COLING 2010, Beijing, China, 2010, pp. 36−44.

[24] D. Zimbra, M. Ghiassi, S. Lee, Brand-related twitter sentiment analysis using feature engineering and the dynamic architecture for artificial neural networks, in: 49th Hawaii International Conference on System Sciences (HICSS), Koloa, HI, 2016, pp. 1930−1938. https://doi.org/10.1109/HICSS.2016.244.

[25] A. Valdivia, M.V. Luzíón, F. Herrera, Neutrality in the sentiment analysis problem based on fuzzy majority, in: Proc. IEEE Int. Conf. on Fuzzy Systems, Naples, Italy, 2017, pp. 1−6.

[26] L. Zhang, S. Wang, B. Liu, Deep learning for sentiment analysis: a survey, WIREs Data Min. Knowl, Discov. 8 (2018).

[27] J. Pennington, R. Socher, C. Manning Glove: Global vectors for word representation. in: Proc. Of the 2014 Conference on Empirical Methods in Natural Language Processing (EMNLP).

[28] T. Mikolov, I. Sutskever, K. Chen, G.S. Corrado, J. Dean, Distributed representations of words and phrases and their compositionality, in: Advances in Neural Information Processing Systems, 2013, pp. 3111–3119.

[29] P. Bojanowski, E. Grave, A. Joulin, T. Mikolov, Enriching word vectors with subword information, Trans. Assoc. Comput. Linguist. 5 (2017) 135–146.

[30] J. Devlin, M.-W. Chang, K. Lee, K. Toutanova, BERT: pre-training of deep bidirectional transformers for language understanding, in: Proceedings of the 2019 Conference of the North American Chapter of the Association for Computational Linguistics: Human Language Technologies vol. 1, 2019, pp. 4171–4186. https://doi.org/10.18653/v1/N19-1423.

[31] M. Arora, V. Kansal, Character level embedding with deep convolutional neural network for text normalization of unstructured data for Twitter sentiment analysis, Soc. Netw. Anal. Min. 9 (1) (2019) 12.

[32] D. Xu, Z. Tian, R. Lai, X. Kong, Z. Tan, W. Shi, Deep learning based emotion analysis of microblog texts, Inf. Fusion 64 (2020) 1–11. https://doi.org/10.1016/j.inffus.2020.06.002.

[33] E. Fersini, Chapter 6: sentiment analysis in social networks: a machine learning perspective, in: F.A. Pozzi, E. Fersini, E. Messina, B. Liu (Eds.), Sentiment Analysis in Social Networks, Morgan Kaufmann, 2017, p. 284. https://doi.org/10.1016/C2015-0-01864-0.

[34] I. Triguero, S. García, F. Herrera, Self-labeled techniques for semi-supervised learning: taxonomy, software and empirical study, Knowl. Inf. Syst. 42 (2013) 245–284.

[35] R. Plutchik, A General Psychoevolutionary Theory of Emotion, Academic press, New York, 1980, pp. 3–33.

[36] J.A. Russell, A circumplex model of affect, J. Pers. Soc. Psychol. 39 (1980) 1161–1178. https://doi.org/10.1037/h0077714.

[37] P. Ekman, Are there basic emotions? Psychol. Rev. 99 (3) (1992) 550–553.

[38] P. Ekman, Facial expression and emotion, Am. Psychol. 48 (4) (1993) 384–392.

[39] H. Scholsberg, Three dimensions of emotions, Psychol. Rev. 61 (1954) 81–88. https://doi.org/10.1037/h0054570.

[40] E. Cambria, A. Hussain, C. Havasi, C. Eckl, Sentic computing: exploitation of common sense for the development of emotion-sensitive systems, in: Development of Multimodal Interfaces: Active Listening and Synchrony, Lecture Notes in Computer Science 5967, 2010, pp. 148–156. Springer.

[41] A. De Luca, S. Termini, A definition of non-probabilistic entropy in the setting of fuzzy sets theory, Inf. Control 20 (1972) 301–312.

[42] A. De Luca, S. Termini, Entropy and energy measures of fuzzy sets. in: M.M. Gupta, R.K. Ragade, R.R. Yager (Ed.), Advances in Fuzzy Set Theory and Applications, North-Holland, Amsterdam, The Netherlands.

[43] J.C. Bezdek, Object Recognition with Fuzzy Objective Function Algorithms, Kluwer Academic Publishers, Norwell, MA, USA, 1981, p. 256. https://doi.org/10.1007/978-1-4757-0450-1.

[44] J.C. Bezdek, R. Ehrlich, W. Full, The Fuzzy C-Means clustering algorithm, Comput. Geosci. 10 (2–3) (1984) 191–203.

[45] T. Velmurugan, T. Santhanam, A survey of partition based clustering algorithms in data mining: an experimental approach, Inf. Technol. J. 10 (3) (2011) 478–484.

[46] S. Gosh, S.K. Dubey, Comparative analysis of k-means and fuzzy c means algorithms, Int. J. Adv. Comput. Sci. Appl. 4 (4) (2013) 35–39.

[47] S. Sivarathri, A. Govardhan, An experiments on hypothesis fuzzy K-means is better than K-means for clustering, Int. J. Data Min. Knowl. Manag. Process 4 (5) (2014) 21–34.

[48] B. Cardone, F. Di Martino, A novel fuzzy entropy-based method to improve the performance of the fuzzy C-means algorithm, Electronics 9 (2020) 554.

[49] F. Di Martino, S. Senatore, S. Sessa, A Lightweight Clustering-Based approach to discover different emotional shades from social message streams, Int. J. Intell. Syst. 34 (7) (2019) 1505–1523.

Fundamental optimization methods for machine learning

Ranjana Dwivedi, Vinay Kumar Srivastava

DEPARTMENT OF ELECTRONICS AND COMMUNICATION ENGINEERING, MOTILAL NEHRU NATIONAL INSTITUTE OF TECHNOLOGY ALLAHABAD, PRAYAGRAJ, UTTAR PRADESH, INDIA

12.1 Introduction

With the rapid growth of machine learning, its popularity among researchers exponentially increases. It plays an important role in numerous applications. Speech recognition, recommendation platforms, image classification, image recognition, search engines, etc. are the several applications of machine learning. One of the main pillar of machine learning is optimization. In case of machine-learning models, optimization involves numerical or analytical computation of parameters in respect of an objective function. For a given learning problem, optimal solution of parameter is tried to achieve through optimization. Success of several optimization methods in machine learning has attracted great number of practitioners and researchers to work on even more challenging machine-learning problems. Based on gradient information, classical optimization method can be classified into three categories, namely, first-order methods, high-order methods, and derivative-free methods.

First-order methods are widely used optimization method particularly Stochastic Gradient Descent (SGD) method [1]. Variants of SGD are also popularly used method. First-order methods are easy to implement and this method also has low computational complexity. This method provides computation with low accuracy and it has slow convergence rate to reach the optimal solution. Various gradient descent (GD) algorithms are discussed here along with their disadvantages and advantages. To overcome the shortcomings of first-order methods, high-order optimization methods [2] attracts attention of researchers. High-order methods captures curvature information along with gradients of objective function. Curvature information helps it to converge faster. Storing the inverse Hessian matrix is one of the main issue of high-order methods. Several algorithms are proposed to solve this issue by approximating the Hessian matrix. Hessian-free Newton method, quasi-Newton method, Gauss-Newton method, and natural gradients are explained here. For the objective functions whose derivative is difficult to compute or its derivative do not exist, derivative-free optimization methods

Statistical Modeling in Machine Learning. https://doi.org/10.1016/B978-0-323-91776-6.00005-1

[3,4] are used to find the optimal solution. Methods for convex objective and methods for stochastic optimization is discussed here. Along with the fundamentals of these optimization methods, challenges for optimization method is also presented.

12.2 First-order optimization methods

One of the most common technique to solve the convex optimization problems is first-order optimization method. In the field of machine learning, imaging, control theory and signal processing, this optimization method primarily based on GD. This method provides low computational complexity with low accuracy. This method is suitable when high accuracy is not crucial. Here the chapter introduce GD of first-order method. Variants of GD along with several algorithms which are based on GD are also introduced.

12.2.1 Gradient descent

In 1847, Augustin-Louis Cauchy proposed the method of GD (also known as Cauchy's method or steepest descent) for unconstrained optimization of a differentiable objective function. It is a fundamental derivative based iterative method. GD method is among the most common algorithm to do optimization. It is one of the popular method to optimize neural networks in machine learning and deep learning. Mostly deep-learning library have numerous algorithms implementation to optimize GD. GD method is used while training a machine learning model. GD method minimizes the objective function $f(\theta)$ in respect of parameter θ. Parameters are updated iteratively in the negative direction of gradient of objective function $\nabla_\theta f(\theta)$. Updating parameters is required to progressively converge to the optimal solution corresponding to objective function [1]. For each iteration, step size is determined by the learning rate η and to reach the (local) minimum value of objective function, η stimuli the number of iterations. The direction of negative gradient created by objective function decreases until the minimum value is obtained.

For a simple function f which is a differentiable function, GD method uses its gradient along with the function itself. Termination criteria will be

$$|d\theta| = |\theta(l+1) - \theta(l)| = |\alpha(l)d(l)| < \varepsilon \tag{12.1}$$

with search direction $d(l) = -\nabla_\theta f(\theta)$ and starting point $\theta(0) \in \Re^n$ at $l = 0$.
Step size α is given by

$$\alpha(l) = \underset{\alpha \in \Re^+}{\arg\min} f(\theta(l) - \alpha(l-1)\nabla_\theta f(\theta) \tag{12.2}$$

is calculated while ensuring $f(\theta(l+1)) < f(\theta(l))$. Position will be updated as,

$$\theta(l+1) = \theta(l) - \left(\underset{\alpha \in \Re^+}{\arg\min} f(\theta(l) - \alpha\nabla_\theta f(\theta)) \right) \nabla_\theta f(\theta) \tag{12.3}$$

GD converges to a local minimum by using optimum $\alpha(l)$.

In machine learning, gradient i.e., the slope of function measures the change in all weights w.r.t. change in error. For a linear regression model, suppose there are N number of training sample, M number of input features, y is the target output. Loss function, $L(\theta)$, is to be minimized with

$$L(\theta) = \frac{1}{2N} \sum_{l=1}^{N} (y(l) - f(\theta(x(l))))^2 \tag{12.4}$$

where $f(\theta(x)) = \sum_{k=1}^{M} \theta(k)x(k)$. Gradient of loss function corresponding to parameter $\theta(k)$ is

$$\frac{\partial L(\theta)}{\partial \theta_k} = \frac{-1}{N} \sum_{l=1}^{N} (y(l) - f(\theta(x(l)))) x_k^l \tag{12.5}$$

Parameter $\theta(k)$ is updated in the direction of negative gradient descent,

$$\theta_{k+1} = \theta_k + \eta \frac{1}{N} \sum_{l=1}^{N} (y(l) - f(\theta(x(l)))) x_k^l \tag{12.6}$$

While determining the direction of movement, GD method utilizes only the function's first-order derivative. For convex objective function, global optimum solution is obtained using this method. GD method usually converges at slower speed and when reaching at global optimum, it converges at much slower speed.

12.2.2 Gradient descent variants

Depending on the amount of data that is used to compute objective function gradient, GD method can be categorized into three variants. GD methods have low computational cost with low accuracy. Batch gradient descent, SGD, and mini-batch gradient descent are the three variants of GD.

12.2.2.1 Batch gradient descent

Batch gradient descent sometimes often called as Vanilla gradient descent. It computes error for an example only after a training epoch. This method calculates the gradient of entire training data set to perform just one update. Gradient of cost function for whole data set is updated with respect to parameter θ.

$$\theta(l+1) = \theta(l) - \eta \nabla_\theta f(\theta) \tag{12.7}$$

As batch gradient descent method computes gradient of entire data set for each update, this method can be very slow. One of its advantage is computational efficiency as it yields a steady error gradient and gives a stable convergence. Sometimes, steady error gradient may result in that state of convergence which is not the optimal that model can accomplish. Batch gradient descent method also does not allow online update. For convex error surface, batch gradient descent method converges to global minimum. For nonconvex surface, this method converges to local minimum. For the N number of samples and M number of features, the computational complexity per iteration of batch gradient descent method will be O(NM).

12.2.2.2   Stochastic gradient descent

In order to overcome the problems of high computational complexity per iteration of batch gradient method, SGD method was proposed. Instead of directly computing the gradient of objective function, SGD method computes the gradient of one random sample. Stochastic gradient provides an impartial estimates of objective function gradient [5]. SGD performs a parameter update of single training example $u^{(i)}$ per iteration instead of objective function gradient of all the training examples.

$$\theta(l+1) = \theta(l) - \eta \nabla_\theta f(\theta)\big(\theta(l); u^{(i)}\big) \tag{12.8}$$

Path taken by SGD method to reach the minima is typically noisy as compared to batch gradient descent method because in SGD method, only one random sample is chosen for updating the gradient per iteration. Fig. 12.1A and B shows the usual path taken to reach the minima by SGD method and by batch gradient descent method. It is clear from the figure that SGD method path to reach the minima is much noisy than batch gradient descent method, but it does not matter. Only criteria are to reach the minima with considerably shorter time is considered. As SGD method is noisier than typical gradient descent, to reach the minima SGD method took higher number of iteration because of its randomness nature to choose the sample. Although SGD method requires higher number of iteration as compared to typical gradient descent method but still SGD method is computationally much less expensive than gradient descent method. Cost of SGD method is independent of number of samples and this method can also achieve sublinear convergence speed [6].

For large data sets, SGD method significantly accelerates the calculation by removing the computational redundancy as it performs one update at a time whereas batch gradient descent performs redundant computation. One major problem of batch gradient descent method is it does not allow online update, SGD method overcomes this disadvantage. Computational complexity per iteration for SGD method is O(M) because this method exploits single sample per iteration, where M is the number of features. In

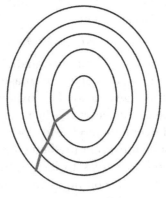

FIGURE 12.1 (A) Path taken by batch gradient descent.

FIGURE 12.1 (B) Path taken by stochastic gradient descent.

case of batch gradient descent, deterministic gradient may lead objective function to converge a local minimum while SGD method have better chance to achieve a global optimal solution. SGD method's fluctuations facilitates it to reach to a possibly better local minimum. But this fluctuation always exists in SGD method, this ultimately complicates convergence to reach global minima. In the existing literature, it has been presented that on slowly decreasing the learning rate, SGD method converges at same speed as batch gradient descent method.

Choice of an appropriate learning rate is still a challenging task while applying SGD method of optimization. A too large learning rate obstructs the convergence of objective function and loss function oscillates to and fro across the ravine. A too small learning rate results in slow convergence speed. One of the solution of the problem of learning rate is to setup a predefined index of learning rates and during the learning process, learning rates are adjusted accordingly [7]. These predefined list of learning rates are needed to be defined in advanced based on the characteristic of data set. Same learning rate for all parameters are not always suitable to use. For less frequently occurring features, a high learning rate is estimated [8]. Objective function in SGD method is trapped in infinite number of local minimum and to avoid this problem is a challenging task. Several works suggest that this problem comes from "saddle point," not from the local minimum values [9]. Saddle points has gradient values as zero in entire directions. As it has positive slope in a specific direction and also has negative slope in opposite direction. It is necessary for SGD to escape from these saddle points. Various works have been developed to escape from these points [10].

12.2.2.3 *Mini-batch gradient descent*

In SGD method, noise causes as a result of random selection causes the gradient descent oscillating. Gradients variance is large and also the direction of movement is biased in SGD method. Mini-batch gradient descent technique was offered [5] to compromise between both methods. Mini-batch gradient descent method updates the parameter per iteration by using n number of samples at a time. n can vary depending on the applications and it ranges between 50 and 256 [1].

$$\theta(l+1) = \theta(l) - \eta \nabla_\theta f(\theta)\big(\theta(l); u^{(ii+n)}\big) \tag{12.9}$$

Mini-batch gradient descent method lessens the gradients variants which led to further steady convergence. This method can also make the use of optimized matrix while calculating the gradients and it makes this method very efficient. Computational complexity is less in mini-batch gradient descent method. For a large neural networks with redundant training sets, it is best to use this method.

12.2.3   Gradient descent optimization algorithms

There are many challenges to SGD method such as how to adjust learning rate, how to escape from being trapped to saddle points during search process and how to increase the convergence rate. To improve SGD, several algorithms are widely used to deal these challenges.

12.2.3.1   Momentum

Momentum method [11] is based on the concept of mechanics, which simulates the inertia of objects. This method helps SGD to accelerate in the appropriate direction and to dampen the oscillations in the direction of high curvature by combining gradients with opposite signs. Momentum stores the impact of preceding updated direction on the subsequent iteration. It makes direction to keep going as in previous direction. Momentum method introduces the variable v which symbolizes the direction and speed of parameter's movement. It accelerates SGD in relevant direction by considering a momentum term γ vary between 0 and 1 along with gradient descent. Momentum term gets multiplied with previous update and provides a friction factor. Mathematically, v is formulated as

$$v_t = \gamma v_{t-1} - \eta \nabla_\theta f(\theta) \tag{12.10}$$

Term γv_{t-1} gives the friction factor, it helps in further increment of velocity as γ is less than 1. Weight change can be expressed as

$$\Delta w_t = \gamma \Delta w_{t-1} - \eta \nabla_\theta f(\theta) \tag{12.11}$$

Parameter will be updated as

$$\theta(l+1) = \theta(l) - v_t \tag{12.12}$$

A suitable momentum term plays an important part in speed up the convergence with low learning rate. Momentum term increases for those gradients which are parallel to previous direction and thus previous velocity can speed up the search. For dimensions whose gradients change direction, momentum term reduces updates and decelerate the search speed. During the training process, it is necessary to get away from the local minimum so that we can gain faster convergence [12]. Momentum term in momentum method also plays a significant part in decreasing the oscillations of convergence.

Selecting a proper momentum term is a challenging task. With small momentum term, it is difficult to achieve faster convergence speed. For large momentum term, recent point might leap out of optimal value. It is verified in various studies that appropriate value of momentum term is 0.9 [1].

12.2.3.2 Nesterov accelerated gradient

In momentum method, gradients at current position are calculated first and then it jumps to the updated accumulated gradient's direction. Nesterov Accelerated Gradient (NAG) [13] method first jumps to the previous accumulated gradient's direction and after that it calculates the gradient and make a correction. Instead of calculating the objective function gradient w.r.t. current parameter θ, in NAG method gradient is calculated with respect to future position of parameter $\theta - \gamma v_{t-1}$. In other words, instead of computing gradient of the current position, gradient of future position is calculated in NAG method. First step in NAG method is to calculate the approximate next position of parameters.

$$\widehat{\theta} = \theta - \gamma v_{t-1} \tag{12.13}$$

Velocity is updated by computing gradients with respect to approximate future position of parameters.

$$v_t = \gamma v_{t-1} - \eta \nabla_\theta f(\theta - \gamma v_{t-1}) \tag{12.14}$$

Now parameter is updated as

$$\theta(l+1) = \theta(l) - v_t \tag{12.15}$$

NAG method have more gradient information as compared to momentum method. One of the main issue is in what manner to decide the learning rate. Oscillations occur when search point is close to optimal point. Decay factor d in learning rate is used in SGD momentum method. With decay factor, learning rate decreases as the number of iteration goes on increasing [14]. Learning rate at i-th iteration is given as,

$$\eta_i = \frac{\eta_0}{1 + d \cdot i} \tag{12.16}$$

where η_0 is initial learning rate and d is decay factor of learning rate and its value ranges between 0 and 1. Learning rate decays faster when d has large value and vice-versa. Adjusting the learning rate manually influences the SGD method. Choosing an appropriate learning rate is a tricky problem [15]. To deal with optimization problems in DNN, various adaptive methods were introduced in which learning rate adjusted automatically. These methods have faster convergence rate and are free of parameter adjustments.

12.2.3.3 AdaGrad

AdaGrad [8] method adapts the learning rate dynamically based on accumulated previous gradient. To deal with sparse data, AdaGrad method is well suited. For frequent parameters, AdaGrad perform smaller updates and for infrequent parameters it performs larger update. In traditional gradient method, learning rate is fixed for updating all parameters, while in AdaGrad method, learning rate is not fixed. It makes use of an individual learning rate for all parameters at each iteration. Gradient of objective function with respect to parameter θ at i-th iteration is given by,

$$g_i = \nabla_\theta f(\theta_i) \tag{12.17}$$

AdaGrad modifies learning rate at each iteration for all parameters based on accumulated previous gradients.

$$\theta_{i+1} = \theta_i - \eta \frac{g_i}{G_i} \tag{12.18}$$

where $G_i = \sqrt{\sum_{k=1}^{i} (g_k)^2 + \varepsilon}$. First term in denominator denotes the accumulated previous gradients. ε is a smoothing term. One of the key benefit of AdaGrad technique is it eliminate the requirement to manually adjust the learning rate. Two main issues of AdaGrad method are:

(i) Although it adapts learning rate dynamically but it still requires to set global learning rate manually.

(ii) Parameter updates have squared gradients in denominator and as it is a positive term, it gets accumulated in the course of training. This bring about learning rate to reduce and ultimately results in an ineffective parameter update.

12.2.3.4   AdaDelta

To solve the issue of monotonically decreasing learning rate, AdaDelta [16] was proposed. Instead of accumulating all previous squared gradients, it focuses on only accumulated previous gradients in a window over a period. Exponentially declining average of accumulated previous gradients are defined as

$$G_i = \sqrt{\beta G_{i-1} + (1-\beta) g_i^2} \tag{12.19}$$

where β as a decay factor similar to momentum term and β is set to around 0.9 [1]. Declining average rely upon only on preceding average and recent gradient.

12.2.3.5   Adaptive moment estimation

Adaptive Moment Estimation (Adam) [15] is an alternative technique which offers adjustable learning rate for each parameters. It incorporates the AdaDelta and momentum approaches. Adam method stores exponentially declining average of accumulated squares of previous gradients G_i like AdaDelta along with exponentially declining average of preceding gradients like momentum method. The estimate of first moment (i.e., mean) of gradient is expressed as m_i. It is expressed as

$$m_i = \beta_1 m_{i-1} + (1-\beta_1) g_i \tag{12.20}$$

G_i is the estimate of the second moment (i.e., variance) of the gradient. Mathematically, it is defined as

$$G_i = \sqrt{\beta_2 G_{i-1} + (1-\beta_2) g_i^2} \tag{12.21}$$

where β_1 and β_2 are the exponential declining rates. It is observed that when exponential declining rates are small (i.e., their value close to 1), during initial iterations method is biased toward zero. To counter these biases, first and second moments with bias corrected are computed.

$$\widetilde{m}_i = \frac{m_i}{1 - \beta_1^i} \tag{12.22}$$

$$\widetilde{G}_i = \frac{G_i}{\sqrt{1 - \beta_2^i}} \tag{12.23}$$

These bias-corrected moments are used to update the parameter. Update formula for Adam method is

$$\theta_{i+1} = \theta_i - \eta \frac{\widetilde{m}_i}{\widetilde{G}_i} \tag{12.24}$$

$$\theta_{i+1} = \theta_i - \eta \frac{\sqrt{1 - \beta_2^i}}{(1 - \beta_1^i)} \frac{m_i}{G_i + \varepsilon} \tag{12.25}$$

Default values of β_1 is 0.9, β_2 is 0.999 and for ε is 10^{-8}.

12.2.3.6   AdaMax
G_i factor in the Adam is oppositely proportional to the l_2 norm of preceding gradients and recent gradients.

$$G_i = \sqrt{\beta_2 G_{i-1} + (1 - \beta_2)|g_i|^2} \tag{12.26}$$

Infinity norm i.e., l_∞ generally exhibits stable behavior. In AdaMax [15] method, G_i with infinity norm converges to more stable value. Infinity norm constrained G_i^∞ is expressed as,

$$G_i^\infty = \sqrt{\beta_2^\infty G_{i-1} + (1 - \beta_2^\infty)|g_i|^\infty} \tag{12.27}$$

AdaMax parameter will be updated using expression,

$$\theta_{i+1} = \theta_i - \eta \frac{\widetilde{m}_i}{G_i} \tag{12.28}$$

Default values of η, β_1, and β_2 are 0.002, 0.9, and 0.999, respectively.

12.2.3.7   Nesterov accelerated adaptive moment estimation
From above, it is clear that Adam can be regarded as grouping of AdaDelta and Momentum method. AdaDelta accounts for the exponentially declining average of preceding squared gradients and momentum contributes the exponentially declining average of preceding gradients. Nesterov Accelerated Adaptive Moment Estimation (Nadam) [17] is the combination of Adam method and NAG method. Look-ahead momentum vector is applied straight to update the present parameter of NAG and bias-

corrected estimate of current momentum is also used for parameter update. Nadam parameter update can be expressed as

$$\theta_{i+1} = \theta_i - \frac{\eta}{\widetilde{G}_i} \left(\beta_i \widetilde{m}_i + \frac{(1-\beta_1) \cdot g_i}{1-\beta_1^i} \right) \tag{12.29}$$

SGD method is quite popular method of optimization but this method has some issues also. One of the issue is sublinear convergence rate of SGD method. Variance of gradients is very large in SGD method. To solve this issue of convergence rate and large variance, there are some variance reduction methods.

Stochastic average gradient [18] method to improve the convergence rate is one of the variance reduction method. In this method, for each update only gradient of one sample needs to be calculated instead of calculating gradients of all samples.

Updated parameter per iteration will be

$$\theta_{k+1} = \theta_k - \frac{\eta}{N} d \tag{12.30}$$

where $d = d - \nabla_\theta f(\theta_i) + \nabla_\theta f(\theta_{k-1})$ and $\nabla_\theta f(\theta_i)$ is the previous accumulated gradient. Memory requirement for this method is much higher than as compared to traditional SGD method.

Stochastic average gradient method achieves linear convergence rate. One of the requirement of this method is that objective function must be smooth and should have convex in nature. Thus for the optimization problems which in nonconvex in nature, Stochastic Variance Reduction Gradient (SVRG) [6] is proposed. Instead of calculating gradients of all samples in each iteration, SVRG computes gradients of sample for every t iteration.

Average gradients of all samples over the interval t is given as

$$\mu = \frac{1}{N} \sum_{i=1}^{N} \nabla_\theta f(\theta_i) \tag{12.31}$$

Next update will be

$$\theta_{k+1} = \theta_k - \eta (\nabla_\theta f(\theta_k) - \nabla_\theta f(\theta_i) + \mu) \tag{12.32}$$

SVRG introduces a concept known as variance reduction in which constant upper bound on gradient variance are assumed. To achieve linear convergence rate, variance of gradients needs to be continuously decreased. SVRG requires less memory because it not store all gradients in memory as compared to stochastic average gradient method.

12.3 High-order optimization method

First-order method of optimization such as SGD and its variant are very popular in machine learning because of their simplicity of applications and low computation cost per iteration. But it has some drawbacks also such as slow convergence rate, low accuracy, sensitivity to learning rate tuning, trouble in get away from saddle points. To overcome these drawbacks, there have been latest attention toward second-order method. Second-order method use gradient information along with capturing curvature information. When second-order method is used in context of machine-learning applications, it provides faster convergence rate. This method also gives stable learning rate tuning. Here the chapter discuss few second-order methods which includes Quasi-Newton method, Gauss-Newton method, and Hessian-free Newton method.

12.3.1 Hessian-free Newton method

For a given unconstrained and smooth convex optimization problem, Newton's method [19] updated direction is based on second-order Taylor series approximation. Newton's method iteration is obtained by

$$\theta_{k+1} = \theta_k - \eta \left(\nabla^2_\theta f(\theta_k)\right)^{-1} \nabla_\theta f(\theta_k) \text{ for k} = 1, 2, 3, \dots. \tag{12.33}$$

where $\left(\nabla^2_\theta f(\theta_k)\right)$ is the Hessian matrix of $f(\theta)$ at $\theta = \theta_k$. Using second-order Taylor series approximation, Objective function can be minimizing as

$$f(\theta) = f(\theta_k) + \nabla_\theta f(\theta_k)^T (\theta - \theta_k) + \frac{1}{2}(\theta - \theta_k)^T \nabla^2_\theta f(\theta_k)(\theta - \theta_k) \tag{12.34}$$

Assuming Hessian is positive definite. Newton's method attains quadratic convergence rate. This technique applies consecutive local rescaling at each iteration. Instead of solving Newton's method employing matrix factorization technique, one can figure out it inexactly using an iterative method. Many iterative linear systems methods require only Hessian-vector product instead of Hessian itself. These type of methods which solves accurate enough are known as Hessian-free Newton method. This method guarantees a faster rate of convergence. With the dimension d and smooth objective function $f(\theta)$, cost for evaluating $\nabla^2_\theta f(\theta)$ is small multiple of computing cost of $\nabla_\theta f(\theta)$. Hessian-free Newton method requires $O(d^2)$ storage without forming the Hessian. While this method save computation cost but it comes at the expense of storage requirement. To yield an effective iteration, it is not necessary Hessian matrix to be as precise as the gradient. In context of large scale data, iteration is much forbearing to noise in case of Hessian-free Newton method as compared to gradient method.

A smaller sample is employed for defining the Hessian in subsample Hessian -ree Newton method. For a subsample size of a_k, stochastic Hessian estimate will be

$$\nabla^2_\theta f_{a_k}(\theta) = \frac{1}{|a_k|} \sum \nabla^2_\theta f(\theta) \tag{12.35}$$

If subsample size of a_k is chosen large, then curvature information captured is productive. Hessian-vector products capture the curvature information. For small subsample size of a_k, cost of each product can be decreased considerably which further reduce the cost per iteration. If achieved properly, Hessian subsampling is robust and efficient [15]. Step computation cost in subsampling Hessian-free Newton method for *max* total number of iteration can be computed as,

$$cost = (max \times factor \times g_{cost}) + g_{cost} \qquad (12.36)$$

where $factor \times g_{cost}$ is the cost of one Hessian-vector product. g_{cost} is the cost of computing gradient estimate $\nabla_\theta f_{a_k}(\theta)$. In Hessian subsampling method, *factor* should be chosen very small such that $max \times factor \approx 1$. This leads to the computation cost per iteration comparative to that of SGD method. It is assumed that chosen subsample size is large enough such that by considering these subsamples, Hessian estimate is sensible. Hessian-free Newton method are usually practiced for solving the nonconvex optimization problems. To ensure positive definite Hessian approximation, Gauss-Newton approximation of Hessian is employed.

$$G_{a_k}(\theta) = \frac{1}{|a_k|} \sum J_h(\theta)^T H_l(\theta) J_h(\theta) \qquad (12.37)$$

where J_h captures the stochastic gradient information and H_l captures loss function second-order information. For analysis the geometry of minimizer Objective function, several numerical tests have been designed and presence of negative curvature is proved. When compared to first-order methods, Newton's method have more memory requirement as for $(n \times n)$ Hessian it requires $O(n^2)$ of storage per iteration while gradient method requires $O(n)$ of storage per iteration for n-dimensional gradient. Newton's method is more sensitive to numerical errors but gradient descent methods are robust against these errors.

12.3.2   Quasi-Newton method

Quasi-Newton method (QNM) is one of the important developments in the field of nonlinear optimization. This method is used when Newton method are difficult to use or when computing Hessian is too expensive per iteration. This method approximates Hessian by using only the gradient information. QNM can be applicable for both convex and nonconvex problems. QNM iteration for minimizing the objective function, which is twice differentiable, is

$$\theta_{k+1} = \theta_k - \eta B_k \nabla_\theta f(\theta_k) \qquad (12.38)$$

where B_k is the symmetric positive definite approximation of $\left(\nabla_\theta^2 f(\theta_k)\right)^{-1}$. Instead of updating each iteration by computing second-order derivative computation, in QNM sequence B_k updated dynamically for each iteration. As B_k already contains information about Hessian, B_k can be updated by using suitable matrix.

$$\nabla_\theta f(\theta_{k+1}) = \nabla_\theta f(\theta_k) + B_{k+1}(\theta_{k+1} - \theta_k) \qquad (12.39)$$

Suppose,

$$y = \nabla_\theta f(\theta_{k+1}) - \nabla_\theta f(\theta_k) \qquad (12.40)$$

and

$$s = \theta_{k+1} - \theta_k \tag{12.41}$$

this gives us Secant equation,

$$B_{k+1} \cdot s = y \tag{12.42}$$

Different QNM such as symmetric rank one (SR1), Davidon-Fletcher-Powell (DFP), Broyden-Fletcher-Goldfarb-Shanno (BFGS), and Broyden class computes B_{k+1} form B_k differently. In case of SR1 update, Hessian approximation is obtained by

$$B_{k+1} = B_k + \frac{(y - B_k s)(y - B_k s)^T}{(y - B_k s)^T s} \tag{12.43}$$

New inverse Hessian approximation is given by expression

$$C_{k+1} = C_k + \frac{(s - C_k \cdot y)(s - C_k \cdot y)^T}{(s - C_k \cdot y)^T y} \tag{12.44}$$

In general, SR1 update method of QNM is simple to implement and has low computation cost. But it has one drawback, this method does not preserve positive definiteness of Hessian matrix approximation. To overcome this shortcoming of SR1 method, BFGS method is proposed. In BFGS method, Hessian approximation is given by

$$B_{k+1} = B_k - \frac{B_k \cdot s \cdot s^T \cdot B_k}{s^T \cdot B_k \cdot s} + \frac{y \cdot y^T}{y^T \cdot s} \tag{12.45}$$

Inverse Hessian approximation of BFGS method is calculated as

$$C_{k+1} = \left(I - \frac{s \cdot y^T}{y^T \cdot s} \right) C_k \left(I - \frac{y \cdot s^T}{y^T \cdot s} \right) + \frac{s \cdot s^T}{y^T \cdot s} \tag{12.46}$$

BFGS update preserve positive definiteness under appropriate conditions and has low computation cost. BFGS update has local superlinear rate of convergence [20] without the need to solve linear systems. However, this method has some issues also. These concerns need to tackle to have an efficient method. Even when the Hessian matrix are sparse, updated inverse Hessian approximation yields dense matrix. This problem restricts BFGS method to use for small scale and midscale data set. This problem can be solved by using limited memory scheme such as L-BFGS method [21] in which B_k not needed be formed explicitly. Individual product can be computed only using current components of series of displacement pairs (s, y)

Development of QNM from deterministic to stochastic form, iteration will take the form as

$$\theta_{k+1} = \theta_k - \eta B_k \nabla_\theta f(\theta_k, \xi_k) \tag{12.47}$$

In spite of convergence rate of stochastic iteration of QNM is always slower than sublinear [22]. But this method could be well prepared to deal with ill-conditioning. For SGD method, constant depends on conditioning of $\{\nabla_\theta^2 f(\theta_k)\}$, whereas in QNM, constant is not dependent on conditioning of Hessian on satisfying $B_k = \left(\nabla_\theta^2 f(\theta_k) \right)^{-1}$. SGD method has low computational complexity as it requires only to compute gradient of objective function $\nabla_\theta f(\theta_k, \xi_k)$ per iteration. In case of QNM, each iteration requires computation of product $B_k \nabla_\theta f(\theta_k, \xi_k)$, this computation makes it more expensive than SGD method. To address this additional per iteration cost, mini-batch gradient estimation is employed. Usually mini-batch of size between 20 and 50 is considered instead of large mini-batch size. When mini-batch gradient estimation is considered, additional per iteration cost is only marginal. In stochastic QNM, gradient $\nabla_\theta f(\theta_k, \xi_k)$ is the noisy estimate of $\nabla_\theta f(\theta_k)$. While updating C_k, it involves difference of gradient estimates y. Updating process may cause to yield poor curvature estimate. Even a single noisy gradient may remain for numerous iteration. To circumvent differencing noisy gradient estimate, one way is to choose identical sample while calculating gradient difference [23].

12.3.3   Gauss-Newton method

It is a traditional approach to solve nonlinear least squares optimization problems. Gauss-Newton method (GNM) minimize the problems in which objective function is sum of squares. One of the pros of using GNM is that it uses only first-order information to construct approximation of Hessian. Though if Hessian may be indefinite, this method guaranteed approximation of Hessian be positive semidefinite. GNM ignores second-order interaction between parameter elements, which could cause loss of curvature information. GNM aims to minimize the objective function which is sum of squares. First step is to define the objective function, which is to be minimized.

$$\|f(\theta_k, \xi_k)\|_2^2 = \sum_{i=1}^{m} f_i(\theta_k, \xi_k)^2 \tag{12.48}$$

Linearize $f(\theta_k, \xi_k)$ around θ_k,

$$f(\theta, \xi) \approx f(\theta_k, \xi_k) + \nabla_\theta f(\theta_k, \xi_k)(\theta - \theta_k) \tag{12.49}$$

Substitute affine transformation for $f(\theta, \xi)$ in least square problem.

$$f(\theta_k, \xi_k) \approx \|f(\theta_k, \xi_k) + \nabla_\theta f(\theta_k, \xi_k)(\theta - \theta_k)\|_2^2 \tag{12.50}$$

If $\nabla_\theta f(\theta_k, \xi_k)$ has full column rank, solution will be

$$\theta_{k+1} = \theta_k - \eta (\nabla_\theta f(\theta_k, \xi_k))^T (\nabla_\theta f(\theta_k, \xi_k))^{-1} (\nabla_\theta f(\theta_k, \xi_k))^T f(\theta_k, \xi_k) \tag{12.51}$$

$$\theta_{k+1} = \theta_k - \eta (\nabla_\theta f(\theta_k, \xi_k))^+ f(\theta_k, \xi_k) \tag{12.52}$$

There are some challenges in applying GNM. Gauss-Newton matrix is generally singular or nearly singular. This issue can be solved by adding a positive multiple of identity matrix in to Gauss-Newton matrix. Hessian-free Newton method and stochastic quasi-Newton method can be applicable with Gauss-Newton approximation for least squared loss functions. Scaling matrices in GNM have guaranteed positive definite. For other loss function also, GNM can be generalized [24]. While training DNNs, generalized

GNM can be applied by considering loss function and prediction function. By using this method, networks computation is performed by loss function rather than prediction function.

12.3.4 Natural gradient method

Newton's method is invariable to linear transformation of parameter θ, while natural gradient method [25] is invariable to differentiable and reversible transformations. Gradient descent algorithms are formulated in space of prediction function instead of parameter. Natural gradient descent method will move parameters quickly in the direction which has less impact of decision function. Before formulating Natural gradient method in prediction function space, geometry of prediction function space must be explained. Amari's technique [26] on information theory gives us an idea about the geometry. Space S of prediction function is a family of densities $j_\theta(x)$, where $\theta \in \Re^n$. Density $j_\theta(x)$ satisfies the normalization condition

$$\int j_\theta(x)dx = 1 \tag{12.53}$$

Derivative of density satisfies the identity for $\forall\ n > 0$,

$$\int \frac{d^n j_\theta(x)}{d\theta^n} dx = \frac{d^n}{d\theta^n} \int j_\theta(x)dx$$

$$= \frac{d^n 1}{d\theta^n}$$

$$= 0 \tag{12.54}$$

By observing Kullback-Leibler (KL) divergence, we can visualize the effect on density by adding a small quantity $\delta\theta$ to parameter.

$$D_{KL}\left(j_\theta \middle\| j_{\theta+\delta\theta}\right) = E_{j_\theta}\left(\log \frac{j_\theta(x)}{j_{\theta+\delta\theta}(x)}\right) \tag{12.55}$$

Approximating divergence with a second-order Taylor expansion,

$$D_{KL}\left(j_\theta \middle\| j_{\theta+\delta\theta}\right) \approx -\delta\theta^T E_{j_\theta}\left[\frac{\partial \log(j_\theta(x))}{\partial\theta}\right] - \frac{1}{2}\delta\theta^T E_{j_\theta}\left[\frac{\partial^2 \log(j_\theta(x))}{\partial\theta^2}\right]\delta\theta \tag{12.56}$$

From above Eq. (12.54), first term in Eq. (12.56) becomes zero.
Fisher information matrix can be defined as

$$G(\theta) = -E_{j_\theta}\left[\frac{\partial^2 \log(j_\theta(x))}{\partial\theta^2}\right] \tag{12.57}$$

Thus Eq. (12.56) becomes

$$D_{KL}\left(j_\theta \middle\| j_{\theta+\delta\theta}\right) \approx \frac{1}{2}\delta\theta^T G(\theta)\delta\theta \tag{12.58}$$

Fisher information matrix is always positive semidefinite and symmetric. Therefore, every small region of space S are similar to small region of Euclidian space. For density estimation, objective function is negative log likelihood,

$$f(\theta) = \frac{1}{k} \sum_{i=1}^{k} -\log(j_\theta(x_i)) \approx D_{KL}(P \,||\, j_\theta) + constant \tag{12.59}$$

where x_i are the independent training samples and P is unknown distribution which training samples have. Hessian matrix, $\nabla_\theta^2 f(\theta)$ will be,

$$\nabla_\theta^2 f(\theta) = - E_P \left[\frac{\partial^2 \log(j_\theta(x))}{\partial \theta^2} \right] \tag{12.60}$$

When optimality is approached, density function approaches to P distribution and thus Fisher information matrix approaches the Hessian matrix. Natural gradient method and Newton method performs in a similar manner when reaching at optimal solution. In large learning systems, numerical computation of Fisher information matrix is difficult task. To overcome this problem, subset of training examples is considered for computing Fisher information matrix [27].

12.4 Derivative-free optimization methods

In many optimizations problems, objective function, and constraints are computed by a "black box," which does not provide derivative information. Objective function computed by black box may also include some noise and calling black box every time may be expensive. These challenges of optimization problems necessitate to use derivative-free optimization (DFO). DFO approximates the objective function explicitly without approximating its derivative. Thus DFO methods are those methods which do not require derivative information for solving optimization problem. Here the chapter mainly focus on DFO methods for convex objective and methods for stochastic optimization.

12.4.1   Methods for convex objective

A function f is said to be convex if its space, S, is convex in nature. For any two point x and y in space S ($S \in \Re^n$) if it satisfies following inequality,

$$f(\alpha x + (1 - \alpha)y) \le \alpha f(x) + (1 - \alpha)f(y) \quad \forall \alpha \in [0, 1] \tag{12.61}$$

Local solution is always a global solution for convex optimization problem, whereas for nonconvex problems, there exists many local solutions. For deterministic optimization problem, objective is the objective function $f(\theta)$ minimization.

$$\underset{\theta}{minimize} \; f(\theta), \tag{12.62}$$

subjected to. $\theta \in \Re^n$

Under appropriate additional assumptions, for convex objective function $f(\theta)$ it satisfies

$$\lim_{k\to\infty} f(\theta_k) - f(\theta_*) = 0 \tag{12.63}$$

where θ_* is global minimizer and is satisfies. $f(\theta_*) \le f(\theta) \ \forall \ \theta \in \Re^n$

For unconstrained convex objective function,

$$f(\theta_k) - f(\theta_*) \le \varepsilon \tag{12.64}$$

R-linear convergence of directional direct search (DDS) method dependence reduced to $\log(\varepsilon^{-1})$ from ε [28]. In DDS method, on restricting the upper limit on step size α_k and for $u > 0$,

$$f(p_i) \le f(\theta_k) - u\alpha_k^2 \tag{12.65}$$

In case of convex stochastic optimization, task is to find the solution of stochastic problem.

$$\underset{\theta}{minimize} \ f(\theta) = E_\xi[f(\theta, \xi)] \tag{12.66}$$

Zeroth order information is referred as bandit feedback. Multiarmed bandit problems can be formulated as a sequential allocation problem. In multiarmed bandit problem [29], a gamester wants to lessen the losses incurred by drawing wrong slot machine arms out of A arms. Total length of discrete sequence is assumed to be T. At time instant k, gamester already knows the losses associated with previous instants. Gamester's loss at k-th instant is $f(\theta_k, \xi_k)$. It is assumed that expectation

$$E_\xi[f(\theta, \xi)] = f(\theta) \quad \forall \ \theta \in (1, 2, \ldots, A) \tag{12.67}$$

To minimize the expected total loss, best long run strategy is to constantly play. Over the course of T drawings, sequence of realized losses contains $\{(f(\theta_1, \xi_1), \ldots \ldots (f(\theta_T, \xi_T))\}$. Cumulative regret contains the metric of gamester's performance. Expected cumulative regret is given as

$$E_\xi[r_T(\theta_1, \ldots \ldots, \theta_T)] = E_\xi\left[\sum_{k=1}^{T} f(\theta_k, \xi_k)\right] - Tf(\theta_*) \tag{12.68}$$

where r_T is cumulative regret, and sequence of ξ_k are identically distributed and are independent to each other. Extending this problem to infinite armed bandits [30] corresponds to DFO. For infinite armed bandit problem, vector ξ defines linear function. In this linear regime, gamester incurs loss.

$$f(\theta_k, \xi_k) = \xi_k^T \theta_k \tag{12.69}$$

Expected cumulative regret is expressed as

$$E_\xi[r_T(\theta_1, \ldots \ldots, \theta_T)] = E_\xi\left[\sum_{k=1}^{T} f(\theta_k, \xi_k)\right] - \underset{\theta}{min} \ E_\xi\left[\sum_{k=1}^{T} f(\theta_k, \xi_k)\right] \tag{12.70}$$

Based on type of bandit feedback, it can be categorized into single-point bandit feedback or multipoint bandit feedback.

12.4.2 Methods for stochastic optimization

To solve the stochastic optimization problem, it is assumed that random variable ξ is independent and identically distributed. Variance of objective function is bounded as

$$E_\xi\left[(f(\theta,\xi)-f(\theta))^2\right] < \sigma^2 < \infty \tag{12.71}$$

If derivative of function exists, under certain regularity condition it follows,

$$\nabla_\theta f(\theta) = E_\xi[\nabla_\theta f(\theta,\xi)] \tag{12.72}$$

Using [31], derivative of objective function is approximated by using central differences. Derivative is approximated as

$$g(\theta_k,\mu_k,\xi_k) = \left[\frac{f\left(\theta_k+\mu_k e_n,\xi_n^+\right) - f\left(\theta_k-\mu_k e_n,\xi_n^-\right)}{2\mu_k}\right] \tag{12.73}$$

To achieve convergence, bounds are applied on objective function, step size. There are always limitations on series of step sizes. Convergence in distribution [32] is given as

$$\frac{1}{k^\gamma}(\theta_k - \theta_*) \to N(0,H) \tag{12.74}$$

where H is a covariance matrix in its elements consists of algorithm parameters and derivative of objective functions. It has been shown that $\gamma = 1/3$ [33]. Gradient estimation in common random number regimes is

$$g(\theta_k,\mu_k,\xi_k) = [\delta_c(f(.,\xi_k);\theta_k;e_n;\mu_k)] \tag{12.75}$$

To compute finite difference approximation, single realization of ξ_k is employed. By considering different realizations of ξ, $f(\theta_k)$ can be computed more accurately. These belongs to sample average approximation methods. Performance of these methods depends on sample size and accuracies used per iteration. These methods shown to achieve convergence for smooth objective for sequence of sample sizes [34]. Sample size can also be dynamically selected from iteration to balance the deterministic and stochastic errors. Stochastic error is given by,

$$\left|f(\theta_k) - \frac{1}{p_k}\sum_{i=1}^{p_k} f(\theta_k,\xi_{k,i})\right| \tag{12.76}$$

Deterministic error using first-order Taylor approximation is given as,

$$\left|f(\theta_k - \eta\nabla_\theta f(\theta_k)) - (f(\theta_k) - \eta\nabla_\theta f(\theta_k)^2)\right| \tag{12.77}$$

12.5 Optimization methods challenges and issues in machine learning

While applying optimization methods to optimize the models in machine learning, some challenges are there. Issues such as insufficient training data, complexity of machine-learning models, nonconvex objective function are very common to occur. Deep neural networks (DNNs) may have sometimes insufficient data to train the model. In the absence of sufficient training data, it may cause high variances and overfitting [35]. Inappropriate selection of learning rate and overfitting may affect the accuracy of model. Number of iteration in SGD method also affect the performance of model which makes it unable to converge.

For the nonconvex optimization problem, it may be possible to attain a local optimal position instead of global optimal position. In case of nonconvex optimization problem, several local optimal solutions are present and choosing global solution out of this local solution is difficult task. There are two approaches to solve the nonconvex problems. First approach transforms the nonconvex problem into convex problem and after that solve it using convex optimization methods. Other approach is to apply specific optimization methods to solve it directly. In sequential models, if the sequence is very large, samples gets truncated and this truncation may lead to deviation. If length of data is not an integer times of sample size, then some previous sampled data are added to sample size while training the model. Complexity of machine-learning models increases. Some of particular optimization problems are developed for specific machine-learning application which cannot be applied to other applications.

12.6 Conclusion

Theoretical basis of fundamental optimization methods is described here. Based on gradient information, Optimization method could be categorized into first-order techniques, high-order techniques and derivative-free optimization techniques. First-order techniques including GD method and its types are presented. Popular algorithms of gradients descent methods are also briefly discussed. SGD method is thoroughly explained along with its advantages and issues. SGD method has low convergence rate and low computational complexity. To improve the convergence rate and to use curvature information, high-order optimization methods were employed in literature. High-order techniques, namely, Newton's method, GNM, QNM, natural gradients methods along with some derivative-free techniques are also explained here. Challenges and issues occurs in optimization problems is also presented here.

References

[1] S. Ruder, An Overview of Gradient Descent Optimization Algorithms, September 15, 2016 arXiv preprint arXiv:1609.04747.

[2] D.F. Shanno, Conditioning of quasi-Newton methods for function minimization, Math. Comput. 24 (111) (1970) 647−656.

[3] A.S. Berahas, R.H. Byrd, J. Nocedal, Derivative-free optimization of noisy functions via quasi-Newton methods, SIAM J. Optim. 29 (2) (2019) 965−993.

[4] L.M. Rios, N.V. Sahinidis, Derivative-free optimization: a review of algorithms and comparison of software implementations, J. Global Optim. 56 (3) (July 2013) 1247–1293.

[5] H. Robbins, S. Monro, A stochastic approximation method, Ann. Math. Stat. (September 1, 1951) 400–407.

[6] R. Johnson, T. Zhang, Accelerating stochastic gradient descent using predictive variance reduction, Adv. Neural Inf. Process. Syst. 26 (2013) 315–323.

[7] C. Darken, J. Chang, J. Moody, Learning rate schedules for faster stochastic gradient search, in: In Neural Networks for Signal Processing 2, August 31, 1992.

[8] J. Duchi, E. Hazan, Y. Singer, Adaptive subgradient methods for online learning and stochastic optimization, J. Mach. Learn. Res. 12 (7) (July 1, 2011).

[9] I. Sutskever, Training Recurrent Neural Networks, University of Toronto, Toronto, Canada, January 1, 2013.

[10] R. Ge, F. Huang, C. Jin, Y. Yuan, Escaping from saddle points—online stochastic gradient for tensor decomposition, in: In Conference on Learning Theory, PMLR, June 26, 2015, pp. 797–842.

[11] N. Qian, On the momentum term in gradient descent learning algorithms, Neural Network. 12 (1) (January 1, 1999) 145–151.

[12] I. Sutskever, J. Martens, G. Dahl, G. Hinton, On the importance of initialization and momentum in deep learning, in: In International Conference on Machine Learning, PMLR, May 26, 2013, pp. 1139–1147.

[13] Y. Nesterov, A method for unconstrained convex minimization problem with the rate of convergence O $(1/k^2)$, in: In Doklady an USSR 269, 1983, pp. 543–547.

[14] L. Baird, A.W. Moore, Gradient descent for general reinforcement learning, Adv. Neural Inf. Process. Syst. (July 20, 1999) 968–974.

[15] D.P. Kingma, J. Ba, Adam: a method for stochastic optimization, arXiv preprint arXiv:1412.6980. 2014 Dec 22.

[16] M.D. Zeiler, Adadelta: an adaptive learning rate method, arXiv preprint arXiv:1212.5701. 2012 Dec 22.

[17] T. Dozat, Incorporating Nesterov Momentum into Adam, 2016.

[18] N.L. Roux, M. Schmidt, F. Bach, A stochastic gradient method with an exponential convergence rate for finite training sets, arXiv preprint arXiv:1202.6258. 2012 Feb 28.

[19] R.H. Byrd, G.M. Chin, W. Neveitt, J. Nocedal, On the use of stochastic hessian information in optimization methods for machine learning, SIAM J. Optim. 21 (3) (July 1, 2011) 977–995.

[20] J.E. Dennis, J.J. Moré, A characterization of superlinear convergence and its application to quasi-Newton methods, Math. Comput. 28 (126) (1974) 549–560.

[21] J. Nocedal, Updating quasi-Newton matrices with limited storage, Math. Comput. 35 (151) (1980) 773–782.

[22] A. Agarwal, P.L. Bartlett, P. Ravikumar, M.J. Wainwright, Information-theoretic lower bounds on the oracle complexity of stochastic convex optimization, IEEE Trans. Inf. Theor. 58 (5) (January 30, 2012) 3235–3249.

[23] A. Bordes, L. Bottou, P. Gallinari, SGD-QN: careful quasi-Newton stochastic gradient descent, J. Mach. Learn. Res. 10 (July 2009) 1737–1754.

[24] N.N. Schraudolph, Fast curvature matrix-vector products, in: In International Conference on Artificial Neural Networks, Springer, Berlin, Heidelberg, August 21, 2001, pp. 19–26.

[25] S.I. Amari, Natural gradient works efficiently in learning, Neural Comput. 10 (2) (February 15, 1998) 251–276.

[26] S.I. Amari, H. Nagaoka, Methods of Information Geometry, American Mathematical Soc., 2000.

[27] H. Park, S.I. Amari, K. Fukumizu, Adaptive natural gradient learning algorithms for various stochastic models, Neural Network 13 (7) (September 1, 2000) 755–764.

[28] M. Dodangeh, L.N. Vicente, Worst case complexity of direct search under convexity, Math. Program. 155 (1–2) (January 1, 2016) 307–332.

[29] H. Robbins, Some aspects of the sequential design of experiments, Bull. Am. Math. Soc. 58 (5) (September 1952) 527–535.

[30] P. Auer, Using confidence bounds for exploitation-exploration trade-offs, J. Mach. Learn. Res. 3 (Nov) (2002) 397–422.

[31] J. Kiefer, J. Wolfowitz, Stochastic estimation of the maximum of a regression function, Ann. Math. Stat. (September 1, 1952) 462–466.

[32] R. Durrett, Probability: Theory and Examples, Cambridge university press, April 18, 2019.

[33] P. L'Ecuyer, G. Yin, Budget-dependent convergence rate of stochastic approximation, SIAM J. Optim. 8 (1) (February 1998) 217–247.

[34] R. Pasupathy, On choosing parameters in retrospective-approximation algorithms for stochastic root finding and simulation optimization, Oper. Res. 58 (4-part-1) (August 2010) 889–901.

[35] N. Srivastava, G. Hinton, A. Krizhevsky, I. Sutskever, R. Salakhutdinov, Dropout: a simple way to prevent neural networks from overfitting, J. Mach. Learn. Res. 15 (1) (January 1, 2014) 1929–1958.

13

Stochastic optimization of industrial grinding operation through data-driven robust optimization

Priyanka D. Pantula, Srinivas Soumitri Miriyala, Kishalay Mitra

GLOBAL OPTIMIZATION AND KNOWLEDGE UNEARTHING LABORATORY, DEPARTMENT OF
CHEMICAL ENGINEERING, INDIAN INSTITUTE OF TECHNOLOGY HYDERABAD, HYDERABAD,
TELANGANA, INDIA

13.1 Introduction

Use of model-based optimization approaches towards improving the design or operational efficiency of engineering processes has become a regular activity in chemical engineering. Recent emphasis on process intensification, approachability of high-end computational power and advancements on numerical algorithms have transformed traditional offerings of these model-based optimization services into highly demanding ones. Depending on the underlying mathematical models' accuracy, which are based on either the underlying physics of the process or the data obtained from the process (surrogate models), the reliability of the optimization study can be established. This is because the models used for mimicking the process behavior are subject to various sources of uncertainties such as variations in kinetic model parameters, process parameter fluctuations, regression errors and so on [1]. Performing optimization exercises assuming a fixed value e.g., nominal value for these uncertain parameters makes the solutions highly sensitive to the parameter variations [1]. This sensitivity may violate the plant safety constraints and thus, may lead to infeasible design of the process or operational conditions. Therefore, optimization under uncertainty (OUU) has emerged as a vital area of research among industries and academicians [2].

Various approaches were reported in the literature in order to handle OUU [1]. Some of the popular methods are as follows: (1) two-stage stochastic programming (TSSP) [3], (2) chance constrained programming (CCP) [4]; (3) robust optimization (RO) [5]; (4) expected value model (EVM) [6]; (5) polynomial chaos expansion (PCE) [7]; (6) fuzzy CCP [4]; (7) fuzzy RO [8]; and (8) fuzzy EVM [6]. The techniques (1)−(5) are usually preferred over the remaining when the uncertain data can be made available and the uncertainty can be measured. Stochastic programming divides the decision variable set into two parts based on their associations with the uncertain parameters. Dependent variables are decided based on several realizations of uncertain parameters (called as "wait and see") whereas the independent variables are allowed to be evolved separately (called as "here and now"). The objective function is set as a combination of these two terms, where deterministic values of the uncertainty independent terms and expected values of the uncertainty associated terms are considered. This technique has been applied in varied fields of engineering applications such as microgrid planning [9], management of distributed energy resources [3], scheduling of the wind-thermal-hydropower-pumped storage system [10], planning of a coal-to-liquids supply chain [11]. Though exhaustively applied on many problems as mentioned above, this approach suffers from the drawback of irrepressible burst in size of the problem due to the rise in uncertain parameters count, apart from practical difficulties involved in decomposition of decision variables in the aforementioned fashion [4].

To facilitate flexibility, a different approach has been adopted in CCP, where violation in constraint satisfaction is allowed to a certain extent by assigning probability with constraints [4]. Assuring more probability of constraint satisfaction, more reliable solutions can be generated and, in this way, CCP generates solutions of different levels of reliability. Recent publications using CCP were presented by the authors in Refs. [11,12] for process optimization under uncertainty, whereas an extensive overview on utilization of CCP in chemical engineering problems can be found in the work of Ref. [13]. In spite of having better control over the size of the problem with the rise in uncertain parameters count, this approach fails to infer much in cases where constraint violation is strictly not allowed. Another popular technique called PCE approximates the relationship among the uncertain parameters and system outputs with the help of orthogonal basis functions [14]. The authors in Ref. [15] presented PCE approach to optimize biological networks under model parameter uncertainty consideration. Some other areas of PCE application include topology optimization [16], analysis of the performance of an H-Darrieus rotor [17], prediction of uncertain frequency response function bounds [18], simulation of probabilistic chemical reactions [19], etc. However, the accuracy of PCE technique depends on the order of expressions considered. The higher the order of expressions in PCE, better is the solution. Nonetheless, higher-order expressions require computation of additional PCE coefficients which makes the approach computationally expensive [20].

As compared to other alternative methods, RO has advantages of handling uncertainties efficiently along-with computational tractability [21]. Some of the RO application areas include process modeling, optimization, planning and scheduling, design of a process, and so on [21]. In general, the parameters in RO are divided into two categories: (1) decision variables, and (2) uncertain parameters. This contrasts with the conventional deterministic optimization technique, where only the decision variables are changed during optimization assuming all other parameters are constant. In a simulation-centered study, the decision variables are optimally determined so that the supremum/infimum values of the stochastic objective functions/constraints, estimated over various uncertain parameter realizations, are optimized [22]. This technique is implemented using a multiloop approach, where the decision variables are varied at the outer loop and for each such decision variable value, the inner loop allows to sample uncertain parameter space several number of times to calculate the statistical terms (e.g., supremum, infimum, mean, standard deviation, etc.) for the objective functions and constraints. While performing sampling, either the distribution information among the uncertain parameters or the data depicting the variation among uncertain parameters should be known. However, as the uncertain data would be coming from realistic sources, it might be very difficult to capture the embedded variation using combinations of several known statistical distributions. During such instances, conducting uncertainty-based analysis assuming availability of uncertain parameter data, not their embedded distribution, is relatively safe. Nonetheless, the solutions of RO are often found to be conservative which might be due to overrepresentation of uncertain parameter space or inefficient sampling in the hyperspace of uncertain parameters [23]. More precisely, when the available uncertain data is scattered, sampling in the places, which may not represent the true nature of uncertainty, might lead to inaccurate solutions. Efficient and adaptive sampling might help in addressing this issue where sampling may be accomplished in the true uncertain space. However, finding such accurate representations, sampling uniformly in those regions and its scalability turn out to be the major challenges toward the proposition of a less conservative and tractable RO approach.

Owing to the increasing applicability of machine learning in various domains of research such as bioinformatics [24], computer vision [25], agriculture [26], time series forecasting [27], classification studies [28], and so on, the combination of RO with machine learning techniques is quite motivating. Therefore, in this chapter, an unsupervised machine learning algorithm [29] is amalgamated with RO to address the issues. This study primarily employs an artificial neural network (ANN) assisted evolutionary Fuzzy C-means clustering algorithm [29] such that the actual regions of uncertainty are recognized. Subsequently, efficient segmentation of uncertain space is performed using cluster boundary detection through Delaunay triangulations [30]. This is followed by an adaptive Sobol sampling method [31] for uniformly sampling the data points in the uncertain space that present significant resemblance to the available/given data. Then, these samples are used for estimating the supremum or infimum of the stochastic/uncertain objective functions and/or constraints in RO, which can be more accurate as compared to the traditional box uncertainty approach [32]. In box uncertainty set, the uncertain space is sampled either uniformly or randomly within the uncertain parameter bounds, regardless of the true representation of uncertain parameters in the entire space. The overall framework is termed as data-driven robust optimization (DDRO), which boosts the RO technique for providing less conservative (more realistic solutions) irrespective of the size of the optimization problem.

The proposed DDRO formulation was demonstrated on an industrial nonlinear grinding model [4] having 10 uncertain parameters and the solutions of DDRO are further compared with those of box uncertainty set based RO. In this model, for ensuring high productivity and significant savings in energy during the optimization of grinding circuits, a three objective optimization problem is formulated. Since the circuits are complex in nature and possess incomplete knowledge, most of the grinding models consist of empirical constants, parameters having physical relevance [4] like rod mill and ball mill grindability indices and hydrocyclones sharpness indices and some others. However, since these parameters are estimated either through rigorous experimentation or regression analysis, it brings in uncertainty in the model and correspondingly they are termed as uncertain parameters [4]. If such kind of uncertainty is ignored and considered as a deterministic optimization study, ignoring their variability, and fixing their values to a specific value, say mean value, then the predictions related to optimization studies and control of industrial grinding circuits turn out to be infeasible or suboptimal [32]. Thus, in this chapter, a three objective OUU problem has been formulated with 3 decision variables and 10 uncertain parameters.

The chapter organization is as follows. Section 13.2 introduces the RO approach and traditional FCM algorithm along-with the existing drawbacks. Section 13.3 illustrates the industrial grinding model followed by the proposition and working of the novel framework, DDRO. Subsequently, Section 13.4 shows DDRO algorithm implementation on the considered grinding circuit model and the results are compared with those of box uncertainty-based RO approach. Finally, the concluding remarks are given in Section 13.5.

13.2 Optimization under uncertainty

13.2.1 Brief overview of robust optimization

Eq. (13.1) presents a multiobjective stochastic optimization problem, where the following notations are used: F: bounded objective functions, ξ: the vector of uncertain variables, $\mathbf{X}$: decision vector and M, I, and E: the number of objectives and the number of inequality and equality constraints respectively.

$$\min_{\mathbf{X}} F = \{f_m(\mathbf{X}, \xi)\} \subset \mathbb{R}^M \; \forall \; m = 1, 2, \dots M \tag{13.1a}$$

$$\text{Constraints}: \; g_i(\mathbf{X}, \xi) \leq 0 \; \forall \; i = 1, 2, \dots I \tag{13.1b}$$

$$h_e(\mathbf{X}, \xi) = 0 \; \forall \; e = 1, 2, \dots E \tag{13.1c}$$

where, $\mathbf{X} = \{x_d\} \subset \mathbb{R}^D | \; x_d^{LB} \leq x_d \leq x_d^{UB} \; \forall \; d = 1 \text{ to } D$

$$\xi = \{\xi_u\} \subset \mathbb{R}^U | \; \xi_u^{LB} \leq \xi_u \leq \xi_u^{UB} \; \forall \; u = 1 \text{ to } U$$

Eq. (13.2) denotes the equivalent deterministic formulation of the above stochastic formulation (Eq. 13.1) using worst-case RO approach, also referred as minmax approach in RO.

$$\min_{X} \left\{ \max_{\xi} [f_m(X, \xi)] \right\} \subset \mathbb{R}^M \qquad (13.2)$$

subject to Eqs. (13.1b) and (13.1c).

Since most often, in real scenario, the probability mass/density function of the uncertain variables (random variables) remains unavailable, the inner optimization loop in Eq. (13.2) is calculated using the realizations of uncertain parameters as shown in Eq. (13.3) (that is supremum value of the objective function(s) over all the instantiations of ξ).

$$\min_{X} \left\{ \max_{\xi} [f_m(X, \xi_n)] \right\} \forall \ n = 1, 2, \dots \ N \qquad (13.3a)$$

$$g_i(X, \xi_n) \leq 0 \ \forall \ n = 1, 2, \dots \ N \qquad (13.3b)$$

$$h_e(X, \xi_n) = 0 \ \forall \ n = 1, 2, \dots \ N \qquad (13.3c)$$

$$\forall \ m = 1, 2, \dots \ M, i = 1, 2, \dots \ I \ \text{and} \ e = 1, 2, \dots \ E$$

The number of realizations of uncertain variables, N (cardinality of given data set {ξ}) needs to be large enough such that the supremum in Eq. (13.3) converges to the true estimates. Nonetheless, in general, N can be quite less in number due to several practical constraints. At the same time, the data can be dispersed in the uncertain space, due to which it might be very difficult to fit any well-behaved statistical distribution function through the data. Under this situation, it is suggested that a large number of uncertain variables are to be sampled from only some specific areas from the overall uncertain space. These regions are called as uncertainty sets and the shape of these uncertainty sets is vital in estimating the final accuracy of RO solutions [32].

13.2.2 Uncertainty set: a paramount element during calculations of statistical terms in RO

The most used uncertainty sets include box, ellipsoidal, budgeted, and polyhedral as presented in Fig. 13.1 [32,33]. Here, the blue colored points represent the data available in the uncertain space while the dotted red lines around each of them indicate the region

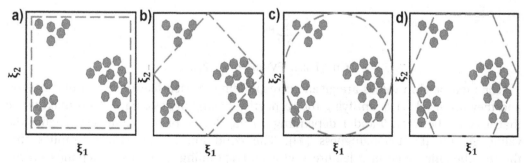

FIGURE 13.1 Uncertainty sets (A) box, (B) budgeted ($\tau = 1$), (C) ellipsoid, and (D) polyhedral.

of sampling approximated by different methods of uncertainty sets. From these uncertainty sets, any number of samples can be drawn, either uniformly or randomly, for evaluating the statistical terms as shown in Eq. (13.3). The larger the number of generated samples, the preferable the convergence of estimated supremum/infimum to the true estimates. However, the box uncertainty set, as shown in Fig. 13.1A, considers all possible combinations of uncertain parameters within the specified bounds. On the other hand, in an effort to accurately transcript the uncertain space, the remaining uncertainty sets (Fig. 13.1B–D) try to eliminate the unnecessary regions of sampling i.e., the blank spaces by varying the geometric shapes. Despite these efforts, it is evident that if these methods are adopted, the chances of samples being picked up from unnecessary regions where the given samples are not present are very high. Hence, the RO solutions thus resulting from such kind of sampling methodologies might deviate drastically from the accurate estimates and lead to conservative RO solutions. To know about more types of sampling in the uncertain space, interested readers may refer to Refs. [32,33]. In contrast to the existing uncertainty sets that are less flexible to capture the true regions of uncertainty, the authors have made an attempt toward designing of a compact uncertainty set using unsupervised machine learning (ML) and boundary detection techniques. Particularly, a well-known unsupervised ML technique, namely, Fuzzy C-means clustering [34] that can deal with interacting data sets, has been implemented for identifying the true regions of uncertainty.

13.2.3 Description of FCM: unsupervised ML approach

Consider the following: Say, $\mathbf{X} = \{\mathbf{x_n}\} | \mathbf{x_n} \subset \mathbb{R}^K \ \forall \ n = 1, 2, \ldots$ FV, presents the feature vectors that are grouped into C clusters ($2 \leq C \leq$ FV, $C \in \mathbb{Z}^+$) and the corresponding membership matrix is indicated by $\mu = \{\mu_{nc}\}$, on which the following three conditions in Eq. (13.4) need to be satisfied.

$$\mu_{nc} \in [0\ 1] \tag{13.4a}$$

$$\sum_{c=1}^{C} \mu_{nc} = 1 \tag{13.4b}$$

$$\sum_{N=1}^{FV} \mu_{nc} > 0 \tag{13.4c}$$

$$\forall \ n = 1, 2, \ldots \text{FV and } c = 1, 2, \ldots C$$

The membership values are given corresponding to each feature vector and cluster, and they are used for quantifying the respective belongingness. Eq. (13.4a) makes sure that μ varies between 0 and 1 depending on the degree of belongingness. Higher the value, greater the belongingness [34]. The condition in Eq. (13.4b) implies that the membership value of a feature vector corresponding to a cluster depends on the presence of other clusters (relative value). Finally, the condition in Eq. (13.4c) makes sure that none of the clusters is empty.

In FCM, each cluster is denoted by a representative called cluster center. Let the cluster centers be denoted by $\Phi = \{\varphi_c\} \subset \mathbb{R}^K \forall \ c = 1, 2, \ldots \ C$ denote and consider $\| \|$ as the Euclidean norm. FCM algorithm primarily works by minimizing a nonlinear objective function as given in Eq. (13.5) (nonlinear programming [NLP] problem).

$$(\mu^* \text{ and } \varphi^*) = \operatorname{argmin}\left(\sum_{c=1}^{C} \sum_{n=1}^{FV} (\|\mu_{nc}\mathbf{x_n} - \varphi_c\|)^2 \right) \tag{13.5}$$

subject to constraints and bounds in Eq. (13.4).

The optimal decision variables (μ^* and φ^*) are usually obtained by solving Eq. (13.5) using Lagrange Multiplier method [34].

13.2.4   Issues in conventional FCM approach

As the decision variable space in FCM algorithm includes membership matrix, the decision variable count depends on the number of (a) feature vectors and (b) cluster centers. On the other hand, the cluster centers in Eq. (13.5) depend only the dimensionality of the given data. Since most often, the number of feature vectors would be too higher than the data dimensionality, μ values alone contribute a considerable size in the decision variable space. Hence, as the number of feature vectors increases, the size of NLP problem increases drastically. This makes the classical optimization algorithms preferable to solve Eq. (13.5) in comparison with evolutionary algorithms. However, since the objective function in Eq. (13.5) is nonconvex in nature, local optimal solutions may be obtained usually on using Lagrange multiplier method and use of evolutionary optimization methods might have additional edge as they are capable of finding global optimal solution in presence of other local optimal solutions.

Some of the researchers in Refs. [35–38] tried to estimate the cluster centers through evolutionary algorithms whereas the μ values (another set of decision variables) were again obtained through Lagrange method. Therefore, despite the cluster representatives being optimally identified, the belongingness values might still be suboptimal. Moreover, in FCM, the identification of optimal cluster number continues to be an age-old problem. Thus, although FCM approach looks encouraging for recognizing the given regions of uncertain space, the above-mentioned issues pertaining with FCM degrade the quality of clustering as well as RO solutions. Moreover, even if the clustering drawbacks are addressed, sampling in the actual uncertain regions presents a much larger challenge since the clusters would no longer follow any specific geometrical shapes. DDRO attempts to address these challenges efficiently as presented in the section below.

13.3 DDRO: data-driven robust optimization for grinding model

13.3.1 Industrial grinding model: description

In the ore beneficiation process of lead-zinc, wet grinding circuit is primarily utilized. The grinding circuit consists of the following basic components: rod mill, ball mill, and a set of hydrocyclones and sumps [4,39]. Prior to passing into the grinding circuit, crushing of the ore is initially performed using primary crusher followed by secondary crusher. Subsequently, the output is directed to the next unit, namely rod mill as a feed together with water. Then, the obtained ore that is grounded, is moved to primary sump and additional water is passed to enable easy flow of material. The slurry that is obtained from the outlet of the sump is correspondingly sent to the hydrocyclones primary bank. The overflow and underflow through the cyclones are sent to the ball mill and secondary sump respectively. Further, the slurry that is attained from secondary sump is passed to the hydrocyclones secondary bank and finally the obtained overflow is considered as the final product. Once again, the underflow of the cyclone is sent back to the ball mill for recycle and the outlet stream is sent to the primary sump, which as a whole, completed the grinding circuit. Fig. 13.2 represents the schematic of the overall grinding circuit.

To process further, the product obtained from secondary cyclone is fed to the flotation circuit and the flotation unit efficiency is largely affected by the particle size

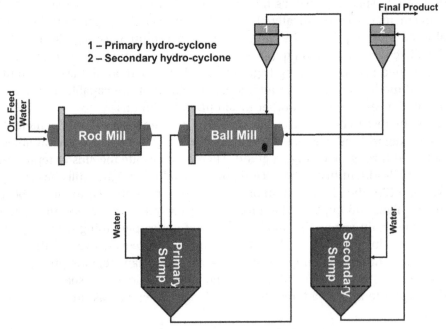

FIGURE 13.2 Representation of Grinding circuit.

distribution of this feed, thereby making grinding operation a vital process in ore beneficiation. Hence, modeling and optimization of industrial grinding circuit is considered to be very significant. Here, to model each unit operation, a hybrid approach has been utilized, which combines population balance with the method of empirical correlation. Through the population balance approach, material balance is performed across all the unit operations, where different major size classes are considered across the circuit. Correspondingly, the empirical correlations are utilized to compute the sharpness indices and breakage functions. Overall, a nonlinear model is developed, since there exists exponential terms along-with power terms in the above correlations [4] and finally, a collection of differential algebraic equations (DAEs) are achieved, which are efficiently solved through DASSL (software found in public repository) [40].

13.3.2 Formulation of optimization problem under uncertainty problem: grinding model

In grinding circuit, since the size of the particle in the product stream is relatively important for enabling the proper working of the succeeding unit, that is flotation, maximization of desired size range in the product stream, also referred as midsize (MS), is considered as an objective for optimal operation. Moreover, for enabling a sustainable operation on a long-term basis and for saving significant amount of energy, the minimization of recirculation load (RCL) is taken as the second objective. Besides these two, the regular objective in any process, that is maximization of productivity, denoted as throughput (TP), is the third objective. Apparently, a trade-off exists between TP and MS, which thereby leads to the formulation of a three objective optimization problem in this study. Moreover, the control variables in the circuit such as classes of coarse size and fine size (CS and FS), percent solids (PS) and recirculation load (RCL) consist of upper bound constraints. In this problem, there are three decision variables, namely, rod mill raw ore feed flowrate (S) and flowrates of water to primary and secondary sump (w_1 and w_2). Overall, for the grinding operation, the deterministic optimization formulation is presented in Eq. (13.6).

$$\max_{\mathbf{X} \in (S,\, w_1,\, w_2)} \mathrm{TP}(\mathbf{X}, \tau, \omega) \tag{13.6a}$$

$$\max_{\mathbf{X} \in S,\, w_1,\, w_2} \mathrm{MS}(\mathbf{X}, \tau, \omega) \tag{13.6b}$$

$$\min_{\mathbf{X} \in S,\, w_1,\, w_2} \mathrm{RCL}(\mathbf{X}, \tau, \omega) \tag{13.6c}$$

subject to:

$$\mathrm{CS}\,(\mathbf{X}, \tau, \omega) \leq \mathrm{CS}^{\mathrm{U}} \tag{13.6d}$$

$$\mathrm{FS}\,(\mathbf{X}, \tau, \omega) \leq \mathrm{FS}^{\mathrm{U}} \tag{13.6e}$$

$$\mathrm{RCL}\,(\mathbf{X}, \tau, \omega) \leq \mathrm{RCL}^{\mathrm{U}} \tag{13.6f}$$

$$PS\ (\mathbf{X}, \tau, \omega) \leq PS^U \tag{13.6g}$$

where,

$$S^L \leq S \leq S^U$$

$$w_1^L \leq w_1 \leq w_1^U$$

$$w_2^L \leq w_2 \leq w_2^U$$

Here, all the objective functions and constraints are not only function of decision variable set $\mathbf{X}$, but also two sets of parameters, denoted by υ, ω. The parameters indicate the following: ball mill and rod mill grindability indices—$\tau_1\ \tau_2$, primary and secondary cyclone sharpness indices—τ_3, τ_4, ball mill and rod mill grindability exponents—τ_5, τ_6, and finally four size fractions present in the feed to rod mill—$\omega_1, \omega_2, \omega_3, \omega_4$. In the input stream, there are basically five size fractions. Since the size fractions sum to 1, four size fractions are sufficient to be determined in order to estimate the fifth size fraction. The vector υ consisting of six parameters (τ_1 to τ_6) is estimated through regression analysis of the experimental data, which thereby leads to uncertainty in the model. Additionally, there exists operational uncertainties (ω_1 to ω_4) because of the disparities in the input feed stream. In general, all these 10 parameters, called as model and operational parameters, are kept at their base values during the deterministic optimization study. Eventhough such kind of analysis is easier to solve, since there are uncertainties involved, it would be realistic to solve the optimization under uncertainty problem. Specifically, in this chapter, RO approach will be used for solving the stochastic optimization problem. The worst-case formulation of RO for this case study is presented in Eq. (13.7).

$$\max_{\mathbf{X} \in (S, w_1, w_2)} \left\{ \min_{\tau, \omega} [TP(\mathbf{X}, \tau_n, \omega_n)] \right\} \quad \forall\ n = 1\ \text{to}\ N \tag{13.7a}$$

$$\max_{\mathbf{X} \in (S, w_1, w_2)} \left\{ \min_{\tau, \omega} [MS(\mathbf{X}, \tau_n, \omega_n)] \right\} \quad \forall\ n = 1\ \text{to}\ N \tag{13.7b}$$

$$\min_{\mathbf{X} \in (S, w_1, w_2)} \left\{ \max_{\tau, \omega} [RCL(\mathbf{X}, \tau_n, \omega_n)] \right\} \quad \forall\ n = 1\ \text{to}\ N \tag{13.7c}$$

subject to constraints (13.6d)−(13.6g) and bounds on decision variables.

Here, it is assumed that the variation in these parameters in the uncertain space is made available through experiments and thus, the bounds on the uncertain parameters are known.

13.3.3   ANN assisted fuzzy C-means clustering technique

Now, prior to solving the formulation in Eq. (13.7), since the conventional uncertainty sets generate conservative solutions, a novel methodology that combines the clustering algorithm with generative modeling framework will be discussed. Primarily, to cluster the uncertain data, an evolutionary ANN-FCM approach, that is suggested by the authors in Ref. [41], has been used, instead of conventional FCM in order to address the drawbacks that are mentioned in Section 13.2.4. In the ANN assisted FCM formulation, a functional map is built using the feature vectors/data to be clustered and the membership grades in FCM using ANNs [42]. The input to the neural network, called as multilayer perceptron (MLP), is the feature vector $\mathbf{x_n}|$ $n \in [1\ FV]$ and the corresponding output is the membership grades (μ_{nc}). Since membership matrix leads to explosion in decision variable space, it is absorbed into ANNs efficiently. Instead, the MLP parameters, namely weights and biases, are tuned in such a way that the resultant membership values enable optimal clustering solution. This tuning exercise is done by minimizing the FCM objective function Eq. (13.5) where μ_{nc} is substituted with the ANN map thus generated. The decision variables now include the parameters of ANN and the cluster centers. Further, to meet the constraints on μ_{nc} (Eqs. 13.4a and 13.4c), log-sigmoid activation function has been implemented in the output layer of ANNs and a normalization function has been utilized in the output layer to satisfy Eq. (13.4b).

Using the ANN based clustering approach, the decision variable space has been significantly reduced in size due to the elimination of number of feature vectors from the decision variable count and constraints. This implies that the large-scale NLP problem (in Eq. 13.5) has been converted into a small-scale NLP, which in turn ensures the usage of genetic algorithm (GA; an evolutionary optimizer) [43] for finding the optimal parameters of MLP and cluster centers, such that hopefully the solutions in the global basin are obtained in FCM. Additionally, in this study, the neural network architecture is optimally identified through maximization of partition coefficient and exponential separation index (PCES) [44]. The hyperparameters of ANN, such as number of nodes, layers, etc. constitute the decision variable set [41]. Since the outputs from MLP constitute the membership values, the estimation of number of nodes in the output layer from the aforementioned optimization formulation enables the detection of optimal cluster number. Here, the validation index, that is, PCES is also maximized using GA [43]. Thus, due to the combination of unsupervised machine learning algorithm with ANN, the usage of evolutionary algorithms in FCM was enabled and further the optimal number of clusters were determined from this study [41]. Now, the uncertain parameter data can be effectively clustered through the ANN reformulated FCM approach which in turn helps in identification of the true regions of uncertainty regardless of the dimensionality of the data along-with the optimal cluster number identification.

13.3.4 Generative modeling framework in the identified clusters

After the identification of true regions of uncertainty using ANN assisted FCM, the next step is to sample the data uniformly in these detected regions such that various statistical modes of objective functions and constraints are estimated more accurately. In case of existing uncertainty sets described in Section 13.2.2, sampling exercise is straightforward since they were constructed based on specific geometrical shapes. On the contrary, the identified uncertain regions here, may not follow any known geometry, thus necessitating the requirement for boundary detection algorithm for each of the associated clusters, prior to sampling. Therefore, the following procedure as represented in Fig. 13.3 has been implemented for cluster boundary detection followed by sampling within the boundaries. Each of the steps presented in Fig. 13.3 must be repeated for each cluster so that all the uncertain regions are considered for generation of new samples from the uncertain space.

The obtained samples vector is denoted by $\{\xi^{DDRO}\}$, where any large number of samples can be generated uniformly within the true regions of uncertainty. Hence, a compact uncertainty set has been designed with the help of given data alone followed by the estimation of the worst-case RO formulation using the generated samples. This is the reason behind terming the algorithm as DDRO approach. Here, let $N_S = \text{cardinality}(\{\xi^{DDRO}\}) \gg N$ and $N_{S,Box}$ represents the number of points considered in box uncertainty set. Algorithm 13.1 presents the pseudocode of DDRO after clustering the given uncertain region using ANN assisted FCM algorithm.

FIGURE 13.3 Sequence of steps for sampling in DDRO.

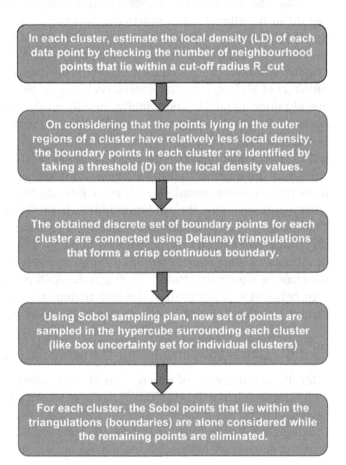

In each cluster, estimate the local density (LD) of each data point by checking the number of neighbourhood points that lie within a cut-off radius R_cut

On considering that the points lying in the outer regions of a cluster have relatively less local density, the boundary points in each cluster are identified by taking a threshold (D) on the local density values.

The obtained discrete set of boundary points for each cluster are connected using Delaunay triangulations that forms a crisp continuous boundary.

Using Sobol sampling plan, new set of points are sampled in the hypercube surrounding each cluster (like box uncertainty set for individual clusters)

For each cluster, the Sobol points that lie within the triangulations (boundaries) are alone considered while the remaining points are eliminated.

13.4 Results and discussions

Since there are 10 uncertain variables in the considered grinding model and the original uncertain data is unavailable from industries, a synthetically generated 10-dimensional (10-D) data set has been assumed as the given uncertain data. The reasons behind generating this particular data are described as follows: (1) it consists of only 150 samples in 10-D space which are assumed to be insufficient for calculating the true statistical supremum/infimum values, (2) it is scattered throughout the uncertain space in a discrete manner (discontinuous data), and (3) it does not follow any known probability distribution. Now, once the uncertain data is provided, ANN assisted evolutionary FCM technique was applied for identifying the true regions of uncertainty where three clusters were optimally identified. Subsequently, the sampling scheme employed in DDRO (refer Section 13.3.4) was applied and thus, a greater number of samples were generated within the identified regions.

Once the supremum/infimum values in Eq. (13.7) (inner-loop) are identified using the generated realizations of uncertain parameters, the 3-objective optimization problem is solved through an evolutionary algorithm, non-dominated sorting genetic algorithm-II (NSGA-II) [43]. In this study, an ideal or desirable Pareto front is generated by solving the DDRO formulation with Monte Carlo Samples that are generated in the identified clusters. Approximately, 100,000 points are generated, and the supremum/infimum values are estimated. The obtained 3-D Pareto fronts thus obtained through DDRO is presented in Fig. 13.4A, for a fixed sample size 500 along-with comparison with ideal solution. Further, in Fig. 13.4B, the box uncertainty set based RO solution is compared with the ideal front. To make the observations clearer, the 3-D front is represented using 2-objectives each as shown in Fig. 13.5. It is found that the solutions that are obtained using DDRO are closer to the ideal solution in comparison with the box uncertainty set based RO and moreover, the solutions are relatively well-spread as seen in Fig. 13.5A and D.

Since any number of samples can be drawn from the designed uncertainty set, Fig. 13.6 demonstrates the effect of varying sample size (from 500 to 3000) on the Pareto front obtained using DDRO and box-based RO. The solutions are analyzed quantitatively using six different convergence and diversity metrics [43] as reported in Table 13.1. It is

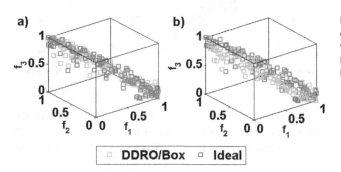

FIGURE 13.4 3-D Pareto fronts obtained on solving the robust optimization formulation of grinding model using NSGA-II (A) DDRO versus Ideal (B) Box uncertainty RO versus Ideal case.

FIGURE 13.5 Comparison of 2-D Pareto fronts obtained using NSGA-II for the grinding model with respect to ideal solutions. (A)–(C) DDRO versus ideal and (D)–(F) Box RO versus Ideal. (In Fig. 13.5A, the DDRO front almost coincides with that of ideal solution).

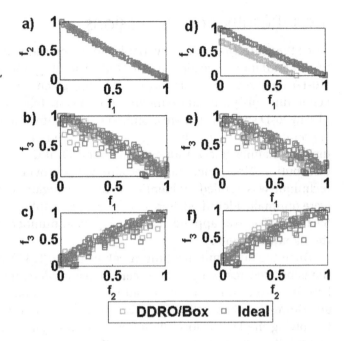

FIGURE 13.6 Comparison of Pareto fronts obtained on solving stochastic optimization problem of the industrial grinding model using DDRO and box uncertainty set based RO with ideal solutions where the number of samples is varying, N equals (A) 500 (B) 1000 (C) 2000 and (D) 3000 (N = N$_S$).

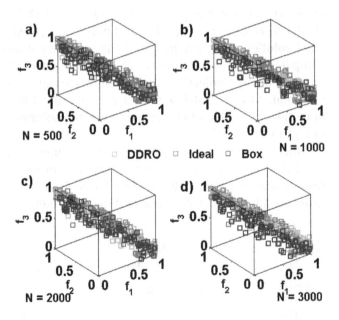

found that most of the metrics for the case of DDRO with 3000 samples are more desirable as compared to the remaining cases. Therefore, for this case study, DDRO outperforms box uncertainty-based RO, which further implies that the conservatism of

Table 13.1 Comparison of 3-D Pareto fronts using quantitative metrics.

| Metric | Desirable | N_S | | | | $N_{S,BOX}$ | | | |
		500	1000	2000	3000	500	1000	2000	3000
Spread	Minimum	**0.54**	0.58	0.58	0.55	0.91	0.91	0.96	0.98
Spacing	Minimum	0.12	0.10	0.12	**0.10**	0.54	0.51	0.55	0.64
Generational distance GD	Minimum	0.02	0.02	0.02	**0.02**	0.15	0.32	0.30	0.29
Inverted GD	Minimum	0.20	0.18	0.18	**0.18**	2.06	2.29	2.29	2.26
Hyper-volume HV	Maximum	7.6e+03	7.7e+03	7.7e+03	**7.9e+03**	6.2e+03	6.8e+03	7.1e+03	7.2e+03
Normalized HV	Maximum	0.98	0.97	0.975	**0.985**	0.61	0.77	0.80	0.85

Algorithm 13.1. DDRO

01.	Inputs: Clustered feature vectors
02.	Steps for DDRO algorithm are listed below.
03.	Inputs: $\{$distance between i and j(d_{ij}), R_cut$\}$ $\forall$ i and j = 1 to $|p_c|$ (No. of points present in one cluster).
04.	Repeat - 1 Loop: i → 1, 2, …, $|p_c|$
05.	flag = 0.
06.	Repeat - 2 Loop 2: j → 1,2, …,$|p_c|$
07.	if $\left(R_{cut} > d_{ij} \right)$
08.	flag = flag + 1
09.	end if
10.	end 2 Loop
11.	LD(i) = flag
12.	end 1 Loop.
13.	Repeat 1 Loop: i → 1,2, …, $|p_c|$
14.	If $(D > LD(i))$
15.	i is a boundary point
16.	end if
17.	end Loop 1.
18.	Repeat steps 4 to 16 for all the clusters.
19.	Perform sampling within the uncertainty set uniformly using Sobol sample plan.
20.	With these samples, evaluate the supremum and infimum values for the stochastic objective functions (inner-loop in equation 13.7).
21.	Using NSGA-II, solve the 3-objective optimization problem and estimate the decision variables.
22.	Output: Decision variables, Pareto Optimal Front

solutions has been relatively reduced. Moreover, for a fixed sample size, say 3000, on fixing the normalized throughput (f_1) to 0.6, it is observed that the midsize fraction objective (f_2) and the recirculation load objective (f_3) have been improved by 30% and 10% respectively, in comparison with the solution obtained using box-based RO. This study shows that the productivity and energy savings in the grinding circuit model are improved on implementing DDRO. Therefore, for this case study, DDRO outperforms box uncertainty-based RO which further implies that the inaccuracy of solutions has been relatively reduced.

Overall, the uncertainty set employed in DDRO algorithm has resulted in better approximations to the desirable solutions in both the case studies considered.

13.5 Conclusion

To handle uncertain optimization problems, robust optimization approach is found to be efficient and tractable. However, the considered uncertainty set is a paramount ingredient in deciding the solution quality in RO. Particularly, if the available uncertain data is scattered, the realizations of samples from the uncertainty set need to be chosen efficiently for evaluating the supremum/infimum of uncertain objectives and constraints in worst-case RO formulation. Generating new points from undesirable regions may overly represent the uncertain space and this in turn leads to undesirable solutions. Therefore, in this chapter, a Data-Driven Robust Optimization, or DDRO, formulation was proposed, which utilizes an unsupervised ML technique and a novel generative modeling framework through which the statistical values in RO formulation (such as, supremum/infimum) were computed. Essentially, DDRO employs an ANN assisted fuzzy clustering mechanism for identifying the true regions of uncertainty followed by cluster boundary detection such that the new samples are generated form within these boundaries. This kind of sampling addresses the issue of overrepresentation of the uncertain parameter space, which in turn reduces the conservatism of RO solutions, making them practically realizable. DDRO was tested on an industrial grinding model, where the stochastic optimization problem was efficiently solved using the constructed uncertainty set, followed by implementation of evolutionary optimization algorithm. In the considered case study, comprehensive comparison was performed using box uncertainty set and varying sample sizes. The less conservative yet more desirable optimization under uncertainty solutions thus obtained, proves the capability of DDRO. Further, $\sim$10%−30% improvement has been observed in the individual objectives as compared to the box uncertainty-based RO, which in turn leads to more energy savings and better productivity in the grinding circuit model.

Acknowledgments

Authors would like to acknowledge Ministry of Human Resources Development (MHRD), Government of India [SPARC/2018−2019/P1084/SL] and National Supercomputing Mission, Department of Science and Technology (DST), Government of India [DST/NSM/R&D_HPC_Applications/2021/23] for providing support to this work.

References

[1] N.V. Sahinidis, Optimization under uncertainty: state-of-the-art and opportunities, Comput. Chem. Eng. 28 (6–7) (June 15, 2004) 971–983.

[2] I.E. Grossmann, R.M. Apap, B.A. Calfa, P. García-Herreros, Q. Zhang, Recent advances in mathematical programming techniques for the optimization of process systems under uncertainty, Comput. Chem. Eng. 91 (August 4, 2016) 3–14.

[3] P. Beraldi, A. Violi, G. Carrozzino, M.E. Bruni, A stochastic programming approach for the optimal management of aggregated distributed energy resources, Comput. Oper. Res. 96 (August 1, 2018) 200–212.

[4] K. Mitra, Multiobjective optimization of an industrial grinding operation under uncertainty, Chem. Eng. Sci. 64 (23) (December 1, 2009) 5043–5056.

[5] X. Xie, R. Schenkendorf, Robust optimization of a pharmaceutical freeze-drying process under non-Gaussian parameter uncertainties, Chem. Eng. Sci. 207 (November 2, 2019) 805–819.

[6] N. Virivinti, K. Mitra, Fuzzy expected value analysis of an industrial grinding process, Powder Technol. 268 (December 1, 2014) 9–18.

[7] D. Kumar, H. Budman, Applications of Polynomial Chaos Expansions in optimization and control of bioreactors based on dynamic metabolic flux balance models, Chem. Eng. Sci. 167 (August 10, 2017) 18–28.

[8] Y. Xu, G. Huang, J. Li, An enhanced fuzzy robust optimization model for regional solid waste management under uncertainty, Eng. Optim. 48 (11) (November 1, 2016) 1869–1886.

[9] A. Narayan, K. Ponnambalam, Risk-averse stochastic programming approach for microgrid planning under uncertainty, Renew. Energy 101 (February 1, 2017) 399–408.

[10] M. Daneshvar, B. Mohammadi-Ivatloo, K. Zare, S. Asadi, Two-stage stochastic programming model for optimal scheduling of the wind-thermal-hydropower-pumped storage system considering the flexibility assessment, Energy 193 (February 15, 2020) 116657.

[11] D. Müller, M. Illner, E. Esche, T. Pogrzeba, M. Schmidt, R. Schomäcker, L.T. Biegler, G. Wozny, R. JU, Dynamic real-time optimization under uncertainty of a hydroformylation mini-plant, Comput. Chem. Eng. 106 (November 2, 2017) 836–848.

[12] G.M. Ostrovsky, T.V. Lapteva, N.N. Ziyatdinov, A.S. Silvestrova, Design of chemical engineering systems with chance constraints, Theor. Found. Chem. Eng. 51 (6) (November 1, 2017) 961–971.

[13] P. Li, H. Arellano-Garcia, G. Wozny, Chance constrained programming approach to process optimization under uncertainty, Comput. Chem. Eng. 32 (1–2) (January 1, 2008) 25–45.

[14] Z.K. Nagy, R.D. Braatz, Distributional uncertainty analysis using power series and polynomial chaos expansions, J. Process Control 17 (3) (March 1, 2007) 229–240.

[15] P. Nimmegeers, D. Telen, F. Logist, J. Van Impe, Dynamic optimization of biological networks under parametric uncertainty, BMC Syst. Biol. 10 (1) (December 1, 2016) 86.

[16] V. Keshavarzzadeh, F. Fernandez, D.A. Tortorelli, Topology optimization under uncertainty via non-intrusive polynomial chaos expansion, Comput. Methods Appl. Mech. Eng. 318 (May 1, 2017) 120–147.

[17] L. Daróczy, G. Janiga, D. Thévenin, Analysis of the performance of a H-Darrieus rotor under uncertainty using Polynomial Chaos Expansion, Energy 113 (October 15, 2016) 399–412.

[18] A. Manan, J.E. Cooper, Prediction of uncertain frequency response function bounds using polynomial chaos expansion, J. Sound Vib. 329 (16) (August 2, 2010) 3348–3358.

[19] M. Villegas, F. Augustin, A. Gilg, A. Hmaidi, U. Wever, Application of the polynomial chaos expansion to the simulation of chemical reactors with uncertainties, Math. Comput. Simulat. 82 (5) (January 1, 2012) 805–817.

[20] T. Crestaux, O. Le Maitre, J.M. Martinez, Polynomial chaos expansion for sensitivity analysis, Reliab. Eng. Syst. Saf. 94 (7) (July 1, 2009) 1161–1172.

[21] A. Ben-Tal, A. Nemirovski, Robust optimization–methodology and applications, Math. Program. 92 (3) (May 1, 2002) 453–480.

[22] H.G. Beyer, B. Sendhoff, Robust optimization–a comprehensive survey, Comput. Methods Appl. Mech. Eng. 196 (33–34) (July 1, 2007) 3190–3218.

[23] D. Bertsimas, E. Litvinov, X.A. Sun, J. Zhao, T. Zheng, Adaptive robust optimization for the security constrained unit commitment problem, IEEE Trans. Power Syst. 28 (1) (July 24, 2012) 52–63.

[24] A. Serra, P. Galdi, R. Tagliaferri, Machine learning for bioinformatics and neuroimaging, Wiley Interdiscip. Rev. Data Min. Knowl. Discov. 8 (5) (September 2018) e1248.

[25] J. Lemley, S. Bazrafkan, P. Corcoran, Deep Learning for Consumer Devices and Services: pushing the limits for machine learning, artificial intelligence, and computer vision, IEEE Consum. Electron. Mag. 6 (2) (March 15, 2017) 48–56.

[26] K.G. Liakos, P. Busato, D. Moshou, S. Pearson, D. Bochtis, Machine learning in agriculture: a review, Sensors 18 (8) (August 2018) 2674.

[27] Z.M. Yaseen, M.F. Allawi, A.A. Yousif, O. Jaafar, F.M. Hamzah, A. El-Shafie, Non-tuned machine learning approach for hydrological time series forecasting, Neural Comput. Appl. 30 (5) (September 1, 2018) 1479–1491.

[28] A. Tripathy, A. Agrawal, S.K. Rath, Classification of sentiment reviews using n-gram machine learning approach, Expert Syst. Appl. 57 (September 15, 2016) 117–126.

[29] C.M. Bishop, Pattern Recognition and Machine Learning, Springer, New York, NY, 2006.

[30] V.T. Rajan, Optimality of the Delaunay triangulation in R d, Discrete Comput. Geom. 12 (2) (December 1, 1994) 189–202.

[31] I.M. Sobol', On the distribution of points in a cube and the approximate evaluation of integrals, Zh. Vychislitel noi Mat. Mat. Fiz. 7 (4) (1967) 784–802.

[32] D. Bertsimas, M. Sim, The price of robustness, Oper. Res. 52 (1) (February 2004) 35–53.

[33] A. Ben-Tal, L. El Ghaoui, A. Nemirovski, Robust Optimization, Princeton University Press, Princeton, New Jersey, August 10, 2009.

[34] J. Nayak, B. Naik, H.S. Behera, Fuzzy C-means (FCM) clustering algorithm: a decade review from 2000 to 2014, in: Computational Intelligence in Data Mining-Volume 2, Springer, New Delhi, 2015, pp. 133–149.

[35] J.C. Bezdek, R.J. Hathaway, Optimization of fuzzy clustering criteria using genetic algorithms, in: Proceedings of the First IEEE Conference on Evolutionary Computation. IEEE World Congress on Computational Intelligence, IEEE, June 27, 1994, pp. 589–594.

[36] S. Bandyopadhyay, Genetic algorithms for clustering and fuzzy clustering, Wiley Interdiscip. Rev. Data Min. Knowl. Discov. 1 (6) (November 2011) 524–531.

[37] O. Castillo, E. Rubio, J. Soria, E. Naredo, Optimization of the fuzzy C-means algorithm using evolutionary methods, Eng. Lett. 20 (1) (March 1, 2012) 61.

[38] G. Balaji, R. Balamurugan, L. Lakshminarasimman, Fuzzy clustered multi objective differential evolution for thermal generator maintenance scheduling, Int. J. Intell. Eng. Syst. 9 (1) (March 2016) 1–3.

[39] S. Sharma, P.D. Pantula, S.S. Miriyala, K. Mitra, A novel data-driven sampling strategy for optimizing industrial grinding operation under uncertainty using chance constrained programming, Powder Technol. 377 (January 2, 2021) 913–923.

[40] L.R. Petzold, Description of DASSL: A Differential/algebraic System Solver, Sandia National Labs., Livermore, CA, September 1, 1982.

[41] P.D. Pantula, S.S. Miriyala, K. Mitra, An evolutionary neuro-fuzzy C-means clustering technique, Eng. Appl. Artif. Intell. 89 (March 1, 2020) 103435.

[42] S. Haykin, N. Network, A comprehensive foundation, Neural Network. 2 (2004) (February 2004) 41.

[43] K. Deb, Multi-objective Optimization Using Evolutionary Algorithms, John Wiley & Sons, Chichester, UK, July 5, 2001.

[44] W. Wang, Y. Zhang, On fuzzy cluster validity indices, Fuzzy Set Syst. 158 (19) (October 1, 2007) 2095–2117.

14 ⠿

Dimensionality reduction using PCAs in feature partitioning framework

Tapan Kumar Sahoo[1], Atul Negi[2], Haider Banka[3]

[1]DEPARTMENT OF COMPUTER SCIENCE AND ENGINEERING, IIIT BHUBANESWAR, BHUBANESWAR, ODISHA, INDIA; [2]SCHOOL OF COMPUTER AND INFORMATION SCIENCES, UNIVERSITY OF HYDERABAD, HYDERABAD, TELANGANA, INDIA; [3]DEPARTMENT OF COMPUTER SCIENCE AND ENGINEERING, IIT (ISM) DHANBAD, DHANBAD, JHARKHAND, INDIA

14.1 Introduction

Dimensionality reduction is an interesting research area in machine learning and predictive data analysis [1−4]. The goal of dimensionality reduction is to transform the data or features from a higher-dimensional space to a lower-dimensional space. Principal component analysis (PCA) is an important unsupervised statistical method which can be used for many purposes, such as finding relationships between observations, extracting essential information from the data, outlier detection and removal, and reducing the dimension of data by keeping only the vital information. PCA technique has a wide range of applications across all fields of engineering, science, medicine, and humanities, such as machine learning [5−11], image processing [12−16], bioinformatics [17−19], information retrieval [20−24], finance [25−27], horticulture [28−30], clinical research [31−33], etc.

There are several variations of PCAs have been developed for unsupervised feature extraction and dimensionality reduction that address the issues, such as locality of feature, summarization of variance, recognition accuracy, recognition time, space and time complexities for one-dimensional and two-dimensional data. The main focus of this chapter is to present and analyze some variations of one-dimensional PCAs in feature partitioning framework.

Statistical Modeling in Machine Learning. https://doi.org/10.1016/B978-0-323-91776-6.00008-7

14.2 Principal component analysis (PCA)

PCA [34] is a subspace method that operates on whole-patterns of a data set and finds a lower-dimensional subspace that is used to transform the data from a higher-dimensional space to a lower-dimensional space. The selected subspace or PCA space represents the direction of the maximum variance of the given data set. PCA is used to find the most accurate data representation in a lower-dimensional space.

Suppose the training set consists of N vectors V_i, $(i = 1, 2, ..., N)$ of each dimension d. The mean vector be represented by $\overline{V}$. Then the covariance matrix (C_{PCA}) is calculated as follows:

$$C_{PCA} = \frac{1}{N} \sum_{i=1}^{N} (V_i - \overline{V})(V_i - \overline{V})^T,$$

$$\text{where } \overline{V} = \frac{1}{N} \sum_{i=1}^{N} V_i.$$

(14.1)

Here, e_{opt} is the optimal projection direction (axis) of C_{PCA} corresponding to the largest eigenvalue. We usually select a set of projection directions (axes) $e_1, e_2, ..., e_{d'}$ to maximize the strength of a feature vector. The optimal projection directions are the orthonormal eigenvectors corresponding to the first d' largest eigenvalues of C_{PCA}. Let $E_{PCA} = \left[e_1^{PCA} \middle| e_2^{PCA} \middle| ... \middle| e_{d'}^{PCA} \right]_{d \times d'}$ be the eigenvectors corresponding to d' largest eigenvalues. Then the family of principal components (projected values) $y_{i1}^{PCA}, y_{i2}^{PCA}, ..., y_{id'}^{PCA}$ for an input vector V_i is obtained as follows:

$$\left[Y_i^{PCA} \right]_{d' \times 1} = \begin{bmatrix} y_{i1}^{PCA} \\ y_{i2}^{PCA} \\ \vdots \\ y_{id'}^{PCA} \end{bmatrix} = [E_{PCA}]_{d' \times d}^T \times (V_i - \overline{V})_{d \times 1}.$$

(14.2)

With this transformation, each feature vector is represented by d' coefficients.

14.3 PCAs in feature partitioning framework

14.3.1 What is feature partitioning framework?

In conventional PCA, the size of a covariance matrix is $d \times d$, where d is the dimension of a vector pattern. In many classification problems, such as face recognition, image compression, neuroimaging based prediction, geographical data analysis, laser data-based classification, etc., the dimensionality of data is very high. In such cases, if we use conventional PCA for dimensionality reduction, then the size of the covariance matrix will be too large to accommodate in memory during execution. Thus, a lot of memory and time are wasted to extract the principal components. Further, sometimes

PCA features may not be effective in terms of recognition when variations are restricted to some parts of the pattern (image), e.g., in face recognition, the variation of the face due to change in expression, pose, etc. In such cases, it is necessary to partition a large sized pattern into subpatterns and apply conventional PCA on subpatterns [35]. If a pattern of dimension d is partitioned into p subpatterns then the size of each sub-patterns will be $\frac{d}{p}$, and hence the size of the subpattern covariance matrix will be $\frac{d}{p} \times \frac{d}{p}$. Thus, it will save a lot of memory and time during execution, and local variations of patterns are effectively handled.

14.3.2 Subpattern principal component analysis (SpPCA)

SpPCA [36] partitions each pattern into nonoverlapping subpatterns, then subpattern groups are formed by selecting the subpatterns belonging to a common region. Standard PCA features are extracted from each subpattern group by independently computing a subpattern covariance matrix for each group. The local subpattern features of a pattern are then synthesized into a single feature vector for the subsequent classification task. It is an effective technique with respect to image classification accuracy, as the image feature is based on local variation of patterns.

Suppose there are N training vectors V_i, $(i = 1, 2, ..., N)$ of each dimension d. Each vector is partitioned into p subvectors of each dimension u_j such that $d = \sum_{j=1}^{p} u_j$, i.e., $V_i = \{V_i^{(1)}, V_i^{(2)}, ..., V_i^{(p)}\}$, where $V_i^{(j)}$ is the jth subvector of ith vector. Since PCA is applied on each subvector group independently, there are p subpattern covariance matrices for p partitions: $C_{PCA}^{(1)}, C_{PCA}^{(2)}, ..., C_{PCA}^{(p)}$. Let $E_{PCA}^{(j)}$ be the selected eigenspace for jth subpattern group. Then, SpPCA feature vector is obtained as follows:

$$\left[Y_i^{SpPCA} \right]_{\left(\sum_{j=1}^{p} r_j \right) \times 1} = \begin{bmatrix} \left[E_{PCA}^{(1)} \right]^T_{r_1 \times u_1} \times \left(V_i^{(1)} - \overline{V^{(1)}} \right) \\ \left[E_{PCA}^{(2)} \right]^T_{r_2 \times u_2} \times \left(V_i^{(2)} - \overline{V^{(2)}} \right) \\ \vdots \\ \left[E_{PCA}^{(p)} \right]^T_{r_p \times u_p} \times \left(V_i^{(p)} - \overline{V^{(p)}} \right) \end{bmatrix}, \tag{14.3}$$

where $r_j (r_j \leq u_j)$ is the number of components selected for the jth subpattern group.

14.3.3 Cross-correlation subpattern principal component analysis (SubXPCA)

SpPCA extracts region-specific local features by computing subpattern covariance matrices. So it ignores the cross-correlations across subpatterns and results in high feature dimensionality. SubXPCA [37] instead reduces the dimensionality by applying standard PCA on SpPCA feature vectors. Thus, it maintains the cross-correlations across subpatterns. The following transformation obtains the SpPCA feature vector:

$$Y_{SubXPCA} = f_{PCA}([Y_{SpPCA}]),$$
(14.4)

where f_{PCA} is the standard PCA algorithm.

14.3.4   Hybrid principal component analysis (HPCA)

Conventional PCA handles global variations of patterns effectively, whereas SpPCA handles local variations of patterns effectively. But, none of them handle global and regional variations of the patterns simultaneously. Although SubXPCA captures hybrid variations of patterns, it still extracts the global features based on locally extracted subpattern features (i.e., cross-correlations across subpatterns are considered only). Especially for face recognition tasks, a face's appearance may vary with illumination, pose, and expression conditions. These factors affect the face locally as well as globally.

The HPCA [38,39] extracts statistical hybrid pattern feature by combining subpattern based SpPCA [36] feature with whole-pattern based standard PCA [34] feature. It can be classified into two types, extended subpattern principal component analysis (ESpPCA) and extended cross-correlation principal component analysis (ESubXPCA). ESpPCA merges local SpPCA features with global PCA features. Thus, it maintains the intra-subpattern and whole-pattern correlations across the vector patterns. However, ESpPCA lacks intersubpattern whole-pattern correlations and which results in high feature dimensionality. ESubXPCA reduces the feature dimensionality of ESpPCA by applying standard PCA on ESpPCA feature vectors, which maintains the intra- and inter-subpattern and whole-pattern correlations. The procedure to extract hybrid PCA (ESpPCA and ESubXPCA) [38] features from 2D images is given below.

14.3.4.1   Procedure to extract hybrid PCA features from 2D face images
Suppose there are N training images, each of dimension $m \times n$. Let the training set be $train = \{A_i : i = 1, 2, ..., N\}$, where A_i is the ith training image.

Step 1: Transform each 2D image A_i of dimension $m \times n$ into 1D vector V_i of dimension $d \times 1$, where $d = m \times n$.

Local pattern feature extraction
Step 2: Partition each vector pattern V_i into p nonoverlapping subpatterns, i.e.,

$$V_i = \left\{ V_i^{(j)} : j = 1, 2, ..., p \right\} \quad such \; that \quad V_i = \begin{bmatrix} V_i^{(1)} \\ V_i^{(2)} \\ \vdots \\ V_i^{(p)} \end{bmatrix}.$$

Step 3: Group statistically similar subpatterns of the training patterns as one group. Since there are p subpatterns per pattern, there will be p subpattern groups is given below:

$$Gr^{(j)} = \left\{ V_i^{(j)} : i = 1, 2, \ldots, N \right\}, j = 1, 2, \ldots, p. \tag{14.5}$$

Step 4: Apply standard PCA [34] on each subpattern vector group $Gr^{(j)}$ independently. The extracted subpattern feature $Y_i^{(j)PCA}$ for ith pattern and jth subpattern is given below:

$$\left[Y_i^{(j)PCA} \right]_{r_j \times 1} = \left[E_{PCA}^{(j)} \right]_{r_j \times u_j}^T \times \left(V_i^{(j)} - \overline{V^{(j)}} \right), j = 1, 2, \ldots, p, \tag{14.6}$$

where $r_j (r_j \le u_j)$ is the number of selected components for the subpattern group $Gr^{(j)}$, u_j is the dimension of *j*th subpattern vector $V_i^{(j)}$, $\overline{V^{(j)}} = \frac{1}{N} \sum_{i=1}^{N} V_i^{(j)}$ is the mean subpattern vector for group $Gr^{(j)}$, $E_{PCA}^{(j)}$ is the eigenspace containing r_j eigenvectors for the subpattern group $Gr^{(j)}$.

Step 5: Obtain SpPCA [36] feature vector by merging each subpattern feature vectors of a pattern. The SpPCA feature vector $\left(Y_{SpPCA}^i \right)$ for ith pattern is given below:

$$\left[Y_i^{SpPCA} \right]_{\left(\sum_{j=1}^{p} r_j \right) \times 1} = \begin{bmatrix} \left[E_{PCA}^{(1)} \right]_{r_1 \times u_1}^T \times \left(V_i^{(1)} - \overline{V^{(1)}} \right) \\ \left[E_{PCA}^{(2)} \right]_{r_2 \times u_2}^T \times \left(V_i^{(2)} - \overline{V^{(2)}} \right) \\ \vdots \\ \left[E_{PCA}^{(p)} \right]_{r_p \times u_p}^T \times \left(V_i^{(p)} - \overline{V^{(p)}} \right) \end{bmatrix}, \tag{14.7}$$

Thus the set of SpPCA feature vectors for all N training images is $Y_{SpPCA} = \left\{ Y_i^{SpPCA} : i = 1, 2, \ldots, N \right\}$.

Global pattern feature extraction
Step 6: Extract global pattern feature using standard PCA on whole-patterns. The global PCA feature $\left(Y_i^{PCA} \right)$ for ith image is given below:

$$\left[Y_i^{PCA} \right]_{d' \times 1} = \left[E_{PCA} \right]_{d' \times d}^T \times \left(V_i - \overline{V} \right)_{d \times 1}, \tag{14.8}$$

where E_{PCA} is the eigenspace containing d' (where $d' \le d$) unit eigenvectors for the whole-patterns, $\overline{V} = \frac{1}{N} \sum_{i=1}^{N} V_i$ is the global pattern mean vector. The set of standard PCA features for all N training images is $Y_{PCA} = \left\{ Y_i^{PCA} : i = 1, 2, \ldots, N \right\}$.

Hybrid pattern feature extraction using ESpPCA
Step 7: The ESpPCA feature is obtained by merging local SpPCA feature $\left(Y_{SpPCA} \right)$ in Eq. (14.7) with standard global PCA feature $\left(Y_{PCA} \right)$ in Eq. (14.8), as they have similar

statistical properties. The dimension of ESpPCA feature is the sum of dimension of SpPCA feature vector and standard PCA feature vector. The ESpPCA feature vector $\left(Y_i^{ESpPCA} \right)$ for ith image is given below:

$$\left[Y_i^{ESpPCA} \right]_{\left(\sum\limits_{j=1}^{p} r_j + d' \right) \times 1} = \begin{bmatrix} \left[E_{PCA}^{(1)} \right]_{r_1 \times u_1}^{T} \times \left(V_i^{(1)} - \overline{V^{(1)}} \right) \\ \left[E_{PCA}^{(2)} \right]_{r_2 \times u_2}^{T} \times \left(V_i^{(2)} - \overline{V^{(2)}} \right) \\ \vdots \\ \left[E_{PCA}^{(p)} \right]_{r_p \times u_p}^{T} \times \left(V_i^{(p)} - \overline{V^{(p)}} \right) \\ \left[E_{PCA} \right]_{d' \times d}^{T} \times \left(V_i - \overline{V} \right) \end{bmatrix}, \tag{14.9}$$

Thus the set of ESpPCA feature for all N training images is $Y_{ESpPCA} = \left\{ Y_i^{ESpPCA} : i = 1, 2, ..., N \right\}$.

Hybrid pattern feature extraction using ESubXPCA

Step 8: The ESpPCA feature obtained from Eq. (14.9) carries hybrid variations of patterns, but it suffers from dimensionality problem. The ESubXPCA calculates hybrid eigenspace using an intersubpattern whole-pattern covariance matrix. The final ESubXPCA feature is obtained by projecting ESpPCA feature on the selective hybrid eigenspace of ESubXPCA. The ESubXPCA feature vector $\left(Y_i^{ESubXPCA} \right)$ for ith image is given below:

$$\left[Y_i^{ESubXPCA} \right]_{d''' \times 1} = \left[E_{ESubXPCA} \right]_{d''' \times \left(\sum\limits_{j=1}^{p} r_j + d' \right)}^{T} \times \left(Y_i^{ESpPCA} - \overline{Y_{ESpPCA}} \right), \tag{14.10}$$

where $E_{ESubXPCA}$ is the selective eigenspace containing d''' (where $d''' \leq \sum\limits_{j=1}^{p} r_j + d'$) unit eigenvectors for ESubXPCA and $\overline{Y_{ESpPCA}} = \frac{1}{N} \sum\limits_{i=1}^{N} Y_i^{ESpPCA}$ is the mean ESpPCA feature. The set of ESpPCA feature for all N training images is $Y_{ESubXPCA} = \left\{ Y_i^{ESubXPCA} : i = 1, 2, ..., N \right\}$.

In the above procedure, the Steps 1 to 5 are for SpPCA-based local feature extraction; Step 6 is for standard PCA-based global feature extraction; Steps 1 to 7 are for ESpPCA-based hybrid feature extraction and Steps 1 to 8 are for ESubXPCA-based hybrid feature extraction. So the standard PCA, SpPCA and ESpPCA are the subprocedure of ESubXPCA. Further, since SubXPCA features can be obtained by applying PCA on SpPCA feature vectors (obtained from Step 5), standard PCA and SpPCA are the subprocedure of SubXPCA. An abstract flow diagram of all the feature partitioning PCAs is depicted in Fig. 14.1.

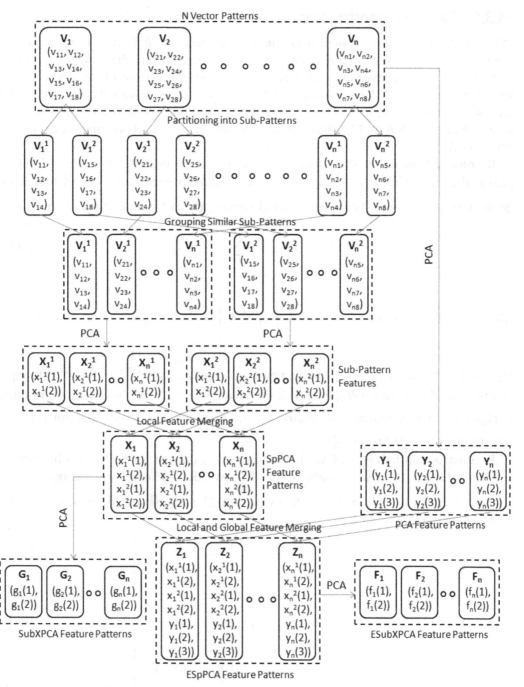

FIGURE 14.1 Feature extraction using various PCAs in feature partitioning framework.

14.3.5   Pattern reconstruction

PCA-based pattern reconstruction is the reverse process of PCA feature extraction [40,41]. Here, the approximate pattern is reconstructed from the selected principal components and eigenvectors. Since SpPCA is the subprocedure of SubXPCA, reconstruction using SubXPCA is dependent on the reconstruction using SpPCA. Further, PCA and SpPCA are the subprocedures of ESpPCA and ESubXPCA, reconstruction using hybrid PCAs (ESpPCA and ESubXPCA) are dependent on the reconstruction using PCA and SpPCA.

Reconstruction using PCA: In PCA, a pattern is reconstructed by combining d' principal components $(y_{i1}, y_{i2}, ..., y_{id'})$ with the eigensubspace consisting of first d' eigenvectors $(e_1, e_2, ..., e_{d'})$. The reconstructed vector $\left(\widetilde{V_i^{PCA}}\right)$ for an image pattern A_i is

$$\widetilde{V_i^{PCA}} = EY_i + \overline{V} = \sum_{l=1}^{d'} e_l y_{il} + \overline{V}, \tag{14.11}$$

where $E = E_{PCA} = [e_1|e_2|...|e_{d'}]_{d \times d'}$, $Y_i = Y_i^{PCA} = \begin{bmatrix} y_{i1} \\ y_{i2} \\ \vdots \\ y_{id'} \end{bmatrix}_{d' \times 1}$, $e_l = e_l^{PCA}$ is the lth best

eigenvector of the covariance matrix using ESubXPCA, $y_{il} = y_{il}^{PCA}$ is the lth principal component of ith image using PCA, $\overline{V}$ is the mean vector, $d' \leq d$, (d is the total number of eigenvectors of the covariance matrix C_{PCA}). Then, the reconstructed vector ($\widetilde{V_i^{PCA}}$) is transformed into 2D image ($\widetilde{A_i^{PCA}}$).

Reconstruction using SpPCA: The reconstructed pattern for SpPCA is obtained by combining reconstructed subpattern of each partition. The reconstruction at each subpattern is similar to PCA-based reconstruction. The reconstructed vector ($\widetilde{V_i^{SpPCA}}$) for an image pattern A_i is

$$\widetilde{V_i^{SpPCA}} = \begin{bmatrix} E^{(1)} Y_i^{(1)} + \overline{V^{(1)}} \\ E^{(2)} Y_i^{(2)} + \overline{V^{(2)}} \\ \vdots \\ E^{(p)} Y_i^{(p)} + \overline{V^{(p)}} \end{bmatrix} = \begin{bmatrix} \sum_{l=1}^{r_1} e_l^{(1)} y_{il}^{(1)} + \overline{V^{(1)}} \\ \sum_{l=1}^{r_2} e_l^{(2)} y_{il}^{(2)} + \overline{V^{(2)}} \\ \vdots \\ \sum_{l=1}^{r_p} e_l^{(p)} y_{il}^{(p)} + \overline{V^{(p)}} \end{bmatrix}, \tag{14.12}$$

where $E^{(j)} = E_{PCA}^{(j)} = \left[e_1^{(j)} \middle| e_2^{(j)} \middle| \cdots \middle| e_{r_j}^{(j)} \right]_{u_j \times r_j}$, $Y_i^{(j)} = Y_i^{(j)PCA} = \begin{bmatrix} y_{i1}^{(j)} \\ y_{i2}^{(j)} \\ \vdots \\ y_{ir_j}^{(j)} \end{bmatrix}_{r_j \times 1}$, $Y_i^{(j)PCA}$ is the jth

subpattern PCA feature vector of ith image, and $E_{PCA}^{(j)}$ is the selected eigenspace of jth subpattern group using PCA, $y_{il}^{(j)} = y_{il}^{(j)PCA}$ is the lth principal component of jth sub-pattern vector of ith image, $e_l^{(j)} = e_l^{(j)PCA}$ is the lth eigenvector of the selected eigenspace of jth subpattern covariance matrix. The reconstructed vector $(V_i^{\widetilde{SpPCA}})$ is then trans-formed into 2D image $(A_i^{\widetilde{SpPCA}})$.

Reconstruction using SubXPCA: Pattern reconstruction for an image A_i using SubXPCA is obtained by the following two steps: (1) reconstructing SpPCA feature vector $\left(Y_i^{\widetilde{SpPCA}} \right)$ from SubXPCA feature vector $(Y_i^{SubXPCA})$,

$$Y_i^{\widetilde{SpPCA}} = EY_i + \overline{Y_{SpPCA}} = \sum_{l=1}^{d''} e_l y_{il} + \overline{Y_{SpPCA}}, \tag{14.13}$$

where $E = E_{SubXPCA}$ is the selected eigenspace consisting of first d'' (where $d'' \leq \sum r_j$) eigenvectors of SubXPCA covariance matrix, $Y_i = Y_i^{SubXPCA}$ is the ith SubXPCA feature vector, $y_{il} = y_{il}^{SubXPCA}$ is the lth principal component of ith image using SubXPCA, $e_l = e_l^{SubXPCA}$ is the lth eigenvector of the selected eigenspace of SubXPCA, and $\overline{Y_{SpPCA}}$ is the mean SpPCA feature vector, and (2) reconstructing the image from SpPCA feature vector using Eq. (14.12).

Reconstruction using ESpPCA: Since ESpPCA combines local SpPCA feature with global PCA feature, the reconstructed pattern (image) for ESpPCA is the reconstructed image for SpPCA $(A_i^{\widetilde{SpPCA}})$ and the reconstructed image for PCA $(A_i^{\widetilde{PCA}})$. It is performed in two steps, (1) separating SpPCA and PCA feature vectors from ESpPCA feature vector, (2) reconstructing images from PCA and SpPCA feature vectors using Eqs. (14.11) and (14.12), respectively.

Reconstruction using ESubXPCA: The reconstructed pattern (image) for ESubXPCA is obtained in three steps, (1) reconstructing ESpPCA feature vector $\left(Y_i^{\widetilde{ESpPCA}} \right)$ from ESubXPCA feature vector $(Y_i^{ESubXPCA})$,

$$Y_i^{\widetilde{ESpPCA}} = EY_i + \overline{Y_{ESpPCA}} = \sum_{l=1}^{d'''} e_l y_{il} + \overline{Y_{ESpPCA}}, \tag{14.14}$$

where $E = E_{ESubXPCA}$ is the selected eigenspace consisting of first d''' (where

$d''' \leq (\sum r_j + d'))$ eigenvectors of ESubXPCA covariance matrix, $Y_i = Y_i^{ESubXPCA}$ is the ith ESubXPCA feature vector, $y_{il} = y_{il}^{ESubXPCA}$ is the lth principal component of ith image using ESubXPCA, $e_l = e_l^{ESubXPCA}$ is the lth eigenvector of the selected eigenspace of ESubXPCA, and $\overline{Y_{ESpPCA}}$ is the mean ESpPCA feature vector; (2) separating SpPCA and PCA feature vector components from ESpPCA feature vector; and (3) reconstructing images from PCA and SpPCA feature vectors using Eqs. (14.11) and (14.12), respectively.

The reconstruction using PCA, SpPCA, SubXPCA, ESpPCA, and ESubXPCA by preserving 90% (approx.) and 95% (approx.) of variances for an image are shown in Fig. 14.2. Since ESpPCA combines SpPCA feature with PCA feature, it's reconstructed image has two components, i.e., reconstructed image using SpPCA and reconstructed image using PCA. Further, since ESubXPCA is an extension of ESpPCA, the reconstruction involves reconstruction from ESubXPCA to ESpPCA and ESpPCA to SpPCA and PCA. It is observed from Fig. 14.2 that the reconstructed images of ESubXPCA (both PCA and SpPCA) and SubXPCA are more blurred as compared to reconstructed images using PCA, SpPCA, and ESpPCA. It is because of the two steps of dimensionality reduction in ESubXPCA. The reconstructed images of ESpPCA are the same as reconstructed images of PCA and SpPCA, as ESpPCA feature vector is the simple concatenation of PCA and SpPCA feature vectors.

14.3.6   Theoretical analysis

In this section, several issues affecting the performance of a feature partitioning PCA such as summarization of variance, time, and space complexities [38,40–43] are briefly discussed as follows.

14.3.6.1   Summarization of variance

It deals with amount of variance carried out by some cumulative principal components [38,40,41]. If λ_i is the ith largest eigenvalue of the covariance matrix of size $d \times d$, then the percentage of variance preserved by the ith principal component is $\frac{\lambda_i}{\sum_{i=1}^{d} \lambda_i} \times 100\%$. The summarization of variance due to d' (where $d' \leq d$) largest principal components is

$\frac{\sum_{i=1}^{d'} \lambda_i}{\sum_{i=1}^{d} \lambda_i} \times 100\%$. A dimensionality reduction is said to be lossless if the summarization of variance is 100%. The principal components corresponding to d' largest eigenvalues are sometimes called as major components and are associated with meaningful information for classification. Whereas, the unused $d - d'$ components are sometimes called as minor components and are meaningless for the classification because there is minimal variance is gained by the eigenvalues corresponding to minor components. The minor components preserve the information due to experimental errors, noises, etc. [44,45].

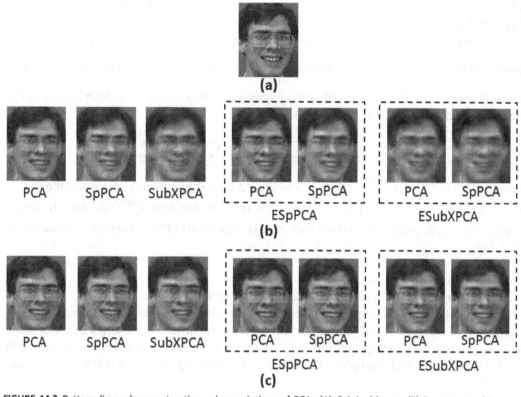

FIGURE 14.2 Pattern (image) reconstruction using variations of PCAs (A) Original image (B) Reconstructed images by preserving 90% (approx.) of variance at each step of dimensionality reduction and (C) Reconstructed images by preserving 95% (approx.) of variance at each step of dimensionality reduction.

If for each step of dimensionality reduction $\alpha\%$ of variance is preserved as major components and remaining $(100 - \alpha)\%$ variance is eliminated as minor components, then the cumulative-variance preserved for a method involving a single level, two levels and three levels of dimensionality reductions are $\alpha\%$, $\alpha\%(\alpha\%(100))\%$, and $\alpha\%(\alpha\%(\alpha\%(100)))\%$ respectively and so on. From Section 14.3.4, it is observed that the methods, such as standard PCA, SpPCA, and ESpPCA require only single step of dimensionality reduction; whereas the methods, such as SubXPCA and ESubXPCA require two steps of dimensionality reduction. The cross-correlation across subpatterns and whole-patterns eliminate the intersubpattern and whole-pattern meaningless information. So, the summarization of variance in ESubXPCA is better than ESpPCA. Further, the smaller number of components sufficiently carry meaningful information for classification. The summarization of variance in ESpPCA is the sum of the summarization of variance of PCA and SpPCA, as the ESpPCA feature vector is obtained by combining the standard PCA and SpPCA feature vectors. For SubXPCA, the feature vectors are obtained by applying PCA on SpPCA feature vectors [37]. The SpPCA has already eliminated the minor components at subpattern level. So, the dimensionality of SubXPCA feature vector is less than those of PCA and SpPCA feature vectors.

Theorem 14.1 If d is the dimension of whole-pattern and each pattern divided into p nonoverlapping subpatterns of dimension u_j such that $d = \sum_{j=1}^{p} u_j$, then the summarization of variance under lossless dimensionality reduction using PCA, SpPCA, SubXPCA, ESpPCA, and ESubXPCA are related as

$$\sum_{i=1}^{d} \lambda_i^{PCA} = \sum_{j=1}^{p}\sum_{i=1}^{u_j} \lambda_{ji}^{SpPCA} = \sum_{i=1}^{d} \lambda_i^{SubXPCA} = \frac{1}{2}\sum_{i=1}^{2d} \lambda_i^{ESpPCA} = \frac{1}{2}\sum_{i=1}^{2d} \lambda_i^{ESubXPCA},$$

where λ_i^{PCA}, $\lambda_i^{SubXPCA}$, λ_i^{ESpPCA}, $\lambda_i^{ESubXPCA}$ are the ith largest eigenvalue of covariance matrix of PCA, SubXPCA, ESpPCA, and ESubXPCA respectively and λ_{ji}^{SpPCA} is the ith largest eigenvalue of jth subpattern covariance matrix. The proof of this theorem is available in Ref. [38].

14.3.6.2 Time complexity analysis

The conventional PCA and its variations spend much time in calculating covariance matrices and relatively insignificant amount of time in reading data, finding eigenvalues, extracting features, etc. [37]. This section focuses on time complexities of calculating covariance matrices using standard *big O* notation [38,46] for the hybrid PCAs (ESpPCA and ESubXPCA), and compare them with the existing standard PCA, SpPCA and SubXPCA.

Let's consider N images, each of dimension $m \times n$ (let the dimension in vector form be $d \times 1$). Then, time complexity of calculating covariance matrix using standard PCA (T_{PCA}) is

$$T_{PCA} = O(Nd^2) = O(Nm^2n^2). \tag{14.15}$$

For SpPCA, the pattern divided into p subpatterns of each dimension $u = \frac{d}{p}$. So the time complexity of calculating subpattern covariance matrices for all subpattern groups (T_{SpPCA}) is

$$T_{SpPCA} = O(Npu^2) = O\left(\frac{Nd^2}{p}\right) = O\left(\frac{Nm^2n^2}{p}\right). \tag{14.16}$$

As the ESpPCA feature is the simple combination of PCA and SpPCA features, so the time complexity to calculate covariance matrix for ESpPCA (T_{ESpPCA}) is

$$T_{ESpPCA} = T_{PCA} + T_{SpPCA} = O\left(N\left(1+\frac{1}{p}\right)m^2n^2\right). \tag{14.17}$$

Let the dimension of each SpPCA feature vector is pr (where $r \le u$). The time complexity to calculate cross subpattern covariance matrix ($T_{PCA-SpPCA}$) is $O(p^2 r^2)$. So, the total time complexity of calculating covariance matrix using SubXPCA ($T_{SubXPCA}$) is

$$T_{SubXPCA} = T_{SpPCA} + T_{PCA-SpPCA} = O\left(N\left(\frac{m^2 n^2}{p} + p^2 r^2\right)\right). \tag{14.18}$$

Let the dimension of ESpPCA feature vector is $(pr + d')$ (where $r \leq u$ and $d' \leq d$). Then, the time complexity to calculate cross subpattern-whole-pattern covariance matrix ($T_{PCA-ESpPCA}$) is $O\left((pr + d')^2\right)$. As ESubXPCA operates on ESpPCA feature vectors, so the total time complexity of calculating the covariance matrix ($T_{ESubXPCA}$) is

$$T_{ESubXPCA} = T_{ESpPCA} + T_{PCA-ESpPCA} = O\left(N\left(\left(1 + \frac{1}{p}\right)m^2 n^2 + (pr + d')^2\right)\right). \tag{14.19}$$

From the above Eqs. (14.15)−(14.19), it is observed that, the order of time complexities of calculating the covariance matrices is

$$T_{SpPCA} \leq T_{PCA} \leq T_{SubXPCA} \leq T_{ESpPCA} \leq T_{ESubXPCA}.$$

The speed of a face recognition system depends on the time required for training and testing related as follows:

Time (Face recognition) = Time (Training) + Time (Testing)

Time (Training) = Time (Image to vector conversion) + Time (Partitioning) + Time (Calculating covariance matrix) + Time (Calculating eigenvectors) + Time (Training feature extraction)

Time (Testing) = Time (Image to vector conversion) + Time (Partitioning) + Time (Test feature extraction) + Time (Recognition)

However, in a practical scenario, the covariance matrix is calculated only once during training. The selected eigenspace of the covariance matrix is stored in memory for subsequent test feature extraction. During testing, the test feature is extracted by projecting a test vector on the selected eigenspace of training vectors. The time required for partitioning, calculating eigenvectors, training feature extraction, test feature extraction are negligible [37]. So, the efficiency of a face recognition system depends on the dimensionality of feature vectors. The order of dimensionality of the variations of PCA is ESubXPCA $\leq$ SubXPCA $<$ PCA $<$ SpPCA $<$ ESpPCA. For this reason, the recognition time using ESubXPCA and SubXPCA is higher than other variations of PCA.

14.3.6.3 *Space complexity analysis*

Space complexity of an algorithm is the amount of space required to execute that algorithm [46]. In a PCA-based face recognition system (FRS), the eigenspace of the covariance matrix and feature vectors are calculated only once for the training set and stored in memory for subsequent test feature extraction and recognition tasks. However, the selected eigenspace requires very less space as compare to the corresponding covariance matrix. The number of components chosen is always less than or equal to the number of rows/columns of the covariance matrix. So the selected eigenspace can be accommodated within the covariance matrix space (if memory overwritten is allowed). Hence, the space complexity of a PCA-based FRS is the amount of space required to store the covariance matrix and feature vectors of the training set [39].

Suppose N training images, each of dimension $m \times n$. Each image is transformed into a vector of dimension d, where $d = m \times n$. For partition-based approaches, each vector is partitioned into p subvectors, each of dimension u. The space complexities for variations of PCAs are given below.

Let the dimension of each PCA feature vector be d' ($d' \leq d$). The space complexity for a PCA-based FRS (S_{PCA}) is the sum of space complexity to store PCA covariance matrices (S^C_{PCA}) and the space complexity to store N PCA training feature vectors (S^F_{PCA}). So

$$S_{PCA} = S^C_{PCA} + S^F_{PCA} = d^2 + Nd' = m^2 n^2 + Nd'. \tag{14.20}$$

The space complexity for a SpPCA-based FRS (S_{SpPCA}) is the space for the corresponding covariance matrices (S^C_{SpPCA}) plus space for N SpPCA training feature vectors (S^F_{SpPCA}). Let the dimension of each subpattern feature vector be r ($r \leq u$). Space needed for subpattern feature vectors per image is pr. So

$$S_{SpPCA} = S^C_{SpPCA} + S^F_{SpPCA} = pu^2 + Npr = \frac{m^2 n^2}{p} + Npr. \tag{14.21}$$

For SubXPCA, the space complexity ($S_{SubXPCA}$) is the sum of space complexities of its covariance matrix ($S^C_{SubXPCA}$) and training feature vectors ($S^F_{SubXPCA}$). Since SubXPCA feature is obtained by applying PCA on SpPCA feature vectors, ($S^C_{SubXPCA}$) is the sum of S^C_{SpPCA} and the covariance matrix ($S^C_{PCA-SpPCA}$) needed to obtain SubXPCA feature vectors from SpPCA feature vectors. Let the dimension of each SubXPCA feature vector is d''. So

$$S_{SubXPCA} = S^C_{SubXPCA} + S^F_{SubXPCA} = \left(S^C_{SpPCA} + S^C_{PCA-SpPCA} \right) + Nd'' = \frac{m^2 n^2}{p} + (pr)^2 + Nd''. \tag{14.22}$$

The space complexity for ESpPCA (S_{ESpPCA}) is the sum of space complexities of its covariance matrix (S^C_{ESpPCA}) and training feature vectors (S^F_{ESpPCA}). Since ESpPCA combines SpPCA feature with and PCA feature, S^C_{ESpPCA} is sum of S^C_{SpPCA} and S^C_{PCA}. So

$$S_{ESpPCA} = S^C_{ESpPCA} + S^F_{ESpPCA} = m^2 n^2 + \frac{m^2 n^2}{p} + N(pr + d') = \left(1 + \frac{1}{p} \right) m^2 n^2 + N(pr + d'). \tag{14.23}$$

For ESubXPCA, the space complexity ($S_{ESubXPCA}$) is the sum of space complexities of its covariance matrix ($S^C_{ESubXPCA}$) and the training feature vectors ($S^F_{ESubXPCA}$). As in ESubXPCA, PCA is applied on ESpPCA feature vectors, $S^C_{ESubXPCA}$ is the sum of S^C_{ESpPCA} and the covariance matrix ($S^C_{PCA-ESpPCA}$) needed to obtain ESubXPCA feature vectors from ESpPCA feature vectors. Let the dimension of each ESubXPCA feature vector be d'''. So

$$S_{ESubXPCA} = S^C_{ESubXPCA} + S^F_{ESubXPCA} = \left(S^C_{ESpPCA} + S^C_{PCA-ESpPCA} \right) + Nd'''$$

$$= \left(1 + \frac{1}{p} \right) m^2 n^2 + (pr + d')^2 + Nd'''. \tag{14.24}$$

From Eqs. (14.20)−(14.24), it is observed that the order of space complexity for the covariance matrices is $S^C_{SpPCA} \le S^C_{SubXPCA} \le S^C_{PCA} \le S^C_{ESpPCA} \le S^C_{ESubXPCA}$. However, for a FRS, both components, i.e., space for covariance matrix and space for feature vectors, are important. The factor N in the feature vector component indicates that the dimension of the feature vector has a high impact on space complexity.

14.4 Summary

This chapter has presented some one-dimensional PCAs (SpPCA, SubXPCA, hybrid PCAs (ESpPCA and ESubXPCA)) in feature partitioning framework. The classical PCA operates on whole-patterns whereas SpPCA operates on subpatterns. Thus, PCA handles the global variations patterns whereas SpPCA handles the local variations of patterns. The SubXPCA effectively handles the cross-correlations across subpatterns and whole-patterns by further applying PCA on SpPCA feature vector set. The hybrid PCAs operate on subpattern vectors of a pattern and the same whole-pattern vector simultaneously and extract the hybrid PCA features. ESpPCA concatenates the local SpPCA feature with the global PCA feature; thus, it lacks the intersubpattern whole-pattern correlations. ESubXPCA further reduces the dimensionality of ESpPCA feature vector by maintaining the cross-correlations across the subpatterns and the whole-pattern of a vector pattern. The critical issues such as summarization of variances, time and space complexities for the feature partitioning framework based PCAs are investigated theoretically. It can be proven experimentally that the ESubXPCA exhibits the superior recognition accuracy and time among other PCAs in feature partitioning framework on a challenging data set (face images with variable illumination, pose, and expression conditions). Although ESubXPCA achieve the minimum feature dimensionality, the total space requirement for the FRS is higher using ESubXPCA than others due to bigger size covariance matrices at multi-levels. Another limitation of this approach is to transform an image into a vector before applying these methods. We can eliminate such drawbacks by extending the above ideas to a two-dimensional feature partitioning pattern framework.

Acknowledgments

The authors wish to acknowledge the editors for encouraging us to contribute a chapter to this book. Further, we are thankful to our previous publishers (Springer Nature and IEEE) for granting permission to include some parts of our previous work in this book.

References

[1] L. Van Der Maaten, E. Postma, J. Van den Herik, Dimensionality reduction: a comparative, J. Mach. Learn. Res. 10 (66−71) (2009) 13.

[2] S. Khalid, T. Khalil, S. Nasreen, A survey of feature selection and feature extraction techniques in machine learning, in: 2014 Science and Information Conference, IEEE, 2014, pp. 372−378.

[3] B. Mwangi, T.S. Tian, J.C. Soares, A review of feature reduction techniques in neuroimaging, Neuroinformatics 12 (2) (2014) 229−244.

[4] M. Belkin, P. Niyogi, Laplacian eigenmaps for dimensionality reduction and data representation, Neural Comput. 15 (6) (2003) 1373−1396.

[5] I. Jolliffe, Principal Component Analysis, Springer, 2011.

[6] T. Metsalu, J. Vilo, ClustVis: a web tool for visualizing clustering of multivariate data using principal component analysis and heatmap, Nucleic Acids Res. 43 (W1) (2015) W566−W570.

[7] J. Chen, Y. Leng, N. Zhang, L. Yu, The Removal of Interference Noise of ICT Using PCA Method.

[8] F. Song, Z. Guo, D. Mei, Feature selection using principal component analysis, in: 2010 International Conference on System Science, Engineering Design and Manufacturing Informatization, Vol. 1, IEEE, 2010, pp. 27−30.

[9] W. Zhao, R. Chellappa, P.J. Phillips, A. Rosenfeld, Face recognition: a literature survey, ACM Comput. Surv. 35 (4) (2003) 399−458.

[10] Y. Fu, G. Guo, T.S. Huang, Age synthesis and estimation via faces: a survey, IEEE Trans. Pattern Anal. Mach. Intell. 32 (11) (2010) 1955−1976.

[11] V. Unnikrishnan, K. Choudhari, S.D. Kulkarni, R. Nayak, V. Kartha, C. Santhosh, Analytical predictive capabilities of laser induced breakdown spectroscopy (LIBS) with principal component analysis (PCA) for plastic classification, RSC Adv. 3 (48) (2013) 25872−25880.

[12] M. Mudrova, A. Procházka, Principal component analysis in image processing, in: Proceedings of the MATLAB Technical Computing Conference, Prague, 2005.

[13] J. Yang, D. Zhang, A.F. Frangi, J.-y. Yang, Two-dimensional PCA: a new approach to appearance-based face representation and recognition, IEEE Trans. Pattern Anal. Mach. Intell. 26 (1) (2004) 131−137.

[14] M. González-Audícana, J.L. Saleta, R.G. Catalán, R. García, Fusion of multispectral and panchromatic images using improved IHS and PCA mergers based on wavelet decomposition, IEEE Trans. Geosci. Rem. Sens. 42 (6) (2004) 1291−1299.

[15] S. Dambreville, Y. Rathi, A. Tannen, Shape-based approach to robust image segmentation using kernel pca, in: 2006 IEEE Computer Society Conference on Computer Vision and Pattern Recognition (CVPR'06), Vol. 1, IEEE, 2006, pp. 977−984.

[16] Y. Ke, R. Sukthankar, et al., PCA-SIFT: a more distinctive representation for local image descriptors, CVPR 4 (2) (2004) 506−513.

[17] W. Stacklies, H. Redestig, M. Scholz, D. Walther, J. Selbig, pcaMethods—a bioconductor package providing pca methods for incomplete data, Bioinformatics 23 (9) (2007) 1164−1167.

[18] S.H. Kim, D. Kang, Z. Huo, Y. Park, G.C. Tseng, Meta-analytic principal component analysis in integrative omics application, Bioinformatics 34 (8) (2017) 1321−1328.

[19] T. Konishi, Principal component analysis for designed experiments, BMC Bioinf. 16 (18) (2015) S7.

[20] T.N. Underhill, An Introduction to Information Retrieval Using Singular Value Decomposition and Principal Component Analysis, 2007.

[21] M.W. Berry, D.I. Martin, Principal component analysis for information retrieval, in: Handbook of Parallel Computing and Statistics, Chapman and Hall/CRC, 2005, pp. 415−430.

[22] V. Vinay, I.J. Cox, K. Wood, N. Milic-Frayling, A comparison of dimensionality reduction techniques for text retrieval, in: Fourth International Conference on Machine Learning and Applications (ICMLA'05), IEEE, 2005, p. 6.

[23] U. Sinha, H. Kangarloo, Principal component analysis for content-based image retrieval, Radiographics 22 (5) (2002) 1271–1289.

[24] F. Cai, H. Chen, Z. Shu, Web document ranking via active learning and kernel principal component analysis, Int. J. Mod. Phys. C 26 (04) (2015) 1550041.

[25] L. Yang, An Application of Principal Component Analysis to Stock Portfolio Management, 2015.

[26] G. Pasini, Principal component analysis for stock portfolio management, Int. J. Pure Appl. Math. 115 (1) (2017) 153–167.

[27] M. Ghorbani, E.K. Chong, Stock price prediction using principal components, PLoS One 15 (3) (2020) e0230124.

[28] I. Gergen, M. Harmanescu, Application of principal component analysis in the pollution assessment with heavy metals of vegetable food chain in the old mining areas, Chem. Cent. J. 6 (1) (2012) 156.

[29] A.F. Iezzoni, M.P. Pritts, Applications of principal component analysis to horticultural research, Hortscience 26 (4) (1991) 334–338.

[30] E. Nevo, D. Zohary, A. Brown, M. Haber, Genetic diversity and environmental associations of wild barley, hordeum spontaneum, in Israel, Evolution (1979) 815–833.

[31] R.J. Martis, U.R. Acharya, K. Mandana, A.K. Ray, C. Chakraborty, Application of principal component analysis to ECG signals for automated diagnosis of cardiac health, Expert Syst. Appl. 39 (14) (2012) 11792–11800.

[32] P. Federolf, K. Boyer, T. Andriacchi, Application of principal component analysis in clinical gait research: identification of systematic differences between healthy and medial knee-osteoarthritic gait, J. Biomech. 46 (13) (2013) 2173–2178.

[33] A. Caprihan, G.D. Pearlson, V.D. Calhoun, Application of principal component analysis to distinguish patients with schizophrenia from healthy controls based on fractional anisotropy measurements, Neuroimage 42 (2) (2008) 675–682.

[34] M. Turk, A. Pentland, Eigenfaces for recognition, J. Cognit. Neurosci. 3 (1) (1991) 71–86.

[35] V. Kadappa, A. Negi, A theoretical investigation of feature partitioning principal component analysis methods, Pattern Anal. Appl. 19 (1) (2016) 79–91.

[36] S. Chen, Y. Zhu, Subpattern-based principle component analysis, Pattern Recogn. 37 (5) (2004) 1081–1083.

[37] K.V. Kumar, A. Negi, SubXPCA and a generalized feature partitioning approach to principal component analysis, Pattern Recogn. 41 (4) (2008) 1398–1409.

[38] T.K. Sahoo, H. Banka, New hybrid PCA-based facial age estimation using inter-age group variation-based hierarchical classifier, Arabian J. Sci. Eng. 42 (8) (2017) 3337–3355.

[39] T.K. Sahoo, H. Banka, A. Negi, Space complexity analysis in hybrid principal component analysis, in: 2020 IEEE-HYDCON, IEEE, 2020, pp. 1–7.

[40] T.K. Sahoo, H. Banka, A. Negi, Novel approaches to one-directional two-dimensional principal component analysis in hybrid pattern framework, Neural Comput. Appl. 32 (9) (2020) 4897–4918.

[41] T.K. Sahoo, H. Banka, A. Negi, Design and analysis of various bidirectional 2dpcas in feature partitioning framework, Multimed. Tool. Appl. (2021) 1–41.

[42] A. Negi, V. Kadappa, An investigation on recent advances in feature partitioning based principal component analysis methods, in: 2010 Second Vaagdevi International Conference on Information Technology for Real World Problems, IEEE, 2010, pp. 90–95.

[43] A. Negi, V.K. Kadappa, SubXPCA versus pca: a theoretical investigation, in: 2010 20th International Conference on Pattern Recognition, IEEE, 2010, pp. 4170–4173.

[44] H.G. Gauch Jr., Noise reduction by eigenvector ordinations, Ecology 63 (6) (1982) 1643–1649.

[45] L. Zhang, W. Dong, D. Zhang, G. Shi, Two-stage image denoising by principal component analysis with local pixel grouping, Pattern Recogn. 43 (4) (2010) 1531–1549.

[46] T.H. Cormen, C.E. Leiserson, R.L. Rivest, C. Stein, Introduction to Algorithms, MIT Press, 2009.

15

Impact of Midday Meal Scheme in primary schools in India using exploratory data analysis and data visualization

Sonal Mobar Roy[1], Tilottama Goswami[2], Charan Kumar Nara[3]

[1]CENTRE FOR POST GRADUATE STUDIES & DISTANCE EDUCATION, NATIONAL INSTITUTE OF RURAL DEVELOPMENT AND PR, HYDERABAD, TELANGANA, INDIA; [2]DEPARTMENT OF INFORMATION TECHNOLOGY, VASAVI COLLEGE OF ENGINEERING, HYDERABAD, TELANGANA, INDIA; [3]COGNIZANT TECHNOLOGY SOLUTIONS INDIA PRIVATE LIMITED

15.1 Introduction and background

India is a developing nation on a roadmap of advancement by using all resources that are available, be it social, economic or human capital. With a good demographic dividend, the country stands to benefit as the youth of the country help in determining the development phase. This makes it pertinent to ensure that the population is healthy and safe. Women in India are anemic, varying in degrees from mild to moderate to severe. This gets transferred to the fetus and consequently this cycle of malnutrition goes on.

The United Nations has hailed the country's nutritional intervention like the midday meal (MDM) scheme as one of the finest approaches to curb malnutrition. All these reports have stressed on the fact that a complete basket of food of balanced diet with adequate micro- as well as macronutrients is required for the upkeep of good health.

Every year since 2018, September is celebrated as the "Poshan Maah" as accorded by the Hon'ble Prime Minister of India. As is seen, malnutrition is the cause of death for 70% of the children under five years of age. Inadequate dietary intake is leading to growth issues such as stunting for as many as 38% of the children. After the initiation of MDM scheme and Poshan Abhiyaan, change in nutritional levels, were expected to raise in populations who were malnourished.

The data collected for such studies in social science domain [1] from various states yearly basis is huge in number. The role of data science and statistical modeling and analysis plays a vital role to investigate this "big data" to draw patterns, outliers and understand the trends. The impact of a factor can be studied objectively and

Statistical Modeling in Machine Learning. https://doi.org/10.1016/B978-0-323-91776-6.00001-4

quantitatively using exploratory data analysis to test the hypothesis. Analysis comprises of basically descriptive statistics in this chapter. The key insights gathered have been graphically represented to make the understanding more intuitive to any stakeholders related to this domain.

The chapter is organized as follows: Section 1.2 describes the conditions of malnutrition and its effects on learning for school children in rural areas and coming from low-income groups. Section 1.3 discusses the MDM scheme. Section 1.4 discusses the exploratory data analysis methodology [2] applied to study the impact of the scheme in the school children from various states of India. Finally, Section 1.5 concludes the chapter summarizing the observations.

15.2 Nutrition in primary schools in rural India

It is pertinent to understand what nutrition is. When we refer to health, it means *physical, mental, social,* and *spiritual health.* The nutritional status is the condition of health of an individual as influenced by the utilization of nutrients. It is assessed by the diet one takes, whether there is any illness or any of the medical tests have shown unwanted results. Nutrition and health are strongly interlinked. Diet and exercise determine the good health of a person. As per the 2020 Global Nutrition Report [3] in the context of COVID-19, the need for more equitable, resilient, and sustainable food and health systems is a priority. It mentions that almost a quarter of all children under five years of age are stunted. It is seen that poor diets that lead to malnutrition is one of the critical challenges that cause health, economic, and environment burden.

15.2.1   Malnutrition

Malnutrition is an impairment of health resulting from a deficiency excess or imbalance of nutrients. It could be both: *under-nutrition* or *over-nutrition.* More so, both would result in ill-health of a person. According to Stratton et al. [4], malnutrition is defined as "a state of nutrition in which a deficiency or excess of energy, protein, and micronutrients causes measurable adverse effects on tissue/body form (body shape, size and composition) and function, and clinical outcome."

In the school-going years, children are growing and need sufficient diet that is nutritious. These are the years that precede adolescence. There are various conditions of malnutrition having different effects on learning. These conditions may include protein energy malnutrition, short term hunger, anemia and so on. The effects on education include restlessness, poor attention, distractedness, and related issues.

The signs of malnutrition can be categorized under three indicators: *physical signs, learning signs,* and *other behavioral signs.* The physical signs may include walking slowly, tiredness, hungry, underweight, swollen throat, and poor eyesight. The learning signs may include lack of concentration, difficulty in cognition, reading and writing and sleeplessness. Last, behavioral signs may include being irritable, aggressiveness, sulkiness and detachment and isolation. As per the 1995 report of World Health Organization [5], nutrition in adults by body mass index is classified as follows: (1) > 18.5 is normal, (2) 17.0–18.49 is mild malnutrition, (3) 16.0–16.99 is moderate malnutrition, and (4) < 16.0 is severe malnutrition.

As Blössner [6] mentions, it is through women that the off-spring gets effected and hence the nutritional status of woman is utmost important.

There are various consequences of malnutrition as seen in children leading to impaired intellectual development and increased lifetime risk of osteoporosis. Micronutrients are an essential part of one's diet. Lesser intake of the same may lead to unwanted consequences. Certain deficiencies that arise from lack of micronutrients are:

  i. Iron-Anemia
  ii. Zinc-Skin rashes and decreased immunity
  iii. Vitamin B_{12}-Anemia
  iv. Vitamin D-Rickets, Osteomalacia
  v. Vitamin C-Scurvy
  vi. Vitamin A-Night blindness

Keeping this relevant issue in highlight, the Government of India started with Integrated Child Development Scheme (ICDS) on October 2, 1975. As per the census (2011) [7], there are 158 million children in the age group of 0–6 years. It is one of the most celebrated flagship program of the country that caters to early childhood care and development. It is in response to the challenge of providing preschool nonformal education as well as to deal with the problems of malnutrition, morbidity, reduced learning capacity, and mortality. The major objectives of the scheme include improving the nutritional and health status of 0–6 years aged children. It focuses on policy implementation for promoting child development and enhancing the health of mothers and children in achieving proper nutrition and health education.

There are six services provided under the ICDS scheme and it is seen that all these services are provided under the gamut of Ministry/Department of Health and Family Welfare through National Rural Health Mission (NRHM). The government has ensured that all the above services are delivered in an integrated manner in all nooks and corners of the country. The Aanganwadi Centers (AWCs) are the platform through which the services are delivered in an integrated manner.

As per the ICDS website, the revised nutritional norms as applicable since February 2009 mention children to consume 500 calories, severely malnourished children to have 800 calories, and pregnant women and lactating mothers to consume 600 calories per day.

As per the table above, the children from 6 to 72 months, i.e., 6 years should be receiving 500 calories wherein the protein content should be 12–15 g; the Severely Malnourished Children (SAM) also in the above-mentioned age group should receive 800 calories and 20–25 g protein, and the pregnant and lactating mothers should receive 600 calories and 18–20 g protein.

As this is a flagship program and is widely delivered across the country, there exists a monitoring system as well for the ICDS scheme. This is mainly done by the Ministry of Women and Child Development. Initially, the Accredited Social Health Activist (ASHA) and Auxiliary Nurse Midwife (ANM) working with the AWCs used to maintain physical registers for all records which were sent to state offices and then to the central government. However, with technological advancement taking place, the ASHAs and ANMs are given trainings and now they update all records on tablets. These tablets store data offline and when connected to the internet, all data gets transferred to the cloud.

Similarly, taking heed of the severe issue of malnutrition, especially in the marginalized sections of the society, the government launched the flagship program called Poshan Abhiyaan or the National Nutrition Mission. The program is delivered under ICDS umbrella and includes the Aanganwadi Service Scheme, the Pradhan Mantri Matru Vandana Yojana (PMMVY), and the Scheme for Adolescent Girls.

The Poshan Abhiyaan was aimed at reaching out to every household with the message of nutrition with the tag line *har ghar Poshan tyohar*. There is an integrated dashboard for monitoring progress. The program is based on Jan Andolan guidelines and is based on eight themes: *Antenatal care, Optimal breastfeeding (early and exclusive), Complementary feeding, Anemia, Growth monitoring, Girls-education, diet, right age of marriage, Hygiene and sanitation* and *Eat healthy food fortification*. There are two categories of activities namely the scheduled activities that include Convergence Action Plan Meeting-state, District, Block; VHSND day; CBE; ECCE day; DAY-NRLM weekly meeting; Home visits by ANM/ASHA/AWW-HBNC and Gram Sabha meeting.

The Poshan Abhiyaan has been put forth as a "jan andolan," meaning people's movements. The main objective is to build recognition across sectors regarding focusing on nutrition and call for action for meeting the challenges. Hence, it is reiterated that malnutrition is a grave matter and needs focused attention (Fig. 15.1).

FIGURE 15.1 Children sitting on floor waiting for MDM and A MDM cook in kitchen shed preparing food. *Source: Photographs are from first author's fieldwork.*

15.3 Midday Meal Scheme

The MDM scheme is a way of moving from food to nutrition security. To date, the government has fed 11 million children through this scheme. There has been considerable change seen in student enrollment and attendance in schools as well.

Though the Government of India launched it, the states play a crucial role in the evolution and improvement of the scheme. The states may use 5% of their annual budget with prior approval from Ministry of Human Resource Development (MHRD). This may be categorized under (1) Infrastructure, (2) Nutrition, and (3) Monitoring. In this chapter, the authors focus on nutrition aspect of MDM scheme.

Nutrition is one of the two primary objectives of the MDM scheme, the other being enrollment. Many states have already started to ensure that nutritional food is delivered in schools for better outcomes. The MDM program, which was commonly called the "noon meal program," commenced in 1962––1963 and was gradually propagated throughout the nation. Initially, it was a centrally sponsored scheme; however, it is implemented by the state government with the assistance provided by the central government. The program aims at retaining the nutritional levels of students in primary schools, so that attendance in schools is increased and dropout rate is decreased. The students coming from the marginalized section are given priority.

15.3.1 Objectives

In this chapter, the authors have made an attempt to explore data from various states across the country to determine the effect/impact of the MDM scheme on three parameters, namely,

  (i) Anemia
 (ii) Stunted Growth
(iii) Enrollment

The authors used Statistical Analysis (EDA) to assess the impact of MDM scheme in India from the period 2015–2020.

15.4 Exploratory data analysis and visualization methodology

Data has been collected from various national surveys, performed on nutrition and education, for the three main indicators namely Anemia, Stunt-Growth, and Enrollment for all states of India yearwise. The table (Table 15.1) below shows how the sample data looks like:

Table 15.1 Sample of data set used in this chapter.

State	isMDM	Anemia 2005–06	Anemia 2015–16	Anemia 2016–18	Stunting 2015–16	Stunting 2016–18	Enrollment 2005–06	Enrollment 2015–16
Goa	Yes	38.2	48.3	22.1	20	19.6	84.8	96.3
Gujarat	Yes	69.7	62.6	38.2	39	39.1	70.8	81.2
Kerala	Yes	44.5	35.7	12.5	20	20.5	89.7	97.4
Madhya Pradesh	Yes	74.1	68.9	53.5	42	39.5	70.5	81.6
Maharashtra	Yes	63.4	53.8	41.6	34	34.1	77.8	87.7

The study on factors such as anemia, stunt-growth and enrollment are collected from the surveys from 29 states of India for years 2005–2006, 2015–2016, and 2016–2018. The data consists of states which have the MDM program and states which do not have MDM. The isMDM attribute is a Boolean flag in the data table as shown above. The flag is set to 'Yes' for MDM states and 'No' for otherwise.

The exploratory data analysis (EDA) explained in Ref. [8] plays an important role in finding out the correlation between the MDM scheme and the factors. EDA is basically performing statistical investigation to test the assumptions. The summary statistics or descriptive statistics such as mean, standard deviation, data quartiles etc. give an overall distribution of data. The outliers can be easily identified. The visualization of the nature of data is more intuitive to humans. Thus EDA along with visualization helps more in understanding and exploring the data points specific to the problem. Various rich libraries are available for drawing plots and graphs. Starting with the simple stacked-bar plot, it gives the information whether the value changes in positive or negative manner. Also, it summarizes the data by visualizing the mean of all the values in a variable. the box plot and violin plot shows the summary of a variable which helps in visualizing the data variability by showing numerical data based on quartiles. A scatter plot is best for regression analysis, as it shows the linear relationship in data and helps in predicting the value for new data. Heatmaps are used to observe the correlations among the feature variables. This is particularly important when we are trying to obtain the feature importance in regression analysis.

The faces called Chernoff Face Visualization techniques [9] represents data points in K-dimensional space and is used in software engineering.

The following types of graphs illustrated in Ref. [8] have been used for plotting using python matplotlib package:

- Stacked-Bar plot
- Box plot
- Violin plot

- Scatter plot
- Heatmap

Various graphs on the three parameters mentioned above are described in detail by the authors.

15.4.1   Stacked-bar plot

Stacked-bar plots exhibit the quantitative relationship that lies between principal category and its subcategories. This section showcases graphs using stacked-bar plot to display the trends on the three parameters for MDM and non-MDM.

15.4.1.1   *Mean percent of anemia in non-MDM versus MDM states*

The graph in Fig. 15.2 compares states where the MDM scheme is available to those states where the scheme is not available. It shows mean percent of MDM states having anemia and mean percent of non-MDM states having anemia for the years 2005, 2015, and 2016. It is inferred from the stacked-bar plot above that every year the prevalence of anemia in MDM states is less when compared to the non-MDM states.

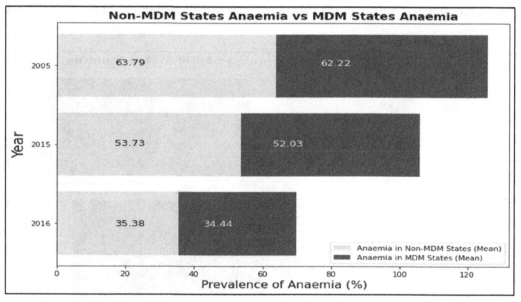

FIGURE 15.2 Mean percent of anemia in non-MDM versus MDM.

15.4.1.2   Mean percent of Stunt-Growth in non-MDM versus MDM states

The graph in Fig. 15.3 above shows the mean stunt-growth in all the MDM states and non-MDM states for anemia in the years 2015 and 2016. It can be observed from the bar plot that the rate of stunt-growth in 2015 and 2016 for MDM and the non-MDM States is approximately the same.

15.4.1.3   Mean percent of enrollment in non-MDM and MDM states

The graph in Fig. 15.4 above shows the mean enrollment rate of all the MDM states and non-MDM states in the years 2005 and 2015. It can be observed from the bar plot that both MDM states mean per cent and non-MDM states per cent are approximately the same for 2005 and 2015.

15.4.2   Box plot

Box plots demonstrates the visualization of data distribution using quartiles. This section showcases graphs using box plot to display the trends on the three parameters for MDM and non-MDM states. Anomaly cases are thus identified in this section, which are rendered as unfilled dots outside the whiskers of the box plot.

15.4.2.1   Quartiles and outliers of anemia in non-MDM and MDM states

From the boxplot in Fig. 15.5, we can observe the median of non-MDM states data in the years 2015 and 2016 is higher than the median of MDM States for the corresponding years. From this, we can conclude that there is an overall decrement in the MDM states

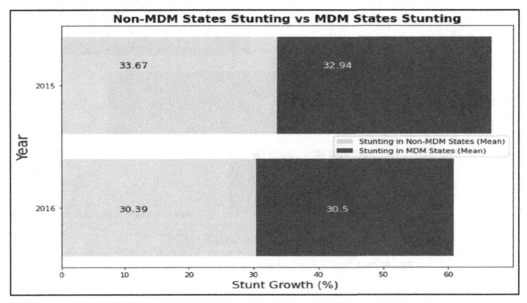

FIGURE 15.3 Mean percent of Stunt-Growth in non-MDM versus MDM states.

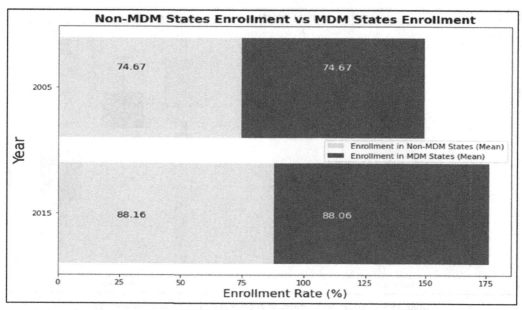

FIGURE 15.4 Mean percent of MDM states enrollment and mean percent of non-MDM states enrollment.

as far as anemia is concerned. As we can point out some outliers in the plot, these are the points which do not lie in the range [(Upper Quartile-1.5 * Inter Quartile Range), (Lower Quartile + 1.5 * Inter Quartile Range)]. From this, the anomaly states are identified as Goa, Mizoram, and Kerala.

15.4.2.2 Quartiles and outliers of Stunt-Growth in non-MDM and MDM states
From the boxplot in Fig. 15.6, we can observe the stunt-growth rate for MDM and non-MDM states in the year 2016 decreased when compared to the year 2015. Also, we can see that median of non-MDM states in 2016 is low than the median of MDM states in 2016. In this plot there are no anomaly cases or outliers present for MDM states.

15.4.2.3 Quartiles and outliers of enrollment in non-MDM and MDM states
From the boxplot in Fig. 15.7, we can observe the median of MDM states in 2005 is comparatively lower than non-MDM states in 2005. In 2015, the median of MDM states has increased approximately equal to the median of the non-MDM state. In this plot there are no anomaly cases or outliers present for MDM states.

15.4.3 Violin plot

Violin plot shows probability density of data. It is a hybrid of two plots: box plot and kernel density plot. This section showcases graphs using violin plot to display the trends on the three parameters for MDM and non-MDM states.

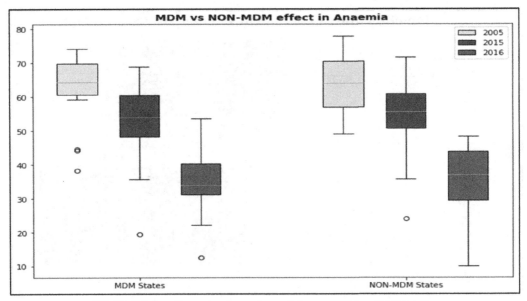

FIGURE 15.5 Quartiles and outliers of anemia in non-MDM and MDM states.

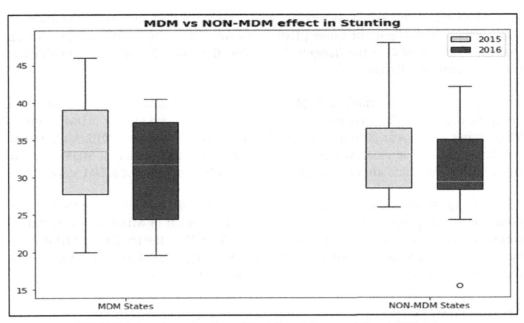

FIGURE 15.6 Quartiles and outliers of Stunt-Growth in non-MDM and MDM states.

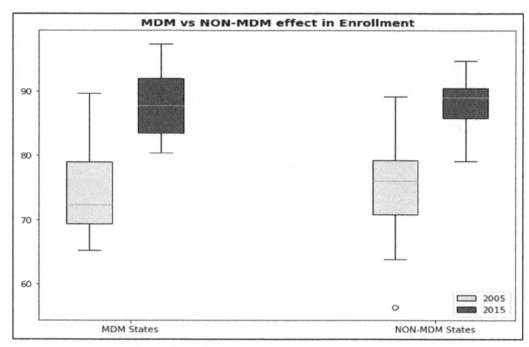

FIGURE 15.7 Quartiles and outliers of enrollment in non-MDM and MDM states.

15.4.3.1   *Density plot distribution of anemia in non-MDM and MDM states*

From the violin plot in Fig. 15.8, we can observe how MDM and non-MDM states with percentages are distributed in the years 2005, 2015 and 2016 for anemia. It also gives an overall range of the data that lies in between. The more states having percentages nearly similar the denser the graph is. In both MDM and non-MDM states the prevalence of Anemia has decreased from 2005 to 2015 and from 2015 to 2016.

15.4.3.2   *Density plot distribution of Stunt-Growth in non-MDM and MDM states*

From the violin plot in Fig. 15.9, we can observe how MDM and non-MDM states with percentages are distributed in the years 2015 and 2016 for Stunt-Growth. It also gives an overall range of the data that lies in between. The more states having percentages nearly similar the denser the graph is. In both MDM and non-MDM states, stunting has decreased at an almost similar rate. Even though in 2016 non-MDM states having very low minimum value, but data is denser at the midregion than in the lower region.

15.4.3.3   *Density plot distribution of enrollment in non-MDM and MDM states*

From the violin plot in Fig. 15.10, we can observe how MDM and non-MDM states with percentages are distributed in the years 2015 and 2016 for enrollment. It also gives an overall range of the data lies between. The more states having percentages nearly similar

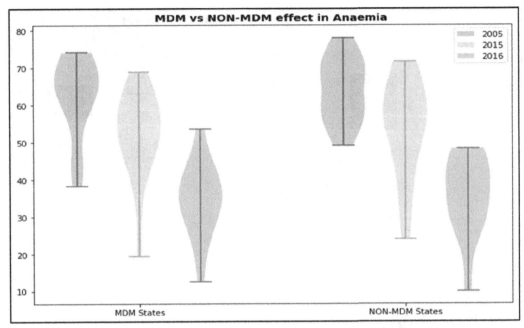

FIGURE 15.8 Density plot distribution of anemia in non-MDM and MDM states.

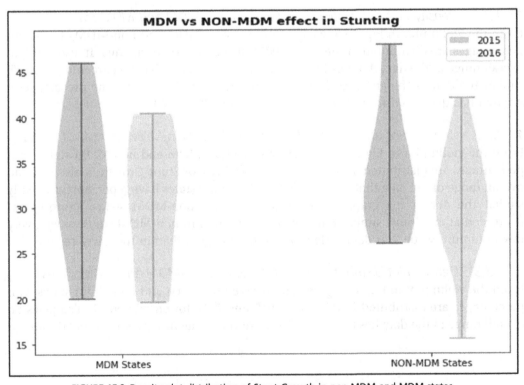

FIGURE 15.9 Density plot distribution of Stunt-Growth in non-MDM and MDM states.

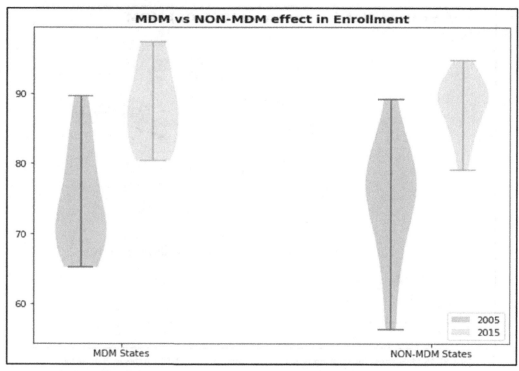

FIGURE 15.10 Density plot distribution of enrollment in non-MDM and MDM states.

the denser the graph is. In both MDM and non-MDM states the enrollment rate has increased. But there is more increase in MDM states than non-MDM states.

15.4.4 Scatter plot

Scatter plots visualize the relationship between two variables. This section showcases five graphs using scatter plot to display the trends on the three parameters for MDM and non-MDM states, displaying the highest and lowest rated state. Hovering over each data point (a state) on the graph will view the rate and name of the state.

15.4.4.1 Prevalence of anemia for all MDM states in the year 2005–2006 (in India)
The scatter plot in Fig. 15.11 depicts the prevalence of Anemia for all the MDM states in the year 2005–2006. Each circle corresponds to a state. Also, the minimum and maximum points are labeled with the state names and percentages. As we can observe, Goa is having the least percentage of 38.2 and Madhya Pradesh is having the highest percentage of 74.1% for Anemia.

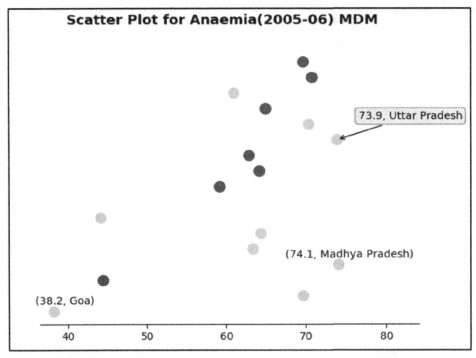

FIGURE 15.11 Scatter plot showing prevalence of anemia for all MDM states in 2005–2006.

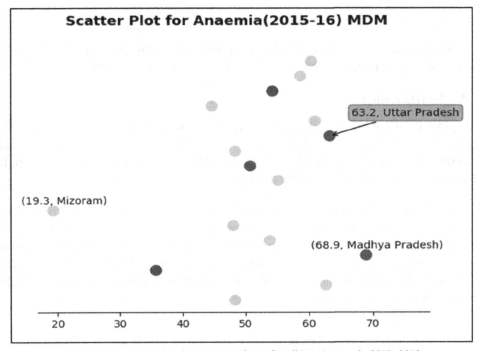

FIGURE 15.12 Scatter plot showing prevalence for all MDM states in 2015–2016.

15.4.4.2   Prevalence of anemia for all MDM states in 2015–2016 (in India)

The scatter plot in Fig. 15.12 depicts the prevalence of Anemia for all the MDM states in the year 2015–2016. Each circle corresponds to a state. By 2015–2016, the anemia percentage has decreased for all the three states: Uttar Pradesh, Madhya Pradesh, and Goa. Obviously, MDM has made an impact from 2005 to 2006 to 2015–2016. Anemia has decreased. Also, the minimum and maximum points are labeled with the state names and percentages. As we can observe, Mizoram is having the least percentage of 19.3% and Uttar Pradesh is having the highest percentage of 68.9% for Anemia.

15.4.4.3   Percentage of Stunt-Growth for all MDM states in the year 2016–2018 (in India)

The scatter plot in Fig. 15.13 depicts the Stunt-Growth for all the Mid-Day-Meal (MDM) states in the year 2016–2018. Each circle corresponds to a state. Also, the minimum and maximum points are labeled with the state names and percentages. As we can observe, Goa is having the least percentage of 19.60% and Meghalaya is having the highest percentage of 40.4% for Stunting.

15.4.4.4   Percentage of enrollment for all MDM states in the year 2005–2006 (in India)

The scatter plot in Fig. 15.14 depicts the Enrollment rate for all the MDM states in the year 2005–2006. Each circle corresponds to a state. Also, the minimum and maximum points are labeled with the state names and percentages. As we can observe, Orissa is having the least percentage of 65.2% and Kerala is having the highest percentage of 89.7% for Enrollment.

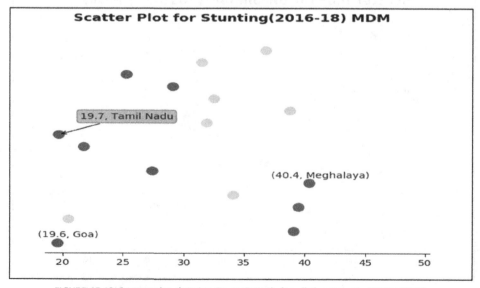

FIGURE 15.13 Scatter plot showing Stunt-Growth for all the states in 2016––2018.

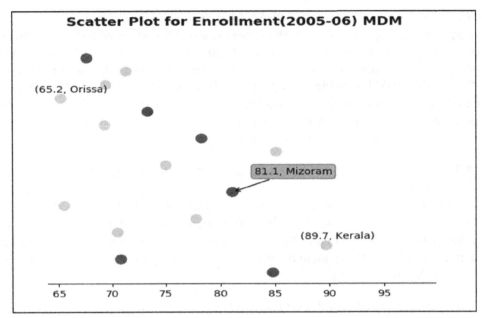

FIGURE 15.14 Scatter plot showing enrollment for the year 2005–2006.

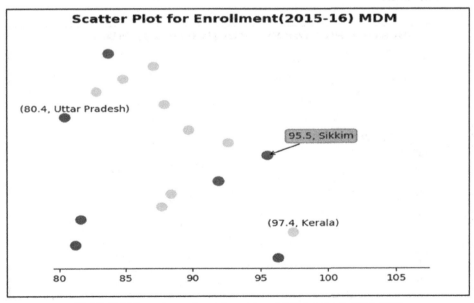

FIGURE 15.15 Percentage of enrollment for all MDM states in 2015–2016.

15.4.4.5   Percentage of enrollment for all MDM states in 2015–2016 (in India)
The scatter plot in Fig. 15.15 depicts the Enrollment rate for all the MDM states in the year 2015–2016. Each circle corresponds to a state. Also, the minimum and maximum points are labeled with the state names and percentages. As we can observe, Uttar Pradesh is having the least percentage of 80.4% and Sikkim is having the highest percentage of 95.5% for Enrollment.

15.4.5   Heatmap

Heatmap is a visualization tool which uses color coding scheme to represent feature values and their correlation. It is represented in a matrix format. Correlation values range from −1 to +1. The closer the cell value to 1, the more positively correlated are the features and the closer the cell value to −1, the features are negatively correlated. A value near to zero means there is no correlation.

From the correlation heatmap in Fig. 15.16, Anemia and Stunt-Growth correlation values are most of the time nearby to 0.7 i.e., darker blue in color. Hence it is positively correlated a rise in anemia shows a rise in stunt-growth. Enrollment and stunt-growth are negatively correlated with values approximately −0.7 and similarly, enrollment and anemia are also negatively correlated. Rise in class enrollment is observed if the students are healthier, nonanemic, and no stunt-growth.

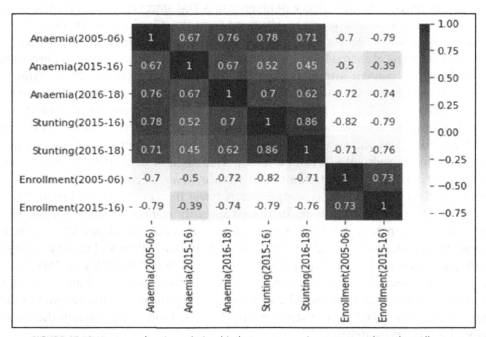

FIGURE 15.16 Heatmap showing relationship between anemia, stunt-growth, and enrollment.

15.5 Data visualization insights on impact of MDM

Malnutrition has emerged as one of the severe problems in India over the past few decades. With various schemes in implementation such as Poshan Abhiyaan and Mid-Day Meal Scheme, various efforts have been made in all across the country to make a dent on this issue and cater better nutritional services to the children, pregnant women and lactating mothers. Mainly it was envisaged that giving nutritious food to the young children at school will help them curb classroom hunger. This will also act as an incentive for ensuring good enrollments in government schools. The children in the rural areas come from families that cannot provide optimum levels of calories required for their growth. Hence, through receiving MDM, the children will at least also achieve physical growth. This chapter discusses the same with three points in focus namely the problem of anemia, stunting and enrollment. The authors have applied statistical analysis especially the EDA on secondary data taken from various national surveys on the above three indicators. The EDA exhibits a pattern that represents a visualization form for the stakeholders [10]. The data was analyzed for states where children were receiving MDM and states where children were not receiving MDM. Various EDA and visualization techniques such as stacked-bar plots, box plots, violin plots, and scatter plots for the three parameters were executed to understand their role in affecting the MDM scheme. From the results achieved, the authors were able to identify patterns between the parameters which have an influence on the performance of children who are participating in MDM and those who do not. It is inferred from the stacked-bar plot above that every year the prevalence of anemia in MDM states is less when compared to non-MDM states. However, it can be observed from the bar plot that the rate of stunt-growth in 2015 and 2016 for MDM and the non-MDM states is approximately the same. Also, it can be observed from the bar plot that both MDM states mean percent and non-MDM states percent are approximately the same for 2005 and 2015. From the boxplot, we can observe the median of non-MDM States data in the years 2015 and 2016 is higher than the median of MDM states for the corresponding years. From this, we can conclude that there is an overall decrement in the MDM states. It is observed the stunt-growth rate for MDM and non-MDM states in the year 2016 decreased when compared to the same in the year 2015. It can be seen that median of non-MDM states in 2016 is low than the median of MDM states in 2016. Moreover, it can be observed that the median of MDM states in 2005 is comparatively lower than non-MDM states in 2005. In 2015, the median of MDM states has increased approximately equal to the median of the non-MDM state. In both MDM and non-MDM states the prevalence of anemia has decreased from 2005 to 2015 and from 2015 to 2016. In both MDM and non-MDM states, stunting has decreased at an almost similar rate. Even though in 2016 non-MDM states had very low minimum value, but data is denser at the midregion than in the lower region. In both MDM and non-MDM states the enrollment rate has increased. But there is more increase in MDM states than non-MDM states. Progressing further, the authors discuss the scenario statewise.

It is observed that Goa is having the least percentage of 38.2 and Madhya Pradesh is having the highest percentage of 74.1% for anemia. As we can observe, Mizoram is having the least percentage of 19.3% and Uttar Pradesh is having the highest percentage of 68.9% for anemia. Odisha is having the least percentage of 65.2% and Kerala is having the highest percentage of 89.7% for enrollment. Also, Goa is having the least percentage of 19.60% and Meghalaya is having the highest percentage of 40.4% for stunting. Uttar Pradesh is having the least percentage of 80.4% and Sikkim is having the highest percentage of 95.5% for enrollment.

15.6 Conclusion

Thus, it can be deduced that MDM scheme has done well in some states whiles others are also catching up. Moreover, MDM does lead to reduction in stunting and increase in enrollment of children in schools. It has been one of the attractive factors for children to come and attend their schools. Thus, MDM has over the period of time, emerged as a successful program. The authors opine that it needs better implementation agency to execute the scheme in a better manner for better results. Certain gap areas need to be addressed. There are rural pockets wherein the staple diet is different and the food delivered through MDM scheme does not match the taste of the children. Same food served everyday results in children wasting food and subsequently missing school. Therefore, the weekly menu should be diverse in nature. Better quality kitchen sheds need to be ensured as these are inside school compounds and also emerge as a safety threat. If these points are taken heed of, the children will be able to achieve better health results and learning outcomes. All human beings deserve a good quality of life including access to healthy, affordable food and nutrition care. It is hoped that through focused interventions and continuous monitoring and evaluation of these schemes, the marginalized people will be able to draw benefits and good health.

References

[1] C.A. Hesse, J.B. Ofosu, Statistical Methods for the Social Sciences, 2017. Retrieved 11 25, 2021.

[2] A.B. Downey, Think Stats: Exploratory Data Analysis, O Reilly, second ed.

[3] Global Nutrition Report, Action on Equity to End Malnutrition, Development Initiatives, Bristol, UK, 2020, 2 September, 2021.

[4] R.J. Stratton, C.J. Green, M. Elia, Disease-Related Malnutrition: An Evidence Based Approach to Treatment, CABI Publishing, Wallingford, 2003.

[5] World Health Organization, . The World Health Report: 1995: Bridging the Gaps Report of the Director-General, World Health Organization, 1995.

[6] M. Blössner, M. De Onis, A. Prüss-Üstün, Malnutrition: quantifying the health impact at national and local levels/Monika Blössner and Mercedes de Onis, World Health Organization, 2005.

[7] Census of India, Census of India, Office of the Registrar General & Census Commissioner, New Delhi, 2011.

[8] https://matplotlib.org/.

[9] S.A. Moiz, R.R. Chillarige, Method level code smells: chernoff face visualization, in: S.C. Satapathy, K.S. Raju, K. Shyamala, D.R. Krishna, M.N. Favorskaya (Eds.), Advances in Decision Sciences, Image Processing, Security and Computer Vision. Learning and Analytics in Intelligent Systems, vol. 3, Springer, Cham, 2020, https://doi.org/10.1007/978-3-030-24322-7_63.

[10] Cole Nussbaumer Knaflic, Storytelling with Data: A Data Visualization Guide for Business Professionals, Wiley, 2015.

Nonlinear system identification of environmental pollutants using recurrent neural networks and Global Sensitivity Analysis

Srinivas Soumitri Miriyala, Ravikiran Inapakurthi, Kishalay Mitra

GLOBAL OPTIMIZATION AND KNOWLEDGE UNEARTHING LABORATORY, DEPARTMENT OF CHEMICAL ENGINEERING, INDIAN INSTITUTE OF TECHNOLOGY HYDERABAD, HYDERABAD, TELANGANA, INDIA

16.1 Introduction

The increase in atmospheric pollution and the consequent adverse effects on the health of organic life-forms on land and in ocean, climate change, economy, infrastructure, and global well-being are prominent issues of universal concern [1]. It all initiated with the advent of industrial revolution, which, although was successful and led to many technological marvels shaping the current society, significant amounts of pollutants were released into the atmosphere without any regulation or monitoring [2]. The effect of industrial revolution can be justified with the rise in global Carbon Dioxide (CO_2) emissions from 1750 (see Fig. 16.1) [3]. These emissions without any check over several decades, imbibed a casual attitude in humans to undermine the importance of environmental pollution and its effects on the society. Today, the growth of industrialization has reached to such proportions that the environmental pollution has started to show its true effects disrupting our daily life-style. Contemporary studies using state-of-the-art instrumentation prove some unprecedented and upsetting facts about the damage incurred by nonhuman life-forms from over the decades due to environmental pollution.

Aerosols, which constitute a majority of environmental pollutants, due to their size, can considerably effect the temperature by dissipating the solar irradiance and thereby directly contribute to the phenomena of climate change [2]. This disrupts the delicate balance in the nature resulting in several life-threatening situations such as the extreme weather events, forest fires, and rise in ocean temperatures causing dangers to the oceanic flora and fauna [4]. The National Interagency Fire Center (NIFC) of the National

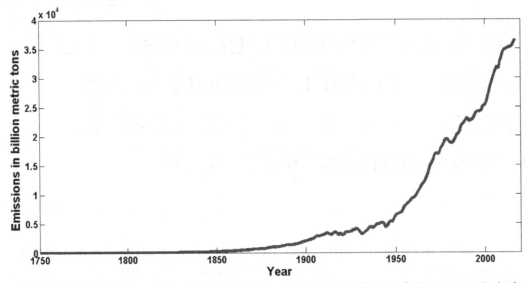

FIGURE 16.1 Carbon dioxide emissions since the beginning of industrial revolution indicating a steep rise in the last five decades.

Oceanic and Atmosphere Administration (NOAA) in USA reported that 7532 fires burned 264,495 acres in March 2021 [5]. The decadal picture of wildfires in USA is presented in Fig. 16.2 [6]. These fires are mainly due to extreme drought events caused by climate change. Similarly, in the last five years, the rising temperatures of ocean are resulting in super cyclones in Arabian sea which used to be calm and witness very few cyclones compared to the Bay of Bengal [7]. Another adverse effect of rising ocean temperatures is

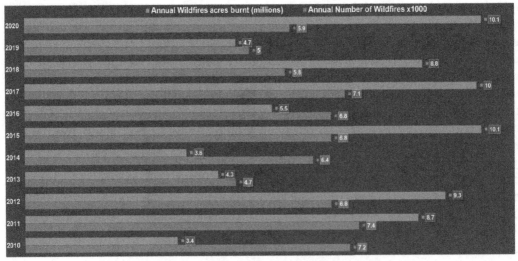

FIGURE 16.2 Wildfire statistics in the United States over the last decade.

the coral bleaching [8]. According to the NOAA, in a short-span of four years (2014−2017) heat-stress was severe enough to trigger bleaching in around 75% of the coral reefs [9]. The heat-stress produced was sufficient to kill 30% of the coral reefs [9]. This was the third mass global bleaching event which was the longest, most widespread, and most destructive on record [9]. Several other related issues stem from climate change and global warming such as the food safety, glacier melting, extinction of flora and fauna. Fig. 16.3 shows the average particulate matter ($PM_{2.5}$) concentrations in 10 most polluted cities in 2019 and 2020 [10]. Anthropogenic pollution due to particulate matter is a major cause of public health hazards resulting in around 9 million deaths per year [2]. Recent studies have reported that short-term exposure to particulate matter is causing pulmonary diseases such as, asthma, shortness of breath and severe cough resulting in high rates of morbidity [2]. Therefore, due to all these dangers posed by environmental pollution, it becomes essential to construct robust, scientific and engineering tools to model, simulate and forecast the future trends of pollutants so that the results can aid in optimal and transparent policy-making to mitigate the issue of environmental pollution.

The models for simulating the environmental data are categorized into physics-based and data-based models. Physics-based models are those which consider various aspects of modeling including the cause and effect relations, source of pollutants, drift of pollutants and interactions between several types of pollutants. These necessitate the implementation of governing equations of the system. On the contrary, data-based models consider only the given data and construct regression based mathematical constructs without involving the physics of the system. While both of them have their own pros and cons, the current situation is advantageous for data-driven models [11]. This is mainly due to the reasons like, (a) science behind the generation, transport and decay of pollutants is not established completely [12], and (b) the easy availability of abundant data obtained from sensors placed in fixed and mobile locations and modeling

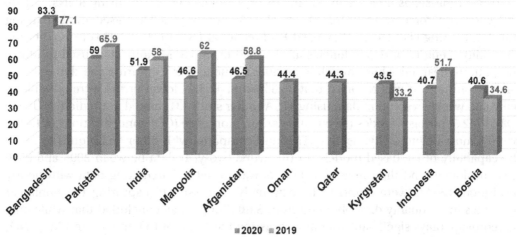

FIGURE 16.3 Average $PM_{2.5}$ concentrations in 10 most polluted countries in 2019 and 2020 in $\mu g/m^3$ of air.

techniques from Data Science (DS) [12]. Chemical Transport Model [13], Atmospheric Dispersion Model [14], Weather Research and Forecast (WRF) coupled with Community Multi-Scale Air Quality model [15] and WRF with Chemistry [16] are some well-known physics-driven models.

Historically, statistical models such as, Auto-Regressive [17], Moving Average [18], Auto-Regressive Integrated Moving Average [19] and their modifications to incorporate the seasonal changes and stationary characteristics were considered as the data-driven models. However, with the popularity of machine learning (ML), much superior models were designed to incorporate the nonlinearities, nonstationarities, and long-term dependencies in the data [12]. For example, researchers in Ref. [20] constructed models for each season using multi-layer perceptron (MLP) Networks for emulating the PM_{10}. In this work, the data with significant correlations was identified using principal component analysis (PCA) and then trained MLP network was implemented to generate 1-day ahead forecasts. Support vector machines (SVMs), MLPs, and boosted regression trees (BRT) were used by Suleiman et al. [21] for modeling PM_{10} and $PM_{2.5}$ generated by vehicular emissions and concentrated on the major roadways. Data from 19 measuring stations was obtained for this study and the results were analyzed using seven statistical measures. BRTs and MLPs' performance was superior to that of SVMs [21]. With the advent of high-performance computers (HPCs) and General Purpose Graphics Processing Units (GPGPUs), Deep Learning (DL) based models are witnessing immense surge in applicability [22]. Apart from the computational aid, the structural modifications enable them to model image based and time-series data more efficiently than their counterparts from ML. Examples for such models include recurrent neural networks (RNNs) [23], Long Short Term Memory Networks (LSTMs) [24], Gated Recurrent Units (GRUs) [25] and convolution-based neural networks, which include the Convolutional Neural Networks (CNNs) [26], Variational Auto-Encoders [27], and generative adversarial networks (GANs) [28].

AirNet time-series data was modeling by researchers in Ref. [29] using RNNs, LSTMs and GRUs for forecasting PM_{10}. Ma et al. [30] proposed a DL based method using transfer learning (TL) where they used Bi-directional LSTM (TL-BLSTM) to predict the air quality. Long-term dependencies are learned using LSTMs and TL is used to utilize the patterns learned while modeling smaller sequences in capturing the dynamics of large resolutions. It was reported that 36.85% and 42.58% lower errors were possible in daily and weekly resolution data using TL. Another study [31] aimed at predicting Ozone (O3), PM2.5, nitrogen oxides (NOx), and carbon monoxide (CO) concentrations in Delhi, world's most polluted city, using LSTMs and compared with other techniques to prove the capability of DL based models. In this work, two-year data between 2008 and 2010 was considered and the capacity of LSTMs was tested for predicting data with hourly resolution. LSTMs were found to be extremely successful in capturing the nonlinear dynamics in air quality data obtained from Safdarjung. They concluded that while only meteorology plays significant role in forecasting the trend of CO, in case of $PM_{2.5}$, NO_x and O_3, all factors including meteorological, vehicular and industrial emissions have strong influence. In another study [32], a spatiotemporal model combining CNNs and LSTMs was presented to predict the 1-day ahead $PM_{2.5}$ concentration. Data collected between 2015 and 2017 were used to train, validate and test proposed technique. Environmental data was obtained from Olympic city in China from 2017 to 2018, is modeled using WRF and RNNs. Readers may refer to Refs. [26,33–39] for other significant and recent works which used CNNs, RNNs, LSTMs, BLSTMs, and other novel DL based methods to model the environmental data.

In the current work, focus is on using RNNs, the basic sequential model, which is the design basis of LSTMs, and GRUs, for modeling environmental data constituting 15 features viz., ambient temperature, CO, methane (CH_4), nitrogen oxide (NO), non-methane hydrocarbons (NMHC), NO_2, O_3, NO_x, PM_{10}, pH rain, $PM_{2.5}$, relative humidity (RH), rainfall, SO_2, and total hydrocarbons (THC), collected from Taiwan over a time frame of 1 year with 1-hour resolution. The time-series data is procured, maintained and monitored by Environmental Protection Agency (EPA) of Taiwan, thus providing the credibility to the open source data set [40]. The RNNs, though proven effective through numerous studies, suffer with several disadvantages, which crop up due to the heuristics involved in designing and training the neural networks. We aim to eliminate these heuristics and automate the process of building RNNs such that the data which needs to be modeled is used as the only basis for determining the optimal RNN with highest accuracy. Our approach based on variance vs. bias trade-off is generic and can be extended to any other neural network and ML models such as SVMs. In this work, we explain the methodology for RNNs.

RNN, as shown and compared with MLP model in Fig. 16.4, is a network of nodes connected in parallel and series similar to MLPs [23]. The nodes in parallel are collectively called a layer, which are in turn connected in series to form the network. In contrast with MLPs, the RNNs, in addition to feed forward connections, have two types of feedback connections originating from each node in the hidden layers: (1) self-feedback, and (2) feedback to the same layer but for other nodes. The nodes in output layer neither have feedback connections nor does there exist any nonlinear activation function. Given a univariate sequential data of length N: $\{X^1, X^2, ...X^t, ...X^N\}$, due to the feedback connections, RNN can be unrolled along the sequence as shown in Fig. 16.5 [23]. Since we will be dealing with time-series data, each step in the sequence is assumed to be a time stamp. Thus, in case of RNNs, the feedforward connections (and the network parameters) are repeated for every time stamp and each of this instance is connected through the feedback loops as can be seen in Fig. 16.5. These connections enable the extraction of temporal features while simultaneously performing the regression, thus enabling the DL. While MLPs deal with time-series by breaking the sequence of time-series data into smaller sequences of predefined length, the RNNs keep the sequence intact while modeling the time-series data [41]. This is the primary reason for the success of RNNs, LSTMs and GRUs when compared with MLPs. The training of weights in RNNs is performed using gradient based optimization techniques. The rate of

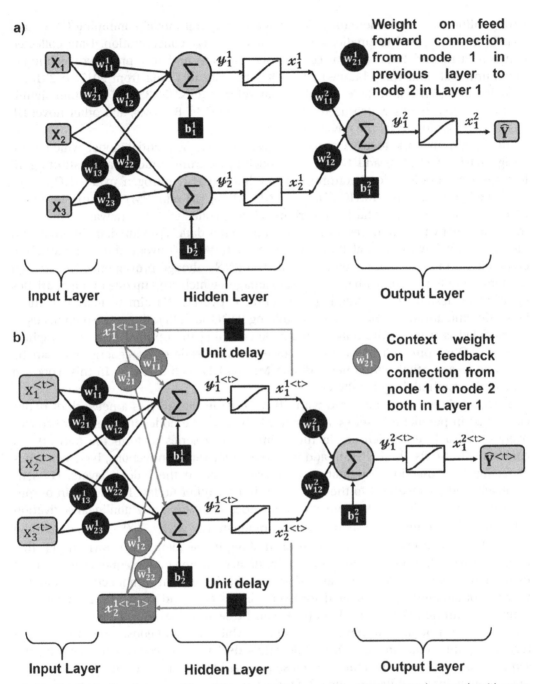

FIGURE 16.4 Comparison between (A) multilayer perceptron and (B) recurrent neural network with same architecture [3-2-1], i.e., three inputs, two nodes in a hidden layer, and one output.

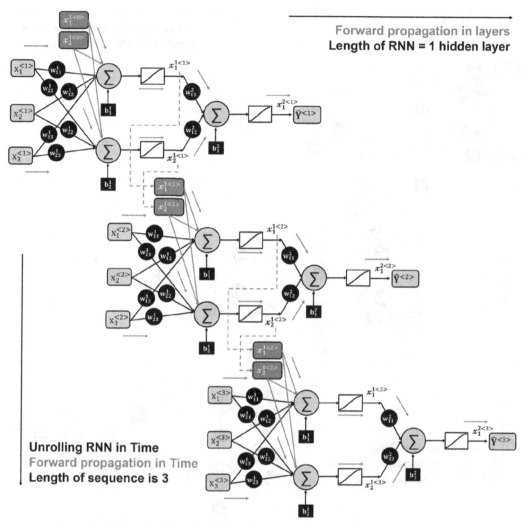

FIGURE 16.5 Unrolling of RNN and forward propagation over a sequence of length N = 3.

change of objective function with the weights are evaluated by back-propagating the sensitivities in length as well as timestamps as shown in Fig. 16.6. The resulting algorithm, called Back Propagation Through Time (BPTT), becomes computationally intensive as the gradients need to be evaluated over the entire sequence. To overcome this issue, Truncated BPTT (T-BPTT) was proposed by authors in Ref. [42]. In T-BPTT, the idea is to forward propagate the information across the full sequence but back propagate the gradients over a truncated sequence of length T as can be seen in Fig. 16.6.

Thus, we observe that, before we model the data using RNN, we need to fix the following hyper-parameters using heuristics which brings in several inefficiencies into RNNs and prohibits them from modeling the data accurately. Therefore it is necessary to

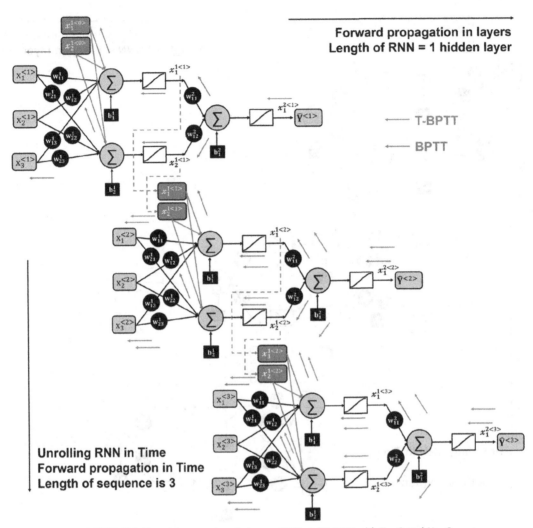

FIGURE 16.6 Pictorial comparison between BPTT and T-BPTT with $T = 2$ and $N = 3$.

eliminate the heuristics and come up with an intelligent way to determine the following hyperparameters in RNNs.

1. Architecture of RNN: Number of hidden layers and nodes in each layer.
2. Nonlinear activation function for hidden layers.
3. Truncation length T in T-BPTT algorithm.

A novel multi-objective optimization problem (MOOP) formulation is presented in this work, which aims at estimation of aforementioned hyper-parameters simultaneously, while maintaining the trade-off between accuracy and parsimony in RNNs. The accuracy of the network is quantified by measuring the correlation coefficient R^2

between the predictions from the trained network and the original test data set. Instead of the training loss, the test set accuracy is made as an objective to prevent the over-fitting. The second objective in the algorithm, i.e., parsimony in RNNs is quantified by measuring total parameters in the network. Aforementioned hyper-parameters are considered as the decision variables in the algorithm. Since each of them are integers and objective functions are nonlinear, the formulation turns out to be an integer non linear programming (INLP) problem. The variance versus bias trade-off in ML forms the basis of this formulation. It states that, bias toward a simpler model (with less param-eters and thus low complexity) might lead to underfitting and thus, more variance in the predictions. On the other hand, a network with large number of parameters might improve the training accuracy but may fail to perform with unseen data set indicating overfitting. This is a clear case of conflicting objectives and thus, the proposed formu-lation is shaped out of it.

The absence of gradient information in the multiobjective framework led to the utilization of population-based algorithm to generate the Pareto solutions. The population-based algorithm selected is Non-Dominated Sorting Genetic Algorithm (NSGA-II) [43]. Each of these solutions will be an RNN model and a single network is selected by implementing a suitable higher-order information (HOI) [43]. In this work, Akaike Information Criterion (AIC) [44], the well-known model evaluation criterion is implemented as the HOI. Thus, using the proposed algorithm optimal RNNs were obtained. For the considered case of modeling environmental parameters using the RNNs, while a univariate optimal RNN model was able to approximate the time-series data of 14 environmental pollutants using nonlinear auto-regression with an accuracy of ~95%, it was realized that univariate model is incapable of emulating pH of rain. For this, hypothesizing that there might be interactions among the environmental parameters, a multi-variate optimal RNN model was trained with pH of rain as output and the remaining environmental parameters as inputs. While the model resulted with credible accuracy, the sensitivity of all the inputs had to be quantified to identify the most important environmental pollutants which strongly influence the pH of rain. For this, Sobol indices-based Global Sensitivity Analysis (GSA) was performed using the predictions from the multivariate RNN model. This analysis concluded that among all the 14 parameters THC, NO_x, and CO, influence the pH of rain significantly. The proposed method for optimal design of RNNs and Monte Carlo−based GSA using the optimally designed networks for modeling, forecasting and quantitatively deter-mining the sensitivities between 15 environmental parameters is a unique and novel study which will help the policymakers to take optimal decisions on environmental regulations. The rest of the manuscript contains the following sections: Formulation section, detailing the data used in this study, novel algorithm for design of RNNs, Monte Carlo−based GSA, section on Results and Discussions, presenting the results of the study and elaborate discussions and finally Conclusions, which summarizes the work.

16.2 Formulation

16.2.1 Environmental pollutants data

Banqiao, Datong, Cailiao, Guanyin, Dayuan, Keelung, Guting, Linkou, Pingzhen, Longtan, Shilin, Sanchong, Tamsui, Songshan, Tucheng, Taoyuan, Wanli, Wanhua, Xinzhuang, Xindian, Yangming, Xizhi, Zhongli, Yonghe, and Zhongshan are the 25 places in Taiwan from where the 15 variable data is collected. As mentioned before, the data is obtained for a time period of one year between January to December 2015 with a resolution of 1 h. Thus, for each of the 15 features, we have a matrix of data with 25 columns and around 8760 rows. To facilitate the modeling using RNNs, which cannot handle the noise in the data, first the data across all measuring stations was averaged, i.e., average was taken across all columns for each row. This was done to deal with loss of information at some measuring stations due to several practical reasons. And then, moving average was performed over the time-series data (i.e., across all the rows in the single column) to filter out the high frequency component from the data which can be classified as measurement noise. The technical information about the sensors used for measuring the 15 environmental parameters is presented in Table 16.1. Readers can refer to Ref. [40] for more elaborate details on the data set.

Table 16.1 Technical details of sensors used for collecting the data [40].

SL No.	Sensor	Feature	Principle
1	Thermometer Metalone T200A	Ambient temperature	
2	Horiba-APHA 360	CH_4	Flame ionization detection (FID) with selective combustion
3	Carbon monoxide analyzer, Horiba, APMA-360	CO	Cross flow modulation, non-dispersive infrared (NDIR) absorption technology
4	Horiba-APHA 360	NMHC	Flame ionization detection (FID) with selective combustion
5	ECOTECH_ML9841	NO	Chemiluminescence
6	ECOTECH_ML9841	NO_2	Chemiluminescence
7	ECOTECH_ML9841	NO_x	Chemiluminescence
8	Ozone analyzer, Ecotech, ML9810B	O_3	Non-dispersive ultraviolet (UV) absorption
9	Rainwater sampler AR-02	pH rain	
10	Aerosol analyzer E-BAM/Met One Instruments, Inc.	PM_{10}	Beta Ray method
11	Aerosol analyzer E-BAM/Met One Instruments, Inc.	$PM_{2.5}$	Beta Ray method
12	Rain gauge, MetOne, 370	Rainfall	
13	Hygrometer, MetOne, 083D	RH	Variance in the capacitance change of dielectric polymer layer.
14	ECOTECH_ML9850	SO_2	Pulsed UV fluorescent radiation technology
15	Horiba-APHA 360	THC	Flame ionization detection (FID) with selective combustion

16.2.2   Algorithm for design of RNNs

$$\text{minimize}\{-R^2 \text{ and } \omega\} \tag{16.1}$$

where,

$$R^2 = \left(\frac{\text{cov}\left(Y, \widehat{Y}\right)}{\sqrt{\text{var}(Y)\,\text{var}\left(\widehat{Y}\right)}} \right)^2$$

$$\text{cov}\left(Y, \widehat{Y}\right) = n\sum_{i=0}^{n} Y_i \widehat{Y}_i - \sum_{i=0}^{n} \widehat{Y}_i \sum_{i=0}^{n} Y_i$$

$$\text{var}(Y) = n\sum_{i=0}^{n} Y_i^2 - \left(\sum_{i=0}^{n} Y_i \right)^2$$

Y is original outputs, $\widehat{Y}$ is the RNN predicted outputs and n is the size of test data set.
ω is the number of weights and biases.
subject to the decision variables.
$\mathscr{H}$ = Number of hidden layers in RNN.
N^h = Number of nodes in hidden layer h in the RNN, where h→1 to $\mathscr{H}$.
$\mathscr{F}$ = Activation function choice for all the nodes in hidden layers. A binary value indicating either log-sigmoid if 0 and tan-sigmoid if 1.
T = Truncation length in T-BPTT algorithm for training RNN.
such that,

$$\left[\mathscr{H}, \mathscr{F}, T \text{ and } N^h \ \forall \ h=1 \text{ to } \mathscr{H}\right] > 0$$

$$\left.\begin{array}{l} 1 \le N^1 \le N^{UB} \\ 0 \le N^h \le N^{UB} \ \forall \ h = 2 \text{ to } \mathscr{H} \end{array}\right\} \text{This is to ensure that there exists at least 1 hidden layer in}$$

the RNNs which are being explored by NSGA – II.

$$1 \le \mathscr{H} \le \mathscr{H}^{UB}$$

$$T^{LB} \le T \le T^{UB}$$

LB and UB stands for lower and upper bounds, respectively.

The INLP formulation of the proposed technique for optimal design of RNNs is presented in Eq. (16.1). All the lower and upper bounds required in the formulation are provided a priori based on the application. Due to the lack of gradient information and presence of multiple objectives, it is solved using NSGA-II as shown in Table 16.2.

In order to ensure that integer values are obtained for decision variables, binary coded NSGA-II was implemented with appropriate bounds. For example, the number of nodes in a hidden layer was denoted using a 4-bit binary representation and the lower bound was set as 0 and upper bound as 15 ($2^4 - 1$). This ensured generation of only

Table 16.2 Pseudocode for the proposed methodology aimed at automated design of RNNs.

Algorithm for optimal design of RNNs using binary coded NSGA-II

Initialize the algorithm with lower and upper bounds for all decision variables and parameters of NSGA-II provided by the user and start the NSGA-II algorithm. There will be a total of $\mathscr{H}^{UB} + 2$ decision variables in the population of which the first $\mathscr{H}^{UB}$ will be the number of nodes in $\mathscr{H}^{UB}$ hidden layers (some of them can be 0), followed by activation function choice $\mathscr{F}$ and truncation length T. It would give a single population of decision variables and the following steps describe the proposed method thereafter. NSGA-II working mechanism is not described in this algorithm to avoid redundancy. Readers can refer [43] for elaborate details on working mechanism of NSGA-II.
Given the population: {$N^h \ \forall \ h = 1$ to $\mathscr{H}^{UB}$, $\mathscr{F}$ and T}, decode the RNN architecture.
Set $\mathscr{H} = 1$
    for $h \rightarrow 2$ to $\mathscr{H}^{UB}$:
      if $\left(N^h > 0\right)$: $\mathscr{H} = \mathscr{H} + 1$, else: break the loop.
    Number of hidden layers in the RNN $= \mathscr{H}$. Let there be I inputs and O outputs.
    RNN architecture is then given by [$I - \left\{N^h : h \rightarrow 2 \text{ to } \mathscr{H}\right\} - O$]
    Activation function: If $\mathscr{F} = 0$, log-sigmoid, else tan-sigmoid.
    Truncation length for T-BPTT $= T$
    Back propagate the gradients and train the RNN using ADAM optimizer [45]
    After training is complete, predict the outputs for test set and evaluate R^2
    For the given RNN, find the total parameters ω.
    Return the objectives $-R^2$ and ω to NSGA-II.
Repeat the same for all populations in the generation.
NSGA-II performs crossover, mutation and selection to create new generation
Repeat for all generations till convergence.
Upon convergence, a Pareto set will be obtained from which one RNN model is selected using AIC [44].

---

integers between the desired bounds. Similarly, 1-bit representation was used for $\mathscr{F}$ and 8-bit representation was used for T. So, if $\mathscr{H}^{UB} = 3$, a chromosome with 21 bits (4 + 4 + 4 + 1 + 8) was created in NSGA-II. Since binary coded NSGA-II was used for solving the INLP formulation, there was a high chance of repetition of decision variables across the generations. Thus, to avoid redundancy, the objective functions corresponding to the architectures which were trained, along with the decision variable information were stored in a database and in all the iterations, first this database was searched for similarity in decision variables.

16.2.3   Global Sensitivity Analysis

GSA is a well-known method for nonlinear sensitivity analysis which ranks the input variables based on their significance with respect to a variable of interest. Consider a square-integrable function $Y = f(X)$ defined on k-dimensional unit hypercube, where X is a vector of k-dimensions and Y is a scalar. The following analysis is written for three dimensions for improving the readability. However, the same can be easily extended to k-dimensions as elucidated in Ref. [46]. The ANOVA decomposition of the function f is given by Eq. (16.2).

$$Y = f(\mathbf{X}) = f_0 + \sum_{i=1}^{k} f_i(x_i) + \sum_{i<j}^{k} f_{ij}(x_i\ x_j) + f_{1,2,\ldots k}(x_1 x_2 \ldots x_k) \tag{16.2}$$

where,

$$\int_0^1 f_{i_1 i_2 i_3 \ldots i_g}(x_{i_1} x_{i_2} \ldots x_{i_g}) dx_j = 0 \ \text{ where } \ j \ \varepsilon \left[i_1 i_2 i_3 \ldots i_g\right] \tag{16.3}$$

For example, when $k = 3$, Eq. (16.2) translates into Eq. (16.4).

$$Y = f(\mathbf{X}) = f_0 + f_1 + f_2 + f_3 + f_{12} + f_{23} + f_{13} + f_{123} \tag{16.4}$$

where, integration of each summand in Eq. (16.4) over any of its independent dimensions is 0.

Due to this property, each summand in Eq. (16.2) can be expressed as a function of $f(\mathbf{X})$. For example, integrating Eq. (16.2) on both sides with respect to $d\mathbf{X}$ in the unit hypercube and applying the condition in Eq. (16.3) leads to the definition of the constant f_0 in Eq. (16.5).

$$f_0 = \int_0^1 Y d\mathbf{X} = \int_0^1 f(\mathbf{X}) d\mathbf{X} \tag{16.5}$$

Similarly, using Eq. (16.3) and Eq. (16.4) f_1 can be written as shown in Eq. (16.6)

$$f_1 = \int_0^1 Y \prod_{j=2}^{k} dx_j = \int_0^1 f(\mathbf{X}) \prod_{j=2}^{k} dx_j - f_0 \tag{16.6}$$

Further, the ANOVA decomposition makes the summands in Eq. (16.2) orthogonal to each other. This kind of decomposition of the function $f(\mathbf{X})$ can be interpreted as follows: If $\mathbf{X}_G$ is group of G inputs from X, then, $f_G(\mathbf{X}_G)$ (a term in the ANOVA decomposition of the function $f(\mathbf{X})$) is the effect of varying all elements in $\mathbf{X}_G$ in addition to the effect of their individual variations on Y.

Subsequently, if X is considered as a random variable, then any function of X and its subsets is also a random variable. Thus, for the function $f(\mathbf{X})$, expectation and variance can be written as in Eq. (16.7) and Eq. (16.8), respectively. It can be seen that f_0 is the expectation of the function $f(\mathbf{X})$.

$$E[f(\mathbf{X})] = \int_0^1 f(\mathbf{X}) d\mathbf{X} \tag{16.7}$$

$$Var[f(\mathbf{X})] = \int_0^1 [f(\mathbf{X})]^2 d\mathbf{X} - [f_0]^2 \tag{16.8}$$

Similarly, for the function $f_G(X_G)$ defined on X_G, the subset of X, the expectation is found to be 0 due to the condition defined in Eq. (16.3). The variance is then given by Eq. (16.9).

$$\text{Var}\big[f_G(X_G)\big] = \int_0^1 \big[f_G(X_G)\big]^2 dX_G \qquad (16.9)$$

The global sensitivity index for the group X_G is called as the Sobol index S_G, defined in Eq. (16.10).

$$S_G = \frac{\text{Var}\big[f_G(X_G)\big]}{\text{Var}[f(X)]} = \frac{\int_0^1 \big[f_G(X_G)\big]^2 dX_G}{\int_0^1 [f(X)]^2 dX - [f_0]^2} \qquad (16.10)$$

16.3 Results and discussions

The preprocessed data was split into a ratio of 70:30 where the first 70% was used for training and the next 30% was used for testing the RNNs. The algorithm for optimal design of RNNs was run with the following parameters: NSGA-II parameters: 100 populations, 100 generations, 0.99 crossover probability, 0.01 mutation probability; RNN parameters: $\mathcal{H}^{UB} = 3$, $N^{UB} = 15$, $T^{LB} = 1$ and $T^{UB} = 64$. Convergence was ensured by running the NSGA-II algorithm for all 100 generations even though no change in Pareto front was reported in much earlier generations and it remained same for remaining generations. The obtained two-dimensional Pareto front for $PM_{2.5}$ is shown in Fig. 16.7.

Similar fronts were obtained for all other 14 pollutants but are not shown to honor the space constraint of the chapter. A single architecture is obtained from the Pareto front by

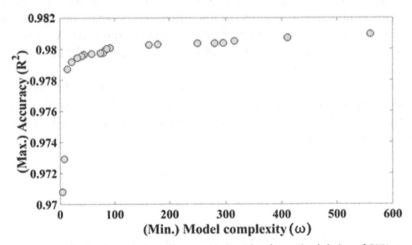

FIGURE 16.7 Pareto front obtained as solution of proposed algorithm for optimal design of RNNs to emulate the univariate data of $PM_{2.5}$.

measuring the AIC and selecting the one with least value. AIC metric penalizes the model for large complexity and thus serves as a robust measure for checking overfitting. The final selected architectures for all 15 pollutants are presented in Table 16.3. Except for pH rain, all other environmental parameters are emulated accurately using univariate optimal RNN models. Thus, multivariate RNN models are implemented, which consider all the 14 environmental parameters as inputs and pH of rain as output. The result of proposed algorithm for the optimal design of such an RNN is shown in Figs. 16.8 and 16.9.

Once again AIC was implemented to select a single architecture from the Pareto front: [14-1-1-2-1], log-sigmoid activation and $T = 22$ and $R^2 = 0.98$. However, to determine the sensitivity of each input parameter, Monte Carlo–based GSA was performed as follows:

Step 1. Define two matrices X_A and X_B with pseudorandom binary signals (PRBS) of 14 input parameter and length N^G. Then, generate the corresponding output sequence from trained network, Y_A and Y_B, respectively.

Step 2. Define a matrix X_{AB}^G such that it's all columns come from X_A except for the column G which come from matrix X_B. Similarly, generate X_{BA}^G.

Step 3. Once again for the input sequences in the matrices X_{AB}^G and X_{BA}^G simulate the trained network and generate the corresponding output sequences for pH rain - Y_{AB}^G and Y_{BA}^G, respectively.

Step 4. According to the work in Ref. [47], the Sobol index S_G is then given by the following approximation (Eq. 16.11) for the integral based equation in Eq. (16.10).

Table 16.3 Selected architectures of RNN for 15 environmental parameters.

Parameters	N^1	N^2	N^3	$\mathcal{F}$	T	R^2
Temp.	8	8	0	1	33	0.996201
CH_4	16	10	0	1	41	0.927706
CO	5	15	0	1	61	0.937332
NMHC	13	10	0	1	41	0.917904
NO	16	10	3	1	41	0.916618
NO_2	2	10	2	1	41	0.952239
NO_x	1	15	6	1	62	0.944167
O_3	6	12	0	1	49	0.971758
pH rain	2	5	0	0	21	0.489056
PM_{10}	4	7	0	1	29	0.985868
$PM_{2.5}$	1	8	0	1	33	0.980065
Rainfall	16	8	4	1	34	0.941299
RH	2	11	0	1	45	0.983110
SO_2	14	9	9	1	39	0.895839
THC	16	9	6	1	38	0.917834

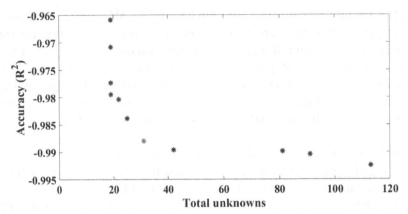

FIGURE 16.8 Pareto front obtained as solution of proposed algorithm for optimal design of multivariate RNNs to emulate the data of pH of rain.

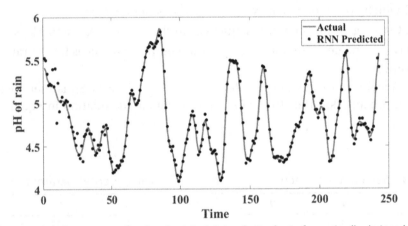

FIGURE 16.9 Comparison between predicted and original data of pH of rain for optimally designed multivariate RNN.

$$S_G = \frac{\left(\frac{1}{N^G}\right)\sum_{j=1}^{N^G}(Y_B)_j\left((Y_{AB}^G)_j - (Y_A)_j\right)}{\left(\frac{1}{N^G}\right)\sum_{j=1}^{N^G}[(Y_A)_j]^2 - \left[\left(\frac{1}{N^G}\right)\sum_{j=1}^{N^G}(Y_A)_j\right]^2} \tag{16.11}$$

Step 5. Vary N^G till the Sobol indices converge.

Thus obtained sensitivity values for all the 14 environmental pollutants with respect to pH of rain are listed in Table 16.4. The convergence graphs for all parameters are presented in Fig. 16.10. Among all the variables, THC, NO_x and CO were found to be most significant, thereby strongly influencing the pH of rain.

Table 16.4 Sobol indices of 14 environmental parameters with respect to pH of rain.

Parameters	Sobol indices
Ambient temperature	0.1092
CH_4	0.1082
CO	0.2197
NMHC	0.1875
NO	0.1979
NO_2	0.1622
NO_x	0.2174
O_3	0.1107
PM_{10}	0.1057
$PM_{2.5}$	0.1080
Rainfall	0.1096
RH	0.1119
SO_2	0.1084
THC	0.7246

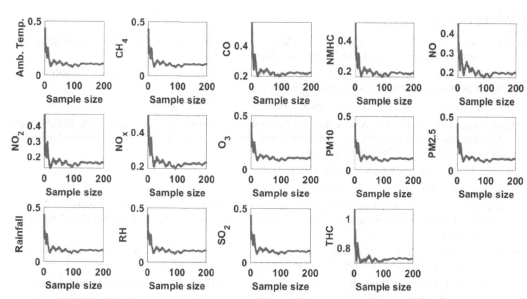

FIGURE 16.10 Convergence characteristics of Sobol indices in the Monte Carlo–based GSA.

16.4 Conclusions

This work proposes a novel MOOP-based neural architecture search strategy for design of RNNs. Pollutants and particulate matter are increasing at a rapid rate in the environment due to industrial development. To tackle the challenges posed by the increasing pollutants, it is necessary to have a representative model, which can predict their dynamic behavior in future. Using a model, decision makers can make better informed decisions which help in tackling the aforementioned issues. Data-based modeling, which has been gaining momentum over the years due to many advantages associated with them, are criticized for following heuristic approaches for selecting the hyperparameters involved with them. A novel algorithm to optimally determine the hyperparameters in an intelligent and scientific way has been proposed in this work and utilized the same to demonstrate its efficacy in modeling the pollutants collected from Taiwan. Further, Monte Carlo–based GSA was performed to quantitatively determine that CH_4, CO, NMHC, and THC are the most significant pollutants influencing the pH of rain. The proposed study is generic and can be implemented to any ML model thus contributing to the field of automated ML.

Acknowledgments

Authors would like to thank Ministry of Human Resources Development (MHRD), Government of India [SPARC/2018–2019/P1084/SL] and Department of Bio-Technology, Government of India [BT/PR34209/AI/133/19/2019] for extending financial support for this work.

References

[1] P. Yadav, J. Singh, D.K. Srivastava, V. Mishra, Environmental pollution and sustainability, in: Environmental Sustainability and Economy, Elsevier, 2021, pp. 111–120.

[2] I. Manisalidis, E. Stavropoulou, A. Stavropoulos, E. Bezirtzoglou, Environmental and health impacts of air pollution: a review, Front. Public Health 8 (2020) 14.

[3] https://www.statista.com/statistics/264699/worldwide-co2-emissions/.

[4] M. Goss, D.L. Swain, J.T. Abatzoglou, A. Sarhadi, C.A. Kolden, A.P. Williams, N.S. Diffenbaugh, Climate change is increasing the likelihood of extreme autumn wildfire conditions across California, Environ. Res. Lett. 15 (9) (2020) 094016.

[5] https://www.ncdc.noaa.gov/sotc/fire/202103.

[6] https://www.nifc.gov/fire-information/statistics.

[7] P.J. Vidya, M. Ravichandran, R. Murtugudde, M.P. Subeesh, S. Chatterjee, S. Neetu, M. Nuncio, Increased cyclone destruction potential in the Southern Indian Ocean, Environ. Res. Lett. 16 (1) (2020) 014027.

[8] S. Sully, D.E. Burkepile, M.K. Donovan, G. Hodgson, R. Van Woesik, A global analysis of coral bleaching over the past two decades, Nat. Commun. 10 (1) (2019) 1–5.

[9] C.M. Eakin, H.P. Sweatman, R.E. Brainard, The 2014–2017 global-scale coral bleaching event: insights and impacts, Coral Reefs 38 (4) (2019) 539–545.

[10] https://www.statista.com/statistics/1135370/most-polluted-capital-cities-in-the-world/.

[11] D. Iskandaryan, F. Ramos, S. Trilles, Air quality prediction in smart cities using machine learning technologies based on sensor data: a review, Appl. Sci. 10 (7) (2020) 2401.

[12] R.K. Inapakurthi, S.S. Miriyala, K. Mitra, Deep learning based dynamic behavior modelling and prediction of particulate matter in air, Chem. Eng. J. 426 (2021) 131221.

[13] F.J. Leij, S.A. Bradford, Combined physical and chemical nonequilibrium transport model: analytical solution, moments, and application to colloids, J. Contam. Hydrol. 110 (3–4) (2009) 87–99.

[14] J. Kukkonen, K. Riikonen, J. Nikmo, A. Jäppinen, K. Nieminen, Modelling aerosol processes related to the atmospheric dispersion of sarin, J. Hazard Mater. 85 (3) (2001) 165–179.

[15] R. Feng, Q. Wang, C.C. Huang, J. Liang, K. Luo, J.R. Fan, H.J. Zheng, Ethylene, xylene, toluene and hexane are major contributors of atmospheric ozone in Hangzhou, China, prior to the 2022 Asian Games, Environ. Chem. Lett. 17 (2) (2019) 1151–1160.

[16] R. Xu, X. Tie, G. Li, S. Zhao, J. Cao, T. Feng, X. Long, Effect of biomass burning on black carbon (BC) in South Asia and Tibetan Plateau: the analysis of WRF-Chem modeling, Sci. Total Environ. 645 (2018) 901–912.

[17] M.I. Rumaling, F.P. Chee, H.W.J. Chang, C.M. Payus, S.K. Kong, J. Dayou, J. Sentian, Forecasting particulate matter concentration using nonlinear autoregression with exogenous input model, Global J. Environ. Sci. Manag. 8 (1) (2022) 27–44.

[18] P. Goyal, A.T. Chan, N. Jaiswal, Statistical models for the prediction of respirable suspended particulate matter in urban cities, Atmos. Environ. 40 (11) (2006) 2068–2077.

[19] L.A. Díaz-Robles, J.C. Ortega, J.S. Fu, G.D. Reed, J.C. Chow, J.G. Watson, J.A. Moncada-Herrera, A hybrid ARIMA and artificial neural networks model to forecast particulate matter in urban areas: the case of Temuco, Chile, Atmos. Environ. 42 (35) (2008) 8331–8340.

[20] F. Taşpınar, Improving artificial neural network model predictions of daily average PM_{10} concentrations by applying principle component analysis and implementing seasonal models, J. Air Waste Manag. Assoc. 65 (7) (2015) 800–809.

[21] A. Suleiman, M.R. Tight, A.D. Quinn, Applying machine learning methods in managing urban concentrations of traffic-related particulate matter (PM_{10} and $PM_{2.5}$), Atmos. Pollut. Res. 10 (1) (2019) 134–144.

[22] L.C. Yan, B. Yoshua, H. Geoffrey, Deep learning, Nature 521 (7553) (2015) 436–444.

[23] A. Sherstinsky, Fundamentals of recurrent neural network (RNN) and long short-term memory (LSTM) network, Phys. Nonlinear Phenom. 404 (2020) 132306.

[24] A. Graves, Supervised sequence labelling, in: Supervised Sequence Labelling with Recurrent Neural Networks, Springer, Berlin, Heidelberg, 2012, pp. 5–13.

[25] J. Chung, C. Gulcehre, K. Cho, Y. Bengio, Empirical Evaluation of Gated Recurrent Neural Networks on Sequence Modeling, 2014 arXiv preprint arXiv:1412.3555.

[26] C.J. Huang, P.H. Kuo, A deep cnn-lstm model for particulate matter ($PM_{2.5}$) forecasting in smart cities, Sensors 18 (7) (2018) 2220.

[27] D.P. Kingma, M. Welling, An Introduction to Variational Autoencoders, 2019 arXiv preprint arXiv: 1906.02691.

[28] A. Makhzani, J. Shlens, N. Jaitly, I. Goodfellow, B. Frey, Adversarial Autoencoders, 2015 arXiv preprint arXiv:1511.05644.

[29] V. Athira, P. Geetha, R. Vinayakumar, K.P. Soman, DeepAirnet: applying recurrent networks for air quality prediction, Procedia Comput. Sci. 132 (2018) 1394–1403.

[30] J. Ma, J.C. Cheng, C. Lin, Y. Tan, J. Zhang, Improving air quality prediction accuracy at larger temporal resolutions using deep learning and transfer learning techniques, Atmos. Environ. 214 (2019) 116885.

[31] M. Krishan, S. Jha, J. Das, A. Singh, M.K. Goyal, C. Sekar, Air quality modelling using long short-term memory (LSTM) over NCT-Delhi, India, Air Quality, Atmosphere & Health 12 (8) (2019) 899–908.

[32] U. Pak, J. Ma, U. Ryu, K. Ryom, U. Juhyok, K. Pak, C. Pak, Deep learning-based $PM_{2.5}$ prediction considering the spatiotemporal correlations: a case study of Beijing, China, Sci. Total Environ. 699 (2020) 133561.

[33] Q. Tao, F. Liu, Y. Li, D. Sidorov, Air pollution forecasting using a deep learning model based on 1D convnets and bidirectional GRU, IEEE Access 7 (2019) 76690−76698.

[34] Y.S. Chang, H.T. Chiao, S. Abimannan, Y.P. Huang, Y.T. Tsai, K.M. Lin, An LSTM-based aggregated model for air pollution forecasting, Atmos. Pollut. Res. 11 (8) (2020) 1451−1463.

[35] W. Tong, L. Li, X. Zhou, A. Hamilton, K. Zhang, Deep learning $PM_{2.5}$ concentrations with bidirectional LSTM RNN, Air Quality, Atmosphere & Health 12 (4) (2019) 411−423.

[36] Q. Zhang, J.C. Lam, V.O. Li, Y. Han, Deep-AIR: A Hybrid CNN-LSTM Framework for Fine-Grained Air Pollution Forecast, 2020 arXiv preprint arXiv:2001.11957.

[37] L. Zhang, D. Li, Q. Guo, Deep learning from spatio-temporal data using orthogonal regularizaion residual CNN for air prediction, IEEE Access 8 (2020) 66037−66047.

[38] D. Qin, J. Yu, G. Zou, R. Yong, Q. Zhao, B. Zhang, A novel combined prediction scheme based on CNN and LSTM for urban $PM_{2.5}$ concentration, IEEE Access 7 (2019) 20050−20059.

[39] H. Gao, W. Yang, J. Wang, X. Zheng, Analysis of the effectiveness of air pollution control policies based on historical evaluation and deep learning forecast: a case study of Chengdu-Chongqing region in China, Sustainability 13 (1) (2021) 206.

[40] https://www.epa.gov.tw/eng/.

[41] S.S. Miriyala, K. Mitra, Deep learning based system identification of industrial integrated grinding circuits, Powder Technol. 360 (2020) 921−936.

[42] I. Sutskever, Training Recurrent Neural Networks, University of Toronto, Toronto, Canada, 2013, pp. 1−101.

[43] K. Deb, A. Pratap, S. Agarwal, T.A.M.T. Meyarivan, A fast and elitist multiobjective genetic algorithm: NSGA-II, IEEE Trans. Evol. Comput. 6 (2) (2002) 182−197.

[44] S. Hu, Akaike Information Criterion, vol 93, Center for Research in Scientific Computation, 2007.

[45] D.P. Kingma, J. Ba, Adam: A Method for Stochastic Optimization, 2014 arXiv preprint arXiv:1412.6980.

[46] A. Saltelli, M. Ratto, T. Andres, F. Campolongo, J. Cariboni, D. Gatelli, et al., Global Sensitivity Analysis: The Primer, John Wiley & Sons, 2008.

[47] Y. Fang, Y. Su, On the use of the global sensitivity analysis in the reliability-based design: insights from a tunnel support case, Comput. Geotech. 117 (2020) 103280.

Comparative study of automated deep learning techniques for wind time-series forecasting

NagaSree Keerthi Pujari[1], Srinivas Soumitri Miriyala[2],
Kishalay Mitra[1]

[1]DEPARTMENT OF CHEMICAL ENGINEERING, INDIAN INSTITUTE OF TECHNOLOGY
HYDERABAD, HYDERABAD, TELANGANA, INDIA; [2]GLOBAL OPTIMIZATION AND KNOWLEDGE
UNEARTHING LABORATORY, DEPARTMENT OF CHEMICAL ENGINEERING, INDIAN INSTITUTE
OF TECHNOLOGY HYDERABAD, HYDERABAD, TELANGANA, INDIA

17.1 Introduction

One of the greatest challenges in 21st century is energy security and environment protection. In view of forthcoming energy shortage, climate change and reduction of fossil fuels, many of the countries are shifting toward cleaner energy sources such as solar, wind, geothermal. In order to reduce carbon emissions, even the national governments have introduced many programs and policies in support to the renewable energy sources. As per the International Renewable Energy Agency (IRENA) Report 2021, the total cumulative installations of renewable energy sources has been increased to 2700 GW [1]. Fig. 17.1 represents the statistics of increase in renewable energy from past decade. Among the available renewable sources, the usage of wind energy has become a promising resource due to its easy availability, cleaner production and affordability. The total wind power capacity across the world rose to 743 GW which helps to avoid 1.1 billion tons of CO_2 [2]. Therefore, the researchers and policy makers turned towards wind energy to explore and discover new strategies for increasing the wind energy production. One of the key challenges in wind energy production is the variability in energy production due to its uncertain nature. Conventionally, the variability in wind is modeled by constructing a wind frequency map (WFM), which is a joint probability distribution over two random variables wind speed and direction. However, due to unavailability of long-term data for wind characteristics, WFM is built with limited amount of data. Since wind is heavily uncertain, usage of WFM, with limited amount of data for constructing a wind farm with long life span, i.e., 25–30 years, is inaccurate. Since we cannot change the nature of wind, an accurate forecasting of wind

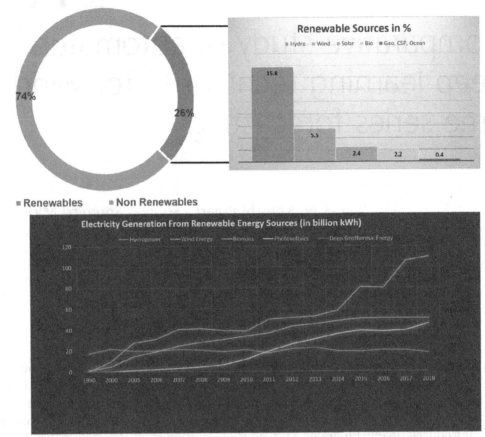

FIGURE 17.1 Pictorial representation of increase in Renewable energy sources (Global Energy Statistical Yearbook 2020 and Federal Ministry for Economic Affairs and Energy based on data from AGEE-Stat and other sources).

characteristics can be a great help for wind farm practitioners in execution of better energy management systems while designing and controlling wind farms.

Wind time-series data being erratic and stochastic, efforts must be made to capture the nonlinear dynamics to enhance the forecasting abilities. Analysis of time-series data (stationarity, nonlinear temporal dynamics and long-term dependencies), therefore, gained importance. To capture the nonlinear dynamics and hidden features in the wind data, examination of the seasonality, trend and randomness is necessary [3,4]. The hidden patterns and seasonality in the wind time-series data can be determined using several decomposition techniques such as classical decomposition, Seasonal and Trend Decomposition using LOESS (STL) and such decompositions contribute to a better depiction of wind time-series data [5,6]. The nonlinear trends of wind characteristics can be summarized to three categories: physical, statistical and data driven models. Generally, physical models are reliable on numerical weather prediction (NWP) [7] and atmospheric data for forecasting. Several works employed NWP models such as Boundary Scaling method and weather research and forecasting models for forecasting wind conditions [8,9]. The complexity of physical models increases due to the fluctuations of atmospheric conditions, which in turn increases the computational cost. Therefore, complex atmospheric conditions, computational cost and lack of professional staff for collection and maintenance of data make the physical models difficult to handle. In contrast to the physics based models, statistical models rely on historic data for modeling and forecasting.

The system identification tools with sufficient historic data have been widely applied for forecasting of wind data with a better forecasting accuracy. From a panoramic view of statistical models, it is evident that the existing models such as Nonlinear Autoregressive Models, AutoRegressive Integrated Moving Average (ARIMA), Wavelet Neural Networks have been widely applied and often results in good accuracies. A hybrid model was proposed with wavelet decomposition and nonlinear autoregressive neural network model (NARX-NN) for forecasting wind power in Ref. [10]. Auto Regressive Moving Average (ARMA) model was employed in Ref. [11] for short-term forecasting of wind data. The four approaches featuring the decomposition of wind speed and direction, i.e., adopting component based ARMA model, traditional ARMA model, vector autoregression (VAR), and restricted version of VAR, were compared in this work using mean absolute error. The results had shown that the component based ARMA model was good at predicting wind speed and the traditional ARMA model was good at predicting wind direction [11]. A hybrid method was proposed utilizing discrete wavelet transform, autoregressive moving average and recurrent neural networks (RNN) for one step ahead univariate time-series prediction where RNN is used as nonlinear autoregressive model with exogenous inputs (NARX) [12]. In recent times, the availability of data turned the researchers towards the machine learning techniques due to their ability in handling extreme nonlinearities. Support vector machine (SVM) combined with improved dragon fly optimization was proposed in Ref. [13]. In this work, dragon Fly optimization was utilized to determine the SVM parameters and the proposed algorithm has shown better results when compared with back propagation neural networks and Gaussian process regression [13]. In Ref. [14], a multilayer perceptron model with multicriterion optimization was proposed for forecasting of wind power for 1 day predictions while in Ref. [15], a hybrid method for hourly predictions using ANFIS combined with a feed forward neural network (FNN) was proposed. A reasonable success has been achieved by all the aforementioned approaches but these are limited to short-term predictions. The deep learning techniques are inherently better to tackle the nonlinear nature of data and long-term dependencies, which encouraged the researchers to use deep learning tools for long-term forecasting. An integrated approach with Bidirectional LSTMs and Discrete Wavelet Packet Transform was proposed in Ref. [16] for dynamic characterizations and predictions of wind speed. Later several techniques were developed to study the long-term forecasts using Deep learning techniques such as RNNs and Long Short-Term Memory Networks (LSTMs) [17,18].

The utilization of system identification as well as machine learning tools has been increasing from past decade due to the availability of voluminous data and the availability of open source software. The construction of these models became tedious and inefficient due to the heuristic estimation of hyperparameters such as architecture of the models, choice of activation function. A resurgence of research in autotuning of these hypermeters has resulted in Automated Machine Learning framework, which helps in reducing the tedious task of heuristic estimation of hyperparameters. A review of algorithms and applications in hyperparameter optimization is presented in Ref. [19]. Recently, Bayesian Optimization and Reinforcement Learning are extensively used in Automated Machine Learning framework [20,21]. From our insights, no work has been reported utilizing the evolutionary multiobjective optimization-based framework for autotuning of hyperparameters in deep learning techniques, which balances accuracy and overfitting.

Therefore, the aim of the current work is to develop automated Machine Learning framework utilizing the evolutionary multiobjective optimization technique, non-dominated sorting genetic algorithm (NSGA-II) [22]. In this work, the nature of time-series is analyzed through nonlinear nature, stationarity and long-term dependencies to decide on appropriate time-series modeling techniques. Since the seasonality has its own effect on time-series data, time-series decomposition was performed to determine the hidden features, trends and sequential patterns in the data. Since the nonlinear nature and long-term dependencies were present in the wind data, two Deep learning techniques, RNNs and long short-term memory networks (LSTMs) were proposed for modeling the wind data. The heuristics associated in these models such as choosing the architecture of the model, time steps to be unrolled, activation function were addressed by considering three conflicting objectives: maximizing of accuracy and minimizing the parameters and unrolling steps. The credibility of these models were validated using the wind characteristics data collected over a span of four years from an open source repository. A comparative study is performed between Automated RNNs and Automated LSTMs in the context of wind time-series forecasting. Fig. 17.2 summarized the present work. The novel contributions of the work are summarized below:

- Hyperparameter optimization of RNNs and LSTMs using evolutionary multi-objective optimization framework.
- Comparison of automated deep learning tools, RNNs, and LSTMs for wind time course modeling.
- Validation of automated models with real data.
- Effective usage of past as well as forecasted data for accurate modeling of wind for wind energy management systems in the aspects of design and control of wind farm.

In the rest of the chapter, Section 17.2 presents time-series analysis and decomposition techniques in brief, followed by the comprehensive details of the proposed holistic algorithm for designing automated deep learning models (including RNNs and LSTMs).

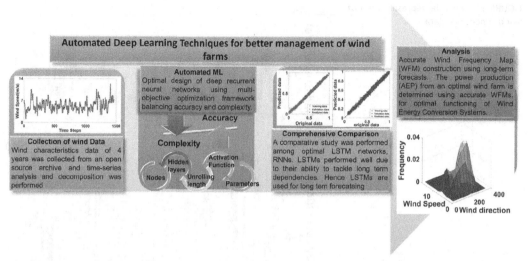

FIGURE 17.2 Overall framework of wind characteristics forecasting and its application in Annual Energy Production from a wind farm.

The results of the proposed work are presented in Section 17.3. The concluded summary and future scope of this work are outlined in Section 17.4.

17.2 Formulation

17.2.1 Data description and analysis

Wind characteristics data of wind speed and wind direction was collected from an open source archive called ENGIE [23]. The data spans over 4 years with 6-hour resolution. The data was collected through sensors from a wind farm with an area of 9 sq. Km in La Haute Barne in France and it comprises four turbines, which are placed at wind farm corners. The measured data is not influenced by wake because of the large distance maintained between the turbines. Since it is a real data obtained from sensors, it is assumed to be corrupted with measurement noise, which is filtered using moving average method. The pictorial representation of wind time-series data is presented in Fig. 17.3 with limited timeframe.

In this work, the wind characteristics data was modeled as multivariate time-series and the time-series analysis on both speed and direction of wind was performed. The time-series analysis includes examining the nonlinearity, stationarity and long-term dependencies. The irregular temporal behavior and a nonlinear relationship between inputs and outputs make the data to be nonlinear and it can be detected using Brock-Dechert-Scheinkman (BDS) test [24]. The stationary time-series is said to have constant statistical properties (mean and variance) and can be discovered using a hypothesis

FIGURE 17.3 Pictorial representation of Wind time-series data.

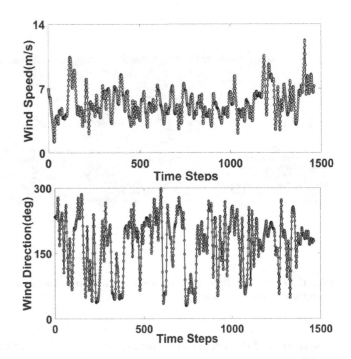

test called Augmented-Dickey-Fuller (ADF) test [25]. The degree of dependency of data at current time stamp on the previous values, known as long-term temporal features, was determined using Hurst exponent [26].

The seasonality component present in the time-series data has sequential influence in the data. Therefore, time-series decomposition is a practical way to determine the hidden patterns and sequential influences in the data. In the literature, there are several ways to decompose the data into trend, cycle and seasonal patterns, such as classical, X11, STL [27]. Among all those, STL decomposition is most robust and efficient [28]. Therefore, in this study STL decomposition is utilized to study the hidden patterns in the data. The results were presented in Section 17.3.

17.2.2 Techniques for modeling the data

Wind characteristics being weather data, it can have daily and yearly periodicity. Therefore, the time steps in seconds is converted to usable signals with daily and yearly periodicity as shown in Eq. (17.1). The collected data can now be modeled as multivariate time-series data, in which the inputs are the four time signals shown in Eq. (17.1) and the outputs are wind characteristics (wind speed and direction).

$$d_s = \sin\left(t * \frac{2\pi}{24 * 60 * 60}\right) \tag{17.1}$$

$$d_c = \cos\left(t * \frac{2\pi}{24 * 60 * 60}\right)$$

$$y_s = \sin\left(t * \frac{2\pi}{365.2425 * 24 * 60 * 60}\right)$$

$$y_c = \cos\left(t * \frac{2\pi}{365.2425 * 24 * 60 * 60}\right)$$

Let $X^t = [X_1^t\ X_2^t\ X_3^t...X_M^t]$ and $Y^t = [Y_1^t\ Y_2^t Y_3^t...Y_N^t]$ be the input and output at time step t, where M and N are dimensions of input and output. For the considered system $M = 4$ and $N = 2$. The conventional way of modeling a time-series is to find an estimate of target at time instance t, $\widehat{Y^t}$, as a function of β_T previous data points and a set of adjusted parameters, w. The error between the target and its estimate is measured using a loss function such as mean square error (MSE) and is minimized by optimizing the parameters, w, which is illustrated in Eqs. (17.2)−(17.4).

$$\widehat{Y^t} = F\left(X^q|_{q=t-\beta_T\ to\ t}, w\right) \tag{17.2}$$

$$L = \frac{1}{T-\beta_T}\sum_{q=t}^{T}\left(Y^q - \widehat{Y^q}\right)^2 \tag{17.3}$$

$$w^* = \text{argmin}(L) \tag{17.4}$$

In Eq. (17.2), F is a functional map between inputs and outputs, T is the sequence length. In this work, two functional maps were considered: one is RNN and the other is LSTMs with weights and biases as the set of adjustable parameters. Unlike state-of-the-art system identification tools, RNNs and LSTMS regress on hidden states rather than regressing over inputs and thus helps in dynamical modeling of multivariate wind data.

In a multivariate time course modeling, both the models take β_T inputs to generate one network predicted output using which the loss function L can be evaluated as per Eq. (17.3) in the manuscript. To train the set of parameters w in these models, the loss function needs to be minimized as shown in Eq. (17.4) in manuscript. Since RNNs and LSTMs used to evaluate the outputs according to Eq. (17.2) are continuous graphs, one can derive gradient of loss function with respect to the decision variables analytically and therefore a suitable gradient based unconstrained optimizer can be implemented to solve the optimization problem in Eq. (17.4). One elegant example of such an optimizer is the gradient descent algorithm, which is an iterative technique that starts with a random initial guess for decision variables and proceeds by updating them using Eq. (17.5) until termination where η is called the learning rate.

$$w^{new} = w^{old} - \eta\left(\frac{\partial L}{\partial w}\right) \tag{17.5}$$

The gradient $\partial L/\partial w$ is evaluated using the chain rule of differentiation. This allows to derive the analytical expression using a recursive relation, which involves the backward propagation of information. Thus, this algorithm is called back propagation through time (BPTT) in case of recurrent networks as the backward flow is not only in layers but also in the other dimension of time [29,30]. However, in case of RNNs, since the network will be unrolled for the entire length of time-series, it becomes computationally intensive to evaluate gradient at every iteration (the entire length of time-series has to be propagated in every iteration of parameter update). A pragmatic approach is to update weights by reducing the extent back propagation in the network. In Truncated-BPTT (t-BPTT), we perform forward and backward propagation only on certain extent of time steps.

Therefore, in recurrent networks, during the training procedure, the network is unrolled only until a prespecified length (T^F) and gradients are back propagated until (β_T), indicating that $T^F \geq \beta_T$. This essentially truncates the length of time-series during BPTT and therefore, this version is called truncated-BPTT (t-BPTT) [31]. In this work, we implemented that $T^F = \beta_T$.

In practice, a small vale of β_T is considered to minimize the cost associated with update of weight. The truncation to β_T helps for practical execution of BPTT algorithm while training the model but it does not imply that the network requires previous β_T data points to predict the current output in validation and testing. Therefore, to decrease the unrolling length to which gradients are back propagated in the network, a truncated-BPTT algorithm (t-BPTT) is used. Further, we use ADAM (adaptive momentum) proposed by Kingma et al. [32] to train all the networks due to its advantage of higher rates of convergence.

17.2.2.1 ADAM optimizer

The loss function in deep neural networks is minimized using the optimization algorithms while training the model. Several optimizers are present in the literature to perform this task. In this work, ADAM (Adaptive moment estimation) optimizer is used for this purpose [32]. Adam uses the squared gradients to enhance the learning rate, η and moving average of gradient, $\partial L/\partial w$ to incorporate the momentum. ADAM uses the estimations of first order and second order moments i.e., mean, μ_t and variance, ϑ_t to adapt the learning rate. Exponential moving averages of gradients are utilized to estimate the first and second order moments. The steps in calculating gradients and parameter update w, using Adam are shown in Eq. (17.6), ω_1 and ω_2 are the parameters of ADAM optimizer. Bias correction factors, $\hat{m}_t$ and $\hat{v}_t$ are estimated for both first two order moments to ensure the initialization at origin.

$$\mu_t = \omega_1 * \mu_{t-1} + (1 - \omega_1) * \left(\partial L/\partial \theta \right) \tag{17.6}$$

$$\vartheta_t = \omega_2 * \vartheta_{t-1} + (1 - \omega_2) * \left(\partial L/\partial \theta \right)^2$$

$$\widehat{m}_t = \frac{\mu_t}{1 - \omega_1^t}$$

$$\widehat{v}_t = \frac{\vartheta_t}{1 - \omega_2^t}$$

$$w^{new} = w^{old} - \eta * \frac{\widehat{m}_t}{\sqrt{\widehat{v}_t} + \varepsilon}$$

17.2.2.2 Recurrent Neural Networks

Recurrent Neural Networks belong to deep neural networks [33]. RNNs model the dynamics in sequential data by regressing over hidden states through a feedback loop on every node in the hidden layer and these hidden layers are connected in a sequence to constitute the deep RNNs. The network has two dimensions: one across the time steps and other across the layers. The dimension across the time is responsible for simulating the feedback loop from previous time step and current time step. The parameters do not change across the dimension of time as the network remains same across all the time steps. Hence, it is able to maintain the sequence in the data while training the model. The RNNs do not have a feedback loop on output layer. The evaluation of activated output of pth node in hidden layer, l at q^{th} time step is determined using Eq. (17.7) and the RNN model is shown in Fig. 17.4.

$$\mathcal{Y}_p^{l<q>} = \sum_{r=1}^{n_{l-1}} w_{pr}^l x_r^{l-1<q>} + \sum_{k=1}^{n_l} \overline{w}_{pk}^l x_k^{l<q-1>} + b_p^l \qquad (17.7)$$

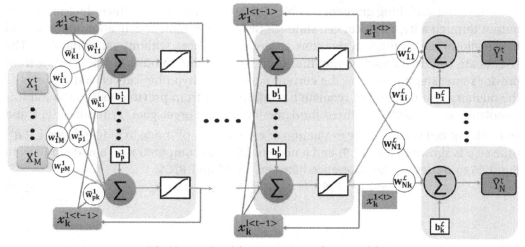

FIGURE 17.4 Pictorial representation of RNN model.

$$x_p^{l<q>} = \sigma\left(\mathcal{Y}_p^{l<q>}\right) \ \forall \ q = t - \beta_T \ \text{to} \ t \ , \ l = 1 \ \text{to} \ \mathcal{L} - 1$$

Here, $\mathcal{Y}_p^{l<q>}$ is the summation of weighted sum of activated outputs in hidden layer, $(l-1)$ at q^{th} time step and weighted sum of activated outputs in hidden layer, l at $(q-1)^{th}$ time step. The additional superscript, q, is added apart from layers, l, to indicate the network variables across the time steps. The activated output, $x_p^{l<q>}$ is known as hidden state and the regression over this hidden state is responsible to model the dynamics in the data. n_l indicates the number of nodes in hidden layer, w_{pr}^l, $\overline{w}_{pk}^l$ and b_p^l are the weights across the dimension of layers, weights across the dimension of time and bias. w_{pr}^l is the weight across the dimension of layers i.e., the weight on feedforward connection from rth node in hidden layer, $(l-1)$ to p^{th} node in hidden layer, l when both nodes are at time step q. $\overline{w}_{pk}^l$ is the weight on feedback connection from k^{th} node in hidden layer, l at $(q-1)^{th}$ time step to p^{th} node in hidden layer, l at q^{th} time step. b_p^l is the bias on p^{th} node in hidden layer,l at q^{th} time step, $\mathcal{L}$ is the maximum number of hidden layers and σ is the activation function. The network output is evaluated at time $q = t$, which is presented in Eq. (17.8). Since the output layer of RNN do not have any feedback connections, it will be the weighted sum of activated outputs from l^{th} hidden layer and has a linear activation function.

$$\widehat{Y^t} = \sum_{r=1}^{n_{\mathcal{L}-1}} w_{pr}^{\mathcal{L}} x_r^{\mathcal{L}-1<q>} + \sum_{k=1}^{n_{\mathcal{L}}} b_k^{\mathcal{L}} \ \text{for} \ q = t \tag{17.8}$$

As the unrolling steps increase in RNNs, gradients will be vanished since the gradients were calculated at every unrolling step and are multiplied to each other. Hence, RNNs will fail numerically due to the problem of vanishing gradients.

17.2.2.3   Long short-term memory networks

To resolve the vanishing gradient trouble, LSTMs are developed, which determines the output through a hypothetical cell state (C) devoid of nonlinear activation function and hence allows the gradient evaluations at larger sequences without vanishing [33]. The LSTMs are trained in such a way that they regulate the amount of information from previous time step to estimate the current output. The hypothetical cell state, C, acts as the memory unit of LSTM to regulate the information from previous cell states and it is calculated as a function of three fundamental units, Forget gate, F, Input gate, I and Intermittent cell value, $\tilde{C}$. The evaluation of cell state for p^{th} node in hidden layer, l at q^{th} time step is shown in Eq. (17.9) and a node in LSTM is compared with a node in RNN in Fig. 17.5 to represent the difference between LSTM and RNN.

$$C_p^{l<q>} = F_p^{l<q>} C_p^{l<q-1>} + I_p^{l<q>} \tilde{C}_p^{l<q-1>} \ \forall \ q = t - \beta_T \ \text{to} \ t \tag{17.9}$$

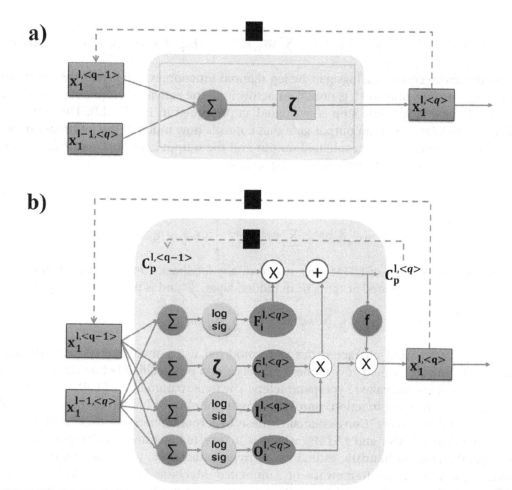

FIGURE 17.5 Pictorial representation of comparison of one LSTM node with RNN node. (A) Represents the RNN node. (B) Represents the LSTM node.

where evaluation of Forget gate of pth node in hidden layer, l at q^{th} time step, $F_p^{l<q>}$ is shown in Eq. (17.10), Input gate of pth node in hidden, l layer at q^{th} time step, $I_p^{l<q>}$ is shown in Eq. (17.11) and Intermittent cell value of pth node in hidden layer, l at q^{th} time step, $\widetilde{C}_p^{l<q-1>}$ is shown in Eq. (17.12).

$$F_p^{l<q>} = \text{logsig}\left[\sum_{r=1}^{n_{l-1}} wf_{pr}^l x_r^{l-1<q>} + \sum_{k=1}^{n_l} \overline{wf}_{pk}^l x_k^{l<q-1>}\right] + bf_p^l \quad \forall \ q = t - \beta_T \text{ to } t \qquad (17.10)$$

$$I_p^{l<q>} = \text{logsig}\left[\sum_{r=1}^{n_{l-1}} wI_{pr}^l x_j^{l-1<q>} + \sum_{k=1}^{n_l} \overline{wI}_{pk}^l x_k^{l<q-1>}\right] + bI_p^l \quad \forall \ q = t - \beta_T \text{ to } t \qquad (17.11)$$

$$\tilde{C}_p^{l<q>} = \zeta \left[\sum_{r=1}^{n_{l-1}} wc_{pr}^l x_j^{l-1<q>} + \sum_{k=1}^{n_l} \overline{wc}_{pk}^l x_k^{l<q-1>} \right] + bc_p^l \ \forall \ q = t - \beta_T \ \text{to} \ t \qquad (17.12)$$

In the above equations, logsig is the log sigmoid function, which results the output in the range of zero–one and ζ is nonlinear activation. The activated output of pth node in hidden layer, l at q^{th} time step is evaluated as presented in Eq. (17.13). The activated output is regulated with an output gate that controls how much of the information can be taken forward from the evaluated output and the output gate of pth node in hidden layer, l at q^{th} time step is presented in Eq. (17.14)

$$x_p^{l<q>} = O_p^{l<q>} \varphi \left(C_p^{l<q>} \right) \qquad (17.13)$$

$$O_p^{l<q>} = \text{logsig} \left[\sum_{r=1}^{n_{l-1}} wo_{pr}^l x_j^{l-1<q>} + \sum_{k=1}^{n_l} \overline{wo}_{pk}^l x_k^{l<q-1>} \right] + bo_p^l \ \forall \ q = t - \beta_T \ \text{to} \ t \qquad (17.14)$$

As the output layer of LSTM does not have any feedback connections, it will be the weighted sum of activated outputs from hidden layer, $\mathscr{L}$ and is presented in Eq. (17.15).

$$\widehat{Y^t} = \sum_{r=1}^{n_{\mathscr{L}-1}} w_{pr}^{\mathscr{L}} x_r^{\mathscr{L}-1<q>} + \sum_{k=1}^{n_{\mathscr{L}}} b_k^{\mathscr{L}} \ \text{for} \ q = t \qquad (17.15)$$

In this work, wind characteristics were modeled using multilayered networks and certain parameters, which govern these models are to be specified before training these models and these are called hyperparameters. The hyperparameters of RNNs and LSTMs include the activation function, unrolling length, number of hidden layers, number of nodes in hidden layers. Conventionally, these hyperparameters are fixed heuristically while modeling RNNs and LSTMs, which is a very tedious and challenging task. To address the issue of heuristic estimation of hyperparameters, we propose a novel, holistic algorithm in the framework of Automated Machine Learning which explores different designs of the model by creating a balance between overfitting and accuracy. This algorithm is generic in nature and can be applied to any sequential data.

17.2.3 Design of optimal networks

The model predictability increases with a greater number of parameters. With increase in parameters, the model complexity increases. Since the wind data is having long-term dependencies, the unrolling length of RNNs and LSTMs also increases, which again increases the model complexity. Hence, to maintain generalization ability of the model, a multiobjective optimization problem with objectives of maximizing accuracy and minimizing the model complexity is formulated. The number of parameters in the model (weights and biases), $\mathscr{N}_p$ and unrolling length, β_T determines the complexity of model while accuracy is measured through correlation coefficient, R^2 on the test set. Hence, the proposed algorithm has three objectives and all the hyperparameters: nodes in hidden layers, number of hidden layers, activation function, unrolling length serves as decision variables. The proposed multiobjective optimization problem is solved using NSGA-II [22]. Since the objectives are linear and decision variables are integers, the proposed formulation is Integer Nonlinear Programming problem and is presented in Eq. (17.16)

$$\min_{\{n_l | l = 1 : \mathscr{L}\}, A^{tf}, \beta_T} - R^2, \mathscr{N}_p, \beta_T \tag{17.16}$$

where,

$$R^2 = \text{mean}\{R_K^2 | K = 1 \text{ to } N\}$$

$$R_K^2 = \frac{\bar{t}\sum_{t=0}^{\bar{t}} Y_K^t \widehat{Y_K^t} - \sum_{t=0}^{\bar{t}} Y_K^t \sum_{t=0}^{\bar{t}} \widehat{Y_K^t}}{\sqrt{v(Y_K)v(\widehat{Y_K})}}$$

$$v(Y_K) = t\sum_{t=0}^{\bar{t}} (Y_K^t)^2 - \left(\sum_{t=0}^{\bar{t}} Y_K^t\right)^2$$

such that

$$\beta_T^{lb} \leq \beta_T \leq \beta_T^{ub}$$

$$N_{lb} \leq n_l \leq N_{ub} \text{ and } N_{lb} = \begin{cases} 1 \text{ if } l = 1 \\ 0 \text{ if } l > 1 \end{cases}$$

$$A^{tf} \in \{1, 2\} | \text{if } A^{tf} = \begin{cases} 1, \text{tansigmoid} \\ 2, \text{logsigmoid} \end{cases}$$

$$\{\beta_T^{lb}, \beta_T^{ub}, N_{ub}\} \in \mathbb{Z}_+$$

where.

$\mathscr{N}_p$: number of weights and biases in the model.
β_T: Back propagation length.
$\mathscr{L}$: Maximum number of hidden layers considered.
n_l: Number of nodes in hidden layer l.
N_{lb} & N_{ub}: Lower and upper bounds of n_l.
$\beta_T^{lb}, \beta_T^{ub}$: Lower and upper bounds of β_T.
A^{tf}: Activation function.
T: Length of training data.
$\bar{t}$: Length of test data.
N: Number of outputs in the data.
M: Number of inputs in the data.
N_{gen}: Number of generations in NSGA-II.
N_{pop}: Population size in NSGA-II.

Table 17.1   Algorithm for Automated Machine Learning framework.

Step 1   Initialize $\mathscr{L}$ +2 binary variables and real variables as 0. $\mathscr{L}$: maximum hidden layers.
Step 2   Set N_{gen} and N_{pop} to start NSGA-II
Step 3   For a given population, build the architecture of the model using first $\mathscr{L}$ decision variables:
      start a loop for I from 1 to $\mathscr{L}$
      if I $= 1$
          set the Ith decision variable as n_I and I $= I + 1$
        else:
          if Ith decision variable is zero,
            set I $= I - 1$, go to step 4,
          else
            set Ith decision variable n_I and let I $= I + 1$
          end
        end
      end loop
Step 4   Assign the architecture as [M, {$n_{1:I}$}, N]. I is the number of hidden layers for the given network.
Step 5   Determine A^{tf} and β_T using last two decision variables
Step 6   If architecture, β_T and A^{tf} are present in database return the objectives from the database else go to step 7
Step 7   Train the model using ADAM and t-BPTT
Step 8   Test the model using test data and evaluate objectives
Step 9   Repeat the steps 3 to 8 for all populations
Step 10 Perform the NSGA-II operations (crossover, mutation, selection and sorting) to produce the new generation and repeat steps 3 to 10 until convergence.

The proposed novel algorithm for optimal RNNs and LSTMs is presented in Table 17.1. Fortran 90 is used to code the NSGA-II optimizer and all the models without the intervention of any open source libraries. Intel Xeon CPU E5-26900 @ 2.90 GHz dual processor 128 GB RAM workstation was utilized to run the simulations.

17.3 Results

Since wind time-series data is collected through anemometers, noise in data needs to be removed. Hence the data is preprocessed with the moving average to filter the noise. The Gaussian shape of histograms for the residuals present in Fig. 17.6A and B proves that it is indeed a white noise. The autocorrelation figures are also plotted for the residuals and the data lies between the 95% confidence intervals which indicates that the residuals are white noise (see Fig. 17.6C and D).

The results of time-series analysis on the 3-year data which proves the existence of nonlinearities and long-term dependencies have been presented in Section 17.3.1 which is followed by the time-series decomposition to remove the seasonality from the data. The results of optimal design of RNNs and LSTMs were presented in Section 17.3.2 and the significance of forecasts in wind farm studies is presented in Section 17.3.3.

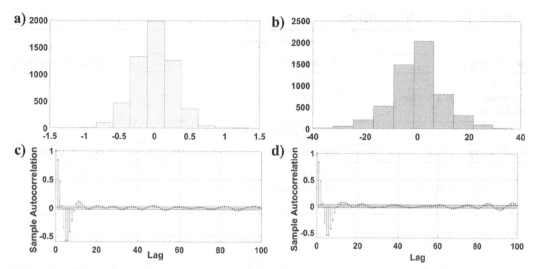

FIGURE 17.6 (A) Represents the Autocorrelation plot for residuals in wind speed data. (B) Represents the Autocorrelation plot for residuals in wind direction data. (C) Represents the white noise of wind speed in Gaussian shape. (D) Represents the white noise of wind direction in Gaussian shape.

17.3.1   Time-series analysis and decomposition

The time-series nature, which can be expressed through nonlinearity, stationarity, long-term dependencies, can be explored through some statistical tests. The nonlinear characteristics in the data were determined through a hypothesis test, BDS test. The time-series data is linear under null hypothesis, H_0 and the time-series data is nonlinear under alternate hypothesis, H_1. A level of significance of 95% is considered to determine the p values. If p value is less than .05 then we reject the null hypothesis. The stationarity nature of the data is determined through ADF test in which the data has a unit root under null hypothesis and data is stationary under alternate hypothesis. A level of significance of 95% is considered to determine the p values. If p value is less than .05, then we reject the null hypothesis. Hurst exponent analysis is used to examine the long-term dependencies. A rescaled range analysis on shorter time spans of the collected data is used to determine the Hurst exponent, H. If H-value lies between 0.5–1, the data is inferred to have long-term dependencies. The results of time-series analysis were present in Table 17.2.

The sequence in the time-series data is influenced by hidden patterns and structures in the data. The behavioral patterns in the time-series data can be determined by time-series data decomposition into seasonal, trend, and reminder components. The seasonal component influences the dynamics in the data. Therefore, STL decomposition is used in this study for filtering the seasonality component. The results of STL decomposition for wind data were presented in Figs. 17.7 and 17.8. After decomposition, the trend and reminder components were modeled together using RNNs and LSTMs. After modeling

Table 17.2 Time-series analysis.

Wind speed		
Nonlinearity	**Stationarity**	**Long-term dependency**
BDS test	ADF test	Hurst exponent
Null hypothesis: data is linear	Null hypothesis: data has a unit root	–
P-value: .001	P-value: .008	H-value: 0.83
Wind direction		
Nonlinearity	**Stationarity**	**Long-term dependency**
BDS test	ADF test	Hurst exponent
Null hypothesis: data is linear	Null hypothesis: data has a unit root	–
P-value: .003	P-value: .01	H-value: 0.79

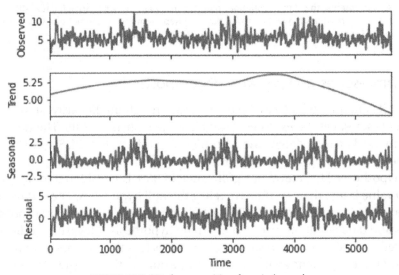

FIGURE 17.7 STL decomposition for wind speed.

and forecasting, the seasonality and trend-reminder components were combined for further wind farm studies such as wind farm micrositing and wind farm control.

17.3.2   Optimal design of RNNs and LSTMs

After extracting the seasonality from the data, the collected 4-year wind time-series data is divided into two parts. The first one contains 3-year data and second contains fourth year data. The 3-year data is used for training. The fourth year data is predicted from the available 3-year data and is validated with the original fourth year data. 70% of data is

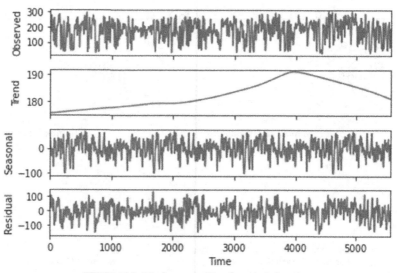

FIGURE 17.8 STL decomposition for wind direction.

Table 17.3 Parameters used in automated RNNs and LSTMs.

S. No	Parameter	Value
1	Number of binary and real variables in NSGA-II	5 and 0
2	Number of population and generations in NSGA-II	200 and 100
3	Mutation and crossover probability in NSGA-II	0.01 and 0.9
4	$\mathscr{L}_{LB}, \mathscr{L}_{UB}$: Bounds on number of hidden layers	1 and 3
5	N_{lb} & N_{ub}: Bounds on nodes in each hidden layer	{1,0,0} and {16, 15, 15}
6	$\beta_T^{lb}, \beta_T^{ub}$: Lower and upper bounds on β_T	2 and 65
7	$\bar{T}$: Number of test data points	1200

utilized for training and 30% for validation. Two simulations were performed with a multivariate data consisting of four inputs (daily and yearly periodicites) and two outputs (wind speed and direction) using the proposed optimal algorithm: one using RNN and other using LSTM. The bounds on the decision variables and parameters of NSGA-II given in Table 17.3. Table 17.3 show that the proposed algorithm obtains the best model from a possibility of $16 \times 15 \times 15 \times 64 \times 2$ alternatives, best of which would be impossible to handle with the heuristic way of search.

The proposed algorithm results in a 3D Pareto front as a solution. The Pareto fronts were obtained within six to eight generations using NSGA-II and did not improve further. The obtained solutions for RNNs and LSTMs were presented in Fig. 17.9. From the results it can be visualized that there lies a trade-off between accuracy and complexity. Every point in the Pareto front is equally important and represents the decision variables. Tables 17.4 and 17.5 represent the embodiment of architectures obtained through Pareto fronts for RNN and LSTM, respectively. Next, one solution from the Pareto set

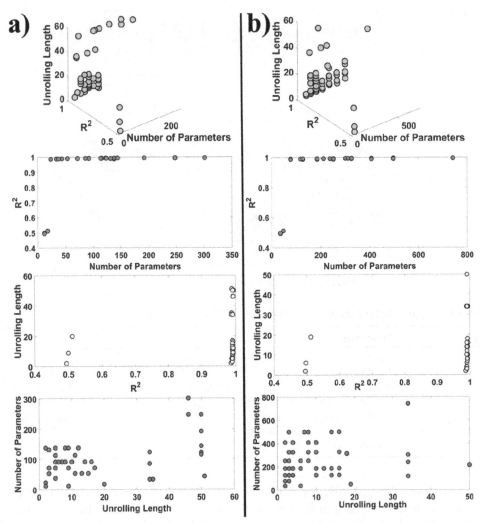

FIGURE 17.9 Converged Pareto fronts for RNN and LSTM respectively. (A) Represents the Pareto fronts obtained through automated RNNs. (B) Represents the Pareto fronts obtained through automated LSTMs.

should be selected. A robust model selection criterion, which penalizes the model for complexity called the Hannan Quinn criterion (HQC) [34] is used as higher order information for selecting a model from the Pareto front. The HQC value is calculated through Eq. (17.17) and the architecture, which gives least value of HQC is selected for further analysis.

$$HQC = \text{Training sample size} * \log(\text{RMSE}^2) + 2 * \text{Number of Parameters} * \log(\log(\text{sample size}))$$

$$(17.17)$$

Table 17.4 List of Pareto solutions for wind data with RNN model.

S. No.	Number of nodes Hidden layer 1	Hidden layer 2	Hidden layer 3	Activation function choice	Unrolling length	RMSE	Number of parameters	R²	HQC
1	1	0	0	0.72	2	0.72	12	0.49	−1957
2	1	0	0	0.71	9	0.71	12	0.49	−2002
3	1	1	1	0.71	20	0.71	18	0.51	−1985
4	2	0	0	0.13	2	0.13	24	0.986	−12,512
5	2	2	0	0.13	34	0.13	34	0.987	−12,235
6	2	2	0	0.14	35	0.14	34	0.988	−11,685
7	2	3	0	0.15	51	0.15	44	0.989	−11,587
8	2	6	0	0.11	34	0.11	86	0.991	−12,956
9	3	0	0	0.12	5	0.12	38	0.988	−12,883
10	3	7	0	0.12	50	0.12	123	0.992	−13,128
11	4	0	0	0.12	3	0.12	54	0.989	−12,977
12	4	0	0	0.11	11	0.11	54	0.99	−13,522
13	4	0	0	0.11	13	0.11	54	0.99	−13,450
14	4	0	0	0.11	15	0.11	54	0.99	−13,472
15	4	6	0	0.111	34	0.11	124	0.992	−12,782
16	4	7	0	0.12	50	0.12	144	0.993	−12,572
17	5	0	0	0.11	3	0.11	72	0.989	−13,106
18	5	0	0	0.11	5	0.11	72	0.99	−13,072
19	5	0	0	0.11	9	0.11	72	0.990	−13,101
20	5	0	0	0.10	17	0.10	72	0.99	−13,614
21	6	0	0	0.10	5	0.10	92	0.99	−13,293
22	6	0	0	0.10	6	0.10	92	0.99	−13,687
23	6	0	0	0.11	7	0.11	92	0.99	−13,160
24	6	0	0	0.09	8	0.09	92	0.99	−13,820
25	6	0	0	0.09	10	0.09	92	0.991	−13,848
26	6	0	0	0.09	14	0.09	92	0.991	−13,882
27	**6**	**0**	**0**	**0.09**	**16**	**0.09**	**92**	**0.991**	**−14,083**
28	6	3	0	0.12	50	0.12	116	0.99	−12,486
29	6	4	0	0.11	3	0.11	132	0.989	−12,720
30	6	6	6	0.16	46	0.16	248	0.992	−10,044
31	6	7	0	0.13	50	0.13	192	0.993	−11,705
32	7	0	0	0.10	5	0.10	114	0.99	−13,409
33	7	0	0	0.11	11	0.11	114	0.99	−12,866
34	8	0	0	0.12	2	0.12	138	0.986	−12,354
35	8	0	0	0.10	5	0.10	138	0.99	−13,334
36	8	0	0	0.10	8	0.10	138	0.991	−13,478
37	8	0	0	0.11	9	0.11	138	0.992	−12,806
38	8	0	0	0.09	12	0.09	138	0.992	−14,078
39	8	6	6	0.12	46	0.12	302	0.993	−11,661
40	8	7	0	0.11	50	0.11	248	0.994	−12,224

Table 17.5   List of Pareto solutions for wind data with LSTM model.

S. No.	Number of nodes Hidden layer 1	Hidden layer 2	Hidden layer 3	Activation function choice	RMSE	Unrolling length	Number of parameters	R^2	HQC
1	1	0	0	1	0.72	2	36	0.49	−1866
2	1	0	0	1	0.73	6	36	0.49	−1753
3	1	1	0	2	0.71	19	48	0.51	−1888
4	2	0	0	1	0.13	2	78	0.986	−12,355
5	2	0	0	2	0.11	3	78	0.989	−13,348
6	2	2	0	1	0.11	34	118	0.989	−12,757
7	2	6	0	1	0.09	34	302	0.992	−13,052
8	3	0	0	1	0.12	2	128	0.987	−12,447
9	3	0	0	2	0.11	3	128	0.989	−12,886
10	3	0	0	1	0.11	4	128	0.989	−12,921
11	3	0	0	1	0.10	10	128	0.99	−13,288
12	3	0	0	1	0.10	14	128	0.99	−13,465
13	3	0	0	1	0.10	16	128	0.99	−13,482
14	3	3	0	1	0.10	50	212	0.99	−13,025
15	3	5	0	1	0.09	18	312	0.99	−12,886
16	4	0	0	1	0.11	2	186	0.987	−12,499
17	4	0	0	2	0.11	3	186	0.989	−12,315
18	4	0	0	1	0.1	4	186	0.99	−13,148
19	4	0	0	1	0.1	6	186	0.99	−13,317
20	4	0	0	1	0.09	10	186	0.99	−13,349
21	4	0	0	1	0.1	12	186	0.991	−13,315
22	4	0	0	1	0.1	14	186	0.991	−13,209
23	4	0	0	1	0.09	16	186	0.991	−13,362
24	4	2	0	1	0.1	34	238	0.991	−13,012
25	5	0	0	1	0.12	2	252	0.988	−12,000
26	5	0	0	2	0.11	3	252	0.989	−12,518
27	5	0	0	1	0.1	6	252	0.991	−12,976
28	5	0	0	1	0.09	8	252	0.991	−13,112
29	6	0	0	2	0.11	3	326	0.99	−12,308
30	6	0	0	1	0.1	4	326	0.99	−12,681
31	6	0	0	1	0.09	8	326	0.991	−13,028
32	6	0	0	1	0.09	10	326	0.992	−13,066
33	**6**	**0**	**0**	**1**	**0.08**	**16**	**326**	**0.992**	**−13,537**
34	7	0	0	1	0.11	2	408	0.988	−11,763
35	7	0	0	1	0.10	4	408	0.99	−11,977
36	7	0	0	1	0.09	8	408	0.991	−12,735
37	7	0	0	1	0.08	10	408	0.993	−13,210
38	7	6	0	1	0.08	34	742	0.994	−12,200
39	8	0	0	2	0.10	3	498	0.99	−11,542
40	8	0	0	2	0.10	7	498	0.99	−11,720
41	8	0	0	1	0.089	8	498	0.992	−12,730
42	8	0	0	1	0.09	14	498	0.993	−12,539
43	8	0	0	1	0.08	16	498	0.994	−13,220

The optimal design of LSTMs has shown an architecture of [4-6-2] with tansigmoid activation function and unrolling length of 16. The number of parameters resulted as 326. The optimal design of RNNs has shown an architecture of [4-6-2] with tansigmoid activation function and unrolling length of 16. The number of parameters for RNN are found as 92. The accuracy of LSTMs and RNNs on validation set is reported as 99.3 and 99.1. The performance of RNNs and LSTMs on wind direction and wind speed is presented in Figs. 17.10 and 17.11 respectively. Though multiple layered architecture didn't emerge as a solution, the proposed algorithm has determined the best model from a total of 460,800 alternatives by only evaluating a maximum of 613 architectures (maximum in case of RNNs). It can be inferred from results that the number of parameters were more in LSTMs as compared to RNNs, while the prediction accuracy is similar on validation data. The increase of parameters in LSTMs are due to the presence of four gates, i.e., input gate, forget gate, intermittent cell value and output gate, all four combined together is similar to one RNN node. This leads to increase in number of parameters in LSTMs.

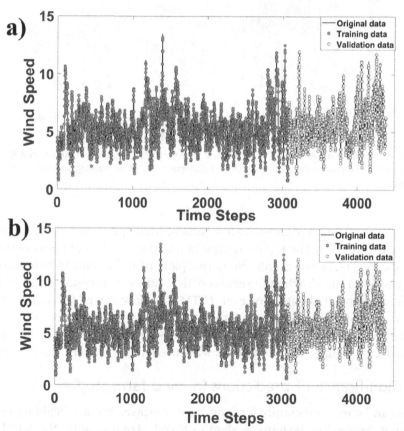

FIGURE 17.10 Performance of RNN and LSTM for Wind Speed. (A) Credibility of Automated RNN on training and validation of wind speed data. (B) Credibility of Automated LSTM on training and validation of wind speed data.

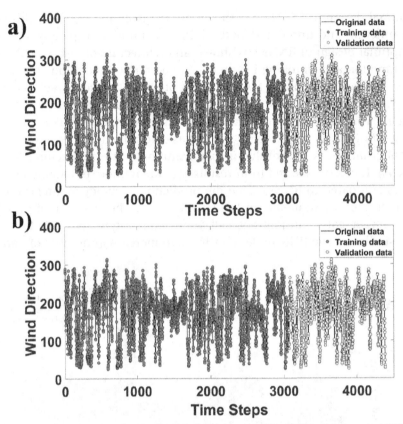

FIGURE 17.11 Performance of RNN and LSTM for Wind Direction. (A) Credibility of Automated RNN on training and validation of wind direction data. (B) Credibility of Automated LSTM on training and validation of wind direction data.

Figs. 17.12 and 17.13 represent the forecasts of wind speed and direction using RNNs and LSTMs, respectively. The figures represent that the LSTMs and RNNs performed well on forecasts. Fig. 17.14 represents the parity plots of RNNs and LSTMs for both wind direction and wind speed. When the extent of dependencies increased further RNNs may fail due to vanishing gradients. However, LSTMs are proven well for long-term forecasts while RNNs work well for short-term forecasts. Thus, for applications such as wind farm control, which requires short-term forecasts, RNNs can be used efficiently while LSTMs can be used in designing the wind farms which requires long-term forecasting.

17.3.3 Significance of predictions in wind farm studies

As LSTMs can be used efficiently for long-term forecasts, the applicability of LSTMs in wind energy conversion systems is demonstrated. Traditionally, the wind energy is extracted by establishing wind turbines in a systematic arrangement which maximizes

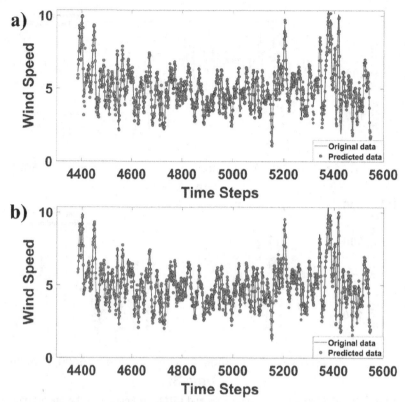

FIGURE 17.12 Forecasting of Wind Speed using RNN and LSTM. (A) Represents the prediction performance of RNN on wind speed data. (B) Represents the prediction performance of LSTM on wind speed data.

the energy production while considering the effects of wake. In any conventional studies of wind farm design, the researchers try to model the wind characteristics, direction and speed as WFM, a wind speed and direction joint probability distribution. Then the effective velocities were determined through the established wake models and the power from each turbines is determined as a function of effective velocities through a power-velocity relationship given by the manufacturer. Thus, the annual energy production is calculated as a function of WFM and expected value of power. Hence, it is important to construct the frequency map accurately. Due to several reasons, only a limited amount of data is available for the construction of wind frequency map. But the standard practice in any wind farm design is to design the wind farm for almost 50 years. Due to the lack of forecasting techniques, wind farm was designed by constructing WFMs with shorter durations which makes the design unrealistic. Therefore, in this study, an effort is made to show the advantages of long-term forecasting using LSTMs with the determination of energy production from four different frequency maps. We first present the construction of frequency maps followed by energy production. Wind Frequency map is constructed through frequentists approach by dividing the wind speed and direction data into several

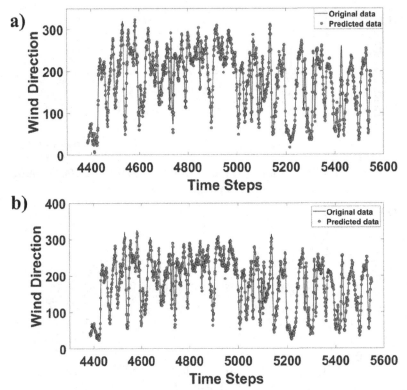

FIGURE 17.13 Forecasting of Wind Direction using RNN and LSTM. (A) Represents the prediction performance of RNN on wind direction data. (B) Represents the prediction performance of LSTM on wind direction data.

intervals (speed bins and direction sectors) and these speed and direction were considered as random variables. The frequentists approach is presented in Eq. (17.18)

$$freq_{mn} = p_{mn}/P \qquad (17.18)$$

where p_{mn} is the number of points in m^{th} speed bin and n^{th} direction sector and P is the total number of data points. In this work, four frequency maps were considered with the collected data. $wfm_{conventional}$ is the frequency map generated with most recent year data, in this work third year data is considered by assuming that we have access to 3-year data. wfm_{avg} is the frequency map generated with consideration all the available data i.e., all the 3 years. $wfm_{realistic}$ is the frequency map obtained by considering the previously available as well as predicted data obtained through the LSTMs. $wfm_{original}$ is the frequency map obtained through the 4-year original data. All these frequency maps are represented in Fig. 17.15. Since we have assumed that we have only 3-year data available with us and modeled the LSTMs, this WFM works as benchmark case to prove the validity of our study. Further we used an optimal layout with 33 turbines spread over a 3000 sq. km (shown in Fig. 17.16), which was obtained by micrositing simulations as described in Ref. [35]. The

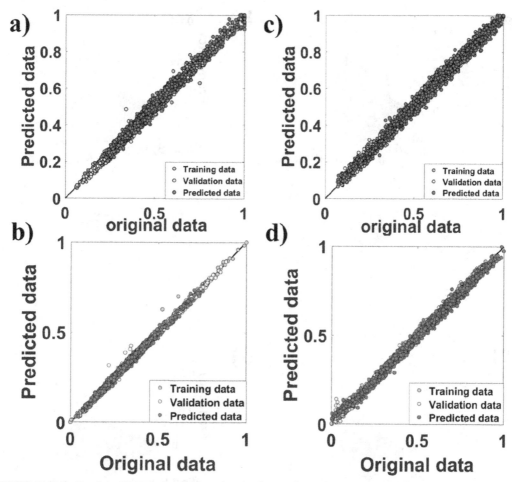

FIGURE 17.14 Parity plots. (A) Represents the parity plot for wind speed using RNN. (B) Represents the parity plot for wind speed using LSTM. (C) Represents the parity plot for wind direction using RNN. (D) Represents the parity plot for wind direction using LSTM.

effective velocities at each turbine were determined by widely used Jensen model [36]. The expected power from the layout is obtained using Eq. (17.19)

$$\text{power} = 8760 \sum_{m=1}^{N} \sum_{r=1}^{D} \sum_{u=1}^{S} \left[P_{\text{curve}} \left(u_{\text{eff}}(d_r, s_u, m) \right) * \text{WFM}(d_r, s_u) \right] \qquad (17.19)$$

where P_{curve} is the power curve, relationship between velocity and power, obtained from turbine manufacturers, D is the number of direction sectors, S is the number of speed bins, s_u, d_r are the values of speed and direction in u^{th} and r^{th} intervals, u_{eff} is the effective velocity determined through wake model for a given turbine, 8760 is the total number of hours in a year. The values of Annual energy obtained through the considered frequency are reported in Table 17.6.

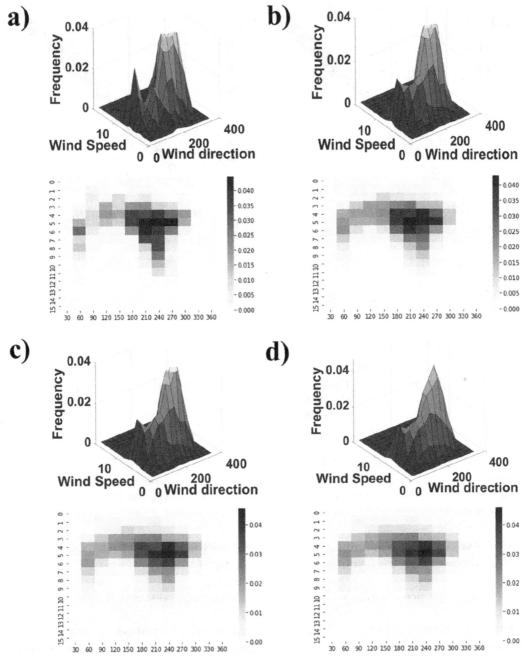

FIGURE 17.15 Wind frequency distribution and corresponding heat maps. (A) Represents wfm$_{conventional}$. (B) Represents wfm$_{avg}$. (C) Represents WFM$_{realistic}$. (D) Represents.wfm$_{original}$.

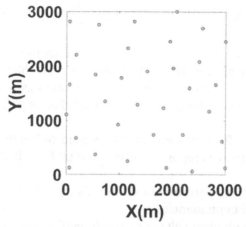

FIGURE 17.16 Optimal wind-farm layout used in this work for analysis using forecasts over long-range of time. The shaded circles indicate turbine locations.

Table 17.6 AEP determined using the frequency maps generated through original data and forecasted data.

S.No	Frequency	AEP(kW)
1	WFM$_{aggressive}$	14,739.92
2	WFM$_{consevative}$	12,273.48
3	WFM$_{realistic}$	11,359.58
4	WFM$_{benchmark}$	11,398.84

The analysis has reported that the WFM constructed by utilizing the original as well as predicted data gives the results close approximation to the benchmark case. The results justify the necessity of long-term forecasting in accurate designing of wind farms.

17.4 Conclusions

In this manuscript, we compared two Deep Learning techniques in terms of their ability in modeling the wind time-series data. The characteristics of wind data were examined through time-series analysis and the behavioral patterns were examined through time-series decomposition and the seasonality component is extracted for better modeling of the data. This provides the justification for selecting RNNs and LSTMs for modeling the data. But the heuristic estimation of hyperparameters made these models tedious and challenging. To mitigate these issues, a novel optimal algorithm is proposed under Automated Machine Learning framework which optimizes the hyperparameters involved in the models. The significance of long-term forecasts is presented using a study on energy production from an optimally designed wind layout. The work is summarized below:

- The efficiency of the models was highly influenced by the hyperparameters.
- A novel algorithm for optimizing the hyperparameters which prevent the heuristics is designed using the evolutionary multiobjective optimization framework.
- A multivariate time-series modeling on real wind data is performed with optimal RNNs and LSTMs.
- The training accuracy for RNNs and LSTMs is reported as 99% for both the models, whereas testing accuracy is reported as 98% and 99% for RNNs and LSTMs, respectively.
- Though both the models performed well on this data, RNNs might suffer in case of increasing long-term dependencies.
- The presence of hypothetical cell state, which regulates the information from previous time steps, makes the LSTMs to perform well with the long-term temporal features.
- The need of long-term forecasts in designing of wind farms was justified with analysis of power using different scenarios of WFMs.
- The advantage of RNNs in short-term forecasts provides a future scope in efficiently forecasting the disturbances during wind-farm control.

Acknowledgments

We would like to acknowledge the support of National Supercomputing Mission (NSM) for the project (DST/NSM/R&D_HPC_Applications/2021/23) sponsored by the Department of Science and Technology (DST), Government of India.

References

[1] https://www.irena.org/Statistics.

[2] https://gwec.net/global-wind-report-2021.

[3] M. Gan, H.X. Li, C.P. Chen, L. Chen, A potential method for determining nonlinearity in wind data, IEEE Power Energy Technol. Syst. J. 2 (2) (2015) 74–81.

[4] A. Sfetsos, A novel approach for the forecasting of mean hourly wind speed time-series, Renew. Energy 27 (2) (2002) 163–174.

[5] V. Prema, K.U. Rao, Development of statistical time-series models for solar power prediction, Renew. Energy 83 (2015) 100–109.

[6] F. Guignard, M. Lovallo, M. Laib, J. Golay, M. Kanevski, N. Helbig, L. Telesca, Investigating the time dynamics of wind speed in complex terrains by using the Fisher− Shannon method, Phys. Stat. Mech. Appl. 523 (2019) 611–621.

[7] X. Zhao, J. Liu, D. Yu, J. Chang, One-day-ahead probabilistic wind speed forecast based on optimized numerical weather prediction data, Energy Convers. Manag. 164 (2018) 560–569.

[8] D.J. Allen, A.S. Tomlin, C.S. Bale, A. Skea, S. Vosper, M.L. Gallani, A boundary layer scaling technique for estimating near-surface wind energy using numerical weather prediction and wind map data, Appl. Energy 208 (2017) 1246–1257.

[9] T.M. Giannaros, D. Melas, I. Ziomas, Performance evaluation of the Weather Research and Forecasting (WRF) model for assessing wind resource in Greece, Renew. Energy 102 (2017) 190–198.

[10] A. Prasetyowati, H. Sudibyo, D. Sudiana, Wind power prediction by using wavelet decomposition mode based NARX-neural network, in: In Proceedings of the 2017 International Conference on Computer Science and Artificial Intelligence, 2017, pp. 275–278.

[11] E. Erdem, J. Shi, ARMA based approaches for forecasting the tuple of wind speed and direction, Appl. Energy 88 (4) (2011) 1405–1414.

[12] H. Nazaripouya, B. Wang, Y. Wang, P. Chu, H.R. Pota, R. Gadh, Univariate time-series prediction of solar power using a hybrid wavelet-ARMA-NARX prediction method, in: 2016 IEEE/PES Transmission and Distribution Conference and Exposition (T&D), IEEE, 2016, pp. 1–5.

[13] L.L. Li, X. Zhao, M.L. Tseng, R.R. Tan, Short-term wind power forecasting based on support vector machine with improved dragonfly algorithm, J. Clean. Prod. 242 (2020) 118447.

[14] J. Wasilewski, D. Baczynski, Short-term electric energy production forecasting at wind power plants in pareto-optimality context, Renew. Sustain. Energy Rev. 69 (2017) 177–187.

[15] I. Okumus, A. Dinler, Current status of wind energy forecasting and a hybrid method for hourly predictions, Energy Convers. Manag. 123 (2016) 362–371.

[16] A. Dolatabadi, H. Abdeltawab, Y.A. Mohamed, Hybrid deep learning-based model for wind speed forecasting based on DWPT and bidirectional LSTM network, IEEE Access 8 (2020) 229219–229232.

[17] H. Liu, X.W. Mi, Y.F. Li, Wind speed forecasting method based on deep learning strategy using empirical wavelet transform, long short-term memory neural network and Elman neural network, Energy Convers. Manag. 156 (2018) 498–514.

[18] H. Liu, C. Chen, Data processing strategies in wind energy forecasting models and applications: a comprehensive review, Appl. Energy 249 (2019) 392–408.

[19] T.Yu, H. Zhu, Hyper-parameter Optimization: A Review of Algorithms and Applications. arXiv preprint arXiv:2003.05689. 2020.

[20] J. Wu, X.Y. Chen, H. Zhang, L.D. Xiong, H. Lei, S.H. Deng, Hyperparameter optimization for machine learning models based on Bayesian optimization, J. Electron. Sci. Technol. 17 (1) (2019) 26–40.

[21] J. Wu, S. Chen, X. Liu, Efficient hyperparameter optimization through model-based reinforcement learning, Neurocomputing 409 (2020) 381–393.

[22] K. Deb, Multi-objective Optimization Using Evolutionary Algorithms, John Wiley & Sons, 2001.

[23] [dataset] Engie. La Haute Borne Data (2017–2020), 2020 [Internet]. Available from: https://opendata-renewables.engie.com/pages/home/.

[24] M.O. Akintunde, J.O. Oyekunle, G.A. Olalude, Detection of non-linearity in the time series using BDS test, Sci. J. Appl. Math. Stat. 3 (4) (2015) 184–192.

[25] D.A. Dickey, W.A. Fuller, Distribution of the estimators for autoregressive time-series with a unit root, J. Am. Stat. Assoc. 74 (366a) (1979) 427–431.

[26] L. Kalo, P. Kamalanathan, H.J. Pant, M.C. Cassanello, R.K. Upadhyay, Mixing and regime transition analysis of liquid-solid conical fluidized bed through RPT technique, Chem. Eng. Sci. 207 (2019) 702–712.

[27] R.J. Hyndman, G. Athanasopoulos, Forecasting: Principles and Practice, OTexts, 2018.

[28] R.B. Cleveland, W.S. Cleveland, J.E. McRae, I. Terpenning, STL: a seasonal-trend decomposition, J. Off. Stat. 6 (1) (1990) 3–73.

[29] T. Hagan Martin, B. Demuth Howard, H. Beale Mark, Neural Network Design, University of Colorado at Boulder, 2002.

[30] A. Graves, Supervised Sequence Labelling with Recurrent Neural Networks, 2012.

[31] I. Sutskever, O. Vinyals, Q.V. Le, Sequence to Sequence Learning with Neural Networks. arXiv preprint arXiv:1409.3215. 2014.

[32] D.P. Kingma, J. Ba, Adam: a method for stochastic optimization. arXiv preprint arXiv:1412.6980. 2014.

[33] A. Sherstinsky, Fundamentals of recurrent neural network (RNN) and long short-term memory (LSTM) network, Phys. Nonlinear Phenom. 404 (2020) 132306.

[34] Y. Miche, A. Lendasse, A faster model selection criterion for OP-ELM and OP-KNN: Hannan-Quinn criterion, InESANN 9 (2009) 177–182.

[35] P. Mittal, K. Mitra, Determining layout of a wind-farm with optimal number of turbines: a decomposition based approach, J. Clean. Prod. 202 (2018) 342–359.

[36] N.O. Jensen, A Note on Wind Generator Interaction, 1983.

Index

Note: 'Page numbers followed by "*f*" indicate figures and "*t*" indicate tables.'

Printed in the United States
by Baker & Taylor Publisher Services